Digital Television

Digital Television

Bandwidth Reduction and Communication Aspects

R. H. Stafford, Ph.D.

A WILEY-INTERSCIENCE PUBLICATION

JOHN WILEY & SONS   New York · Chichester · Brisbane · Toronto

Library of Congress Cataloging in Publication Data:
Stafford, Richard H.   1914–
  Digital television.

  "A Wiley-Interscience publication."
  A revision of the author's thesis, California
Western University.
  Bibliography: p.
  Includes index.
  1. Digital television. I. Title.
TK6678.S7 1980   621.388   80-17542
ISBN 0-471-07857-3

Printed in the United States of America

10 9 8 7 6 5 4 3 2 1

Preface

The field of digital image processing has grown considerably during the past ten years. The decrease in size and cost of discrete components has made it possible to achieve practical hardware implementation of many new types of image processors. The same can be said of digital TV in the area of real time usage.

Most of the books presently available are devoted to non-real time digital image processing, with very little mention of the degrading effects of channel noise in the transmission systems, usually the binary symmetrical channel type. Therefore, this book is devoted to discussing state-of-the-art digital image processing systems with an emphasis on real time digital TV systems in the real world of noisy communications channels.

A new technique of pseudorandom sampling is introduced for use with various types of pulse modulation and coding systems. Six chapters are devoted to practical applications of pulse modulation transmission systems, two of which are types of convolutional and Reed-Solomon error detection and correction systems. One chapter is devoted to a comprehensive coverage of state-of-the-art digital image processing methods as they pertain to band-width reduction.

The first three chapters are devoted to a tutorial prerequisite type of discussion.

The reader will find the book reasonably comprehensive in coverage of the field with some new and challenging ideas for the future.

R. H. STAFFORD

Citrus Heights, California
August 1980

Acknowledgments

This book is an expansion of my doctoral dissertation, presented to California Western University, and prepared while I was working as an engineer for Collins Radio/Rockwell International Corp. at Newport Beach, California.

I would like to thankfully acknowledge permission to use my office at Collins while preparing the thesis, which constitutes about 80% of this book. My use of the office during early-morning and late-night hours probably inconvenienced to some extent the security guards on duty; I wish to thank them for their cooperation. While at Collins I also discussed some of the aspects of the thesis with my colleagues. Two fellow engineers who were most helpful to me in the area of computer programming were Mr. Fred Sendra and Mr. Jay Robertson.

I reformatted and expanded my doctoral thesis into this book in my spare time while working at Calspan Advanced Technology Corp., at their White Sands Missile Range office. I wish to thank them for use of their facilities.

R. H. S.

Contents

Digital Television

chapter 1

Introduction

1.1 REASONS FOR USING DIGITAL TV

The field of digital television communications is a very broad one. There are many reasons for its existence. The digital form of communications in general has several advantages over the analog form. These can be listed as follows:

1 Reliability.
2 More efficient multiplexing.
3 Miniaturization.
4 Data handling precision.
5 Less calibration trouble.
6 Greater complexity handling capability.
7 More versatile.
8 Greater channel capacity.

The modern trend in communications is to use all solid-state components in the equipment. Digital circuitry has grown up with advances in the solid-state art. Solid-state components leave much greater reliability than many of the components used in analog systems.

Time division multiplexing is based on digitization, whereas analog systems must resort to frequency domain multiplexing. Time division multiplexing is much more efficient than frequency domain multiplexing.

Due to the use of the modern solid-state components, such as the integrated circuits, package size of the equipment has been tremendously reduced.

For the reasons outlined above, data handling is done more efficiently with digital techniques, and there are fewer calibration problems with digital circuitry.

Modern trends are towards more and more complex systems, which are handled with greater ease by digitization. This is partly due to the greater versatility of digital systems.

1

Furthermore, digital communications systems come closer to achieving maximum channel capacity. The Shannon theory of channel capacity is based on the assumptions of continuous Gaussian channel noise, correlation detection, and a signal resembling noise. His criterion is an error-free data rate.

The video television signal has a probability distribution that is very close to being Laplacian in nature, and a shade less to being Gaussian. The digitized television signal therefore approaches the fulfillment of Shannon's requirements for optimum channel capacity.

1.2 DIGITAL TV APPLICATIONS

There are many applications for a digital TV communication system. It is not used at present in the commercial television field, because of certain bandwidth problems, but these problems are being overcome. Eventually, it is probable that commercial TV will become all digital.

Digital TV is a must in our space probes to the moon and the planets. It has already been proven successful in this field.

Armed Services TV surveillance systems have used digital TV successfully, and the experiments with videophone systems have shown that such use is just around the corner. Wherever analog TV systems are in use, they can be replaced by digital systems.

1.3 SCOPE OF THE BOOK

The main body of the book includes a general discussion of various digital TV communications modulation techniques, with a performance comparison, a discussion of bandwidth requirements, and the need for bandwidth compression.

A large section is devoted to the large variety of means of achieving bandwidth reduction.

Most digital TV communications systems in the available literature are based on the use of pulse-code modulation (PCM) in one of its forms. A discussion of these techniques, of course, is included in what follows. However, there are other efficient coding forms besides PCM that could be used for digital TV, one of which is pulse-position modulation (PPM). This system is more efficient in many ways than is PCM, and is discussed here. A new method for implementing PPM systems for use with digital TV is presented.

One of the needs of digital TV communications is the use of error correction coding. This subject has been mentioned very briefly in the available literature.

Two forms of error correction coding are discussed in detail; the topics covered include performance comparisons and detailed implementation diagrams. The two coding systems are Convolutional coding and Reed-Solomon coding. A new algorithm for Reed-Solomon decoding is also presented in detail.

Performance comparison curves are presented for PCM/FM, DPCM/FM, PRN/PPM/FM, convolutional-FM, and Reed-Solomon-FM digital TV communication systems, where PRN is an abbreviation for pseudorandom noise coding.

Where it is considered wise to do so, some of the chapters conclude with a short summary section. Chapter 11 contains a general summary, a comparison of performance characteristics for eight types of digital TV communications applications, and a section on conclusions.

Bibliography and References for this book are found at the end of the book; not at the end of chapters. References are designated simply by the author's last name and the year of publication, for example, Viterbi (1964). This portion of the book is divided into two parts: Digital TV Communication Systems and Error Correction Coding.

A list of symbols and abbreviations appears at the end of the book. Even though references may point to a specific author and a specific paper, there are several symbols that have multiple meanings. Various authors do not standardize upon a particular meaning for a given symbol. Therefore, it is necessary to list page numbers along with the symbols and abbreviations. Not all symbols are listed; only those that might cause confusion.

Pulse Modulation Systems Applicable to TV Transmission

introduction

In considering the types of feasible and applicable modulation systems for use in a TV communications link, three types were studied: pulse-amplitude modulation/FM (PAM/FM), pulse-duration modulation/FM (PDM/FM), and PCM/FM. A PAM/FM system samples the video waveform with a train of pulses. The result of the sampling process is a train of pulses having an amplitude variation, the envelope of which matches the sampled video waveform. The PAM train of pulses is then used to frequency modulate a carrier to give a PAM/FM type of modulation.

In PDM/FM the pulse amplitude remains constant, but the trailing edge of the pulse is caused to vary in accordance with the sampled video waveform.

In a PCM/FM system the pulse amplitude remains constant, but the samples are quantized and the binary output represents various video signal levels.

2.1 THEORETICAL COMPARISON

The pulse systems are usually compared on the basis of performance in the face of interfering Gaussian noise. Signal-to-noise ratios (SNRs) are measured or calculated versus some point of reference common to all systems. The best common reference point to use is the rms error rate probability expressed in percent $(\overline{e_n^2})$. Another reference that is used is the rms signal-to-error ratio $(\overline{s_n^2}/\overline{e_n^2})$.

Viterbi (1962) has stated mathematically the mean-square error for Gaussian type noise by

$$\overline{e_n^2} = \lim_{N\to\infty} \frac{1}{N} \sum_{n=1}^{N} e_n^2 \qquad (2.1)$$

where N is the number of samples taken per second. He has also expressed the mean-square signal as

$$\overline{s_n^2} = \lim_{N\to\infty} \frac{1}{N} \sum_{n=1}^{N} s_n^2. \qquad (2.2)$$

2.2 EXPERIMENTAL COMPARISON

Aeronutronic Inc. has made a systems evaluation study of the three pulse-modulation systems mentioned above: PAM/FM, PDM/FM, and PCM/FM. In their Report U-743 of December 18, 1959 (AD 234 960), vols. II and III, they discussed the experimental performance data and the testing methods used in their evaluation study program. Their test setups and procedures were discussed in vol. III.

Basically, the test setups for all three of the pulse systems tested were similar to each other. The setups provided for an assembly of the telemetering system to be tested, a source of interference, monitoring equipments, and error comparators designed specifically for the system under test.

2.3 THE PAM/FM SYSTEM

Viterbi (1962) has also shown that PAM/FM signals can be detected by a pulsed integrator, which is synchronous with the sampled transmission time, and that the output signal-to-mean-square-error ratio is a linear function of the channel signal-to-noise ratio (SNR). PAM consists simply of extending the sample amplitude to last over the allotted transmission time of T seconds and multiplying the carrier by this waveform.

The multiplier output over the nth sample is

$$\frac{(3S)^{1/2}}{A} x_n(1 - \cos 2\omega_c t) \qquad (2.3)$$

When average power is assumed, the PAM output signal is s_n^2/S, but when peak power is assumed as a basis for comparing PAM and PDM performance, the $s_n^2 = S/3$ and

$$\frac{\overline{s_n^2}}{\overline{e_n^2}} = \frac{2ST}{3N/B} \qquad (2.4)$$

where S is the received signal power, T the time of the sample transmission period, N the noise power measured at the output of a band-pass filter, and B the filter bandwidth. This assumes that the PAM/FM energy per sample transmission period does not exceed that for PDM/FM. When this limiting condition does not apply, PAM/FM is superior to PDM/FM in rms–SNR performance by a ratio of $3:1$.

2.4 THE PDM/FM SYSTEM

Viterbi (1962) has shown that PDM signals can be detected by a pulsed integrator, which is synchronous with the sampled transmission time, and that the output signal-to-mean-square-error ratio is a linear function of the channel signal-to-noise ratio. In PDM a pulse is generated with width proportional to the amplitude of the sample. For this purpose the sample must be amplitude-limited, say between A and $-A$. Then a sample x_n (where $-A < x_n < A$) will produce a pulse duration $1 + (x_n/A)T/2$ seconds. He has also stated that the output signal power or mean-square signal at the end of a sample transmission time T will be

$$\overline{s_n^2} = \int_{-A}^{A} \frac{x_n^2}{2A} \, dx_n = \frac{S}{3} \tag{2.5}$$

and

$$\frac{\overline{s_n^2}}{\overline{e_n^2}} = \frac{2ST}{3N/B},$$

as was previously mentioned.

2.5 THE PCM/FM SYSTEM

The PCM/FM system transmits the binary equivalent of the numerical value for the given data level, by transmitting the carrier $+f$ to represent a "one" and the carrier $-f$ to represent a "zero". Thus an L-level sample is represented by a binary code of length $\log_2 L$ bits. The detector is a synchronous integrator that operates on each bit at a time for a period of $T/\log_2 L$ seconds, and then decides whether a zero or a one was sent. The variance is expressed by

$$\sigma^2 = \frac{(\log_2 L)^2}{T^2} \iint_0^{T/\log_2 L} \frac{N}{B} \delta(T - W)\left(\sqrt{2}\,\sin\omega_0 t\right)\left(\sqrt{2}\,\sin\omega_0 W\right) dt\, dw$$

$$= \frac{\log^2 L}{T} \frac{N}{2B}. \tag{2.6}$$

Thus the signal-to-rms error at time T when a one was sent is

$$\frac{\overline{s_n^2}}{\overline{e_n^2}} = \frac{2ST}{\log_2 L(N/B)}. \tag{2.7}$$

The probability of bit error when a one is sent is the probability that the noise contribution at time T is less than $-S^{1/2}$ or

$$P_b = \int_{-\infty}^{-(2ST/\log_2 L(N/B))^{1/2}} \left[e^{-x^2/2}(2\pi)^{1/2} \right] dx. \tag{2.8}$$

Viterbi has determined the PCM mean-square error for a four-level code to be

$$\overline{e_n^2} = \int x^2 P(X) dx$$

$$= \frac{A^2}{3(4)^2} \left[P(o) + 2 \sum_{i=1}^{3} p(i)(12_i^2 + 1) \right]$$

$$= \frac{(1 + 60P_B)}{48} A^2 \tag{2.9}$$

where $P(i)$ is the probability of an i-level error. The mean-square signal is $A^2/3$.

So, for a four-level coded PCM

$$\frac{\overline{s_n^2}}{\overline{e_n^2}} = \frac{16}{1 + 60P_B} \qquad \text{for } L = 4. \tag{2.10}$$

By definition a four-level code is that produced by a two-binary bits per word code, such as produced by the use of two cascaded flip-flops. A 2-bit code system can sense four different word levels. A 3-bit/word code can sense eight different word levels.

Other signal-to-mean-square error equations were also developed:

$$\frac{\overline{s_n^2}}{\overline{e_n^2}} = \frac{64}{1 + 252P_B} \qquad \text{for } L = 8, \tag{2.11}$$

$$\frac{\overline{s_n^2}}{\overline{e_n^2}} = \frac{256}{1 + 1020P_B} \qquad \text{for } L = 16, \tag{2.12}$$

$$\frac{\overline{s_n^2}}{\overline{e_n^2}} = \frac{1024}{1 + 4092P_B} \qquad \text{for } L = 32, \tag{2.13}$$

$$\frac{\overline{s_n^2}}{\overline{e_n^2}} = \frac{4096}{1 + 16,380P_B} \qquad \text{for } L = 64. \tag{2.14}$$

From these equations it is possible to compute the $\overline{e_n^2}$ value for any given P_B value.

The author has also worked out the mean-square error formulas for several word levels:

$$\overline{e_n^2} = \frac{1 + 252 P_B}{192} A^2 \qquad \text{for } L = 8, \qquad (2.15)$$

$$\overline{e_n^2} = \frac{1 + 4092 P_B}{768} A^2 \qquad \text{for } L = 16, \qquad (2.16)$$

$$\overline{e_n^2} = \frac{1 + 4092 P_B}{3072} A^2 \qquad \text{for } L = 32, \qquad (2.17)$$

and

$$\overline{e_n^2} = \frac{1 + 16{,}380 P_B}{12{,}288} A^2 \qquad \text{for } L = 64. \qquad (2.18)$$

2.6 COMPARISON OF SYSTEM PERFORMANCE

With the information given above, it is possible to compute theoretical performance comparison curves versus SNR's. Furthermore, the Aeronutronic data as previously mentioned can be used to form comparison curves. The test results for the PAM/FM and the PDM/FM systems can be used directly, because they are already in the form of rms errors. However, the Viterbi equations must be used to convert the PCM/FM bit error rate data into percent rms errors.

The Viterbi equations were used to form the curves shown in Figs. 2.1 and 2.2. Figure 2.1 shows a plot of mean-square-signal-to-error ratio versus SNR in the carrier. Figure 2.2 shows a plot of rms errors versus the carrier SNR. From Fig. 2.2 it can be seen that theoretically PCM/FM is several decibels better in performance against SNR than either PAM/FM or PDM/FM, and that PAM/FM and PDM/FM are equal in performance. These curves were prepared using the assumption that the PAM/FM pulse energy levels were limited to that of the PDM/FM pulses. If this limit is not applied, PAM/FM is theoretically better than PDM/FM. However, in practice, such limiting is usually applied to PAM/FM pulses because of system component limitations such as the maximum pulse peak power restrictions.

2.7 TEST RESULTS

Volume II of the Aeronutronic Report U-743 gives the test performance result for the three systems under discussion. Mostly normalized dimensionless ratios were used in presenting the test data. One such parameter was the

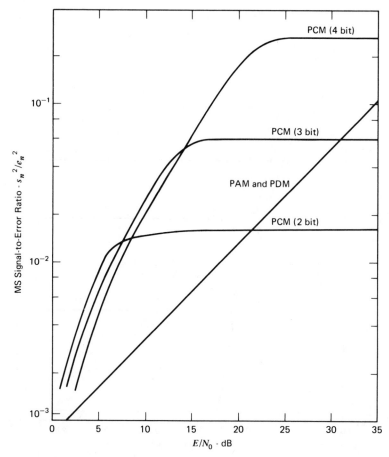

Figure 2.1. Mean-square signal-to-error ratio/signal-to-noise ratio.

normalized SNR, in the form of $S/K_1 f_S^{1/2}$. This bears the following relationship to the normally used S/N ratio:

$$\frac{S}{K_1 f_S^{1/2}} \, dB = \frac{S}{N} \, dB + \frac{10 \log B}{f_s} \qquad (2.19)$$

where B is the IF bandwidth and f_S is the sampling frequency in pulses per second (PPS).

Figure II-3-9 of Report U-743, of the PCM/FM test result, shows a graph of the bit error probability versus $S/K_1 f_B^{1/2}$, where f_B is the bit rate in pps. In this case both the bit rate and the IF bandwidth were equal, so $S/K_1 f_B^{1/2}$ would equal S/N.

Figure II-4-13 of Report U-743, of the PDM/FM test results, shows a graph of the rms errors in percent versus $S/K_1 f_S^{1/2}$.

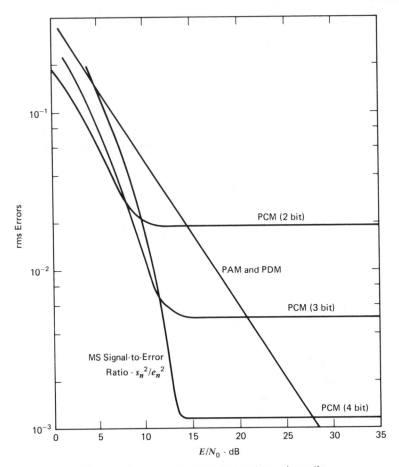

Figure 2.2. rms errors versus signal-to-noise ratio.

Figure II-6-17 of Report U-743, of the PAM/FM test results, shows a graph of the rms errors in percent versus $S/K_1 f_S^{1/2}$.

In order to compare the three systems on one graph the bit rate errors of the PCM/FM data must be converted to rms errors. This was done for three separate code formats, 2 bits/word, 3 bits/word, and 4 bits/word. The resultant rms error curves are plotted in Fig. 2.3 along with PDM/FM and PAM/FM curves. The rms errors are plotted versus the signal-to-noise parameter S/N.

Representative curves from Report U-743 for the PDM and PAM curves are shown in Figs. II-4-13 and II-6-17. The curve of Fig. II-6-17 (PAM/FM) for a B/f_S ratio of 5 was chosen, in which the IF bandwidth was 10 kHz ($f_S = 2$ kHz). The curve of Fig. II-4-13 for B/f_S of 10 was chosen, in which the IF bandwidth of 10 kHz was used ($f_S = 1$ kHz). Thus the two curves shown in Fig. II-6-17 are for the same IF bandwidth. In this dimensionless

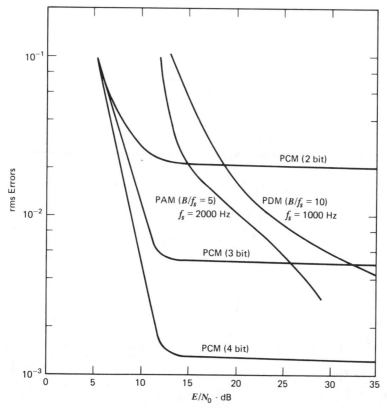

Figure 2.3. System performance comparison: PCM/FM, PDM/FM, and PAM/FM for PAM ($B/f_s = 5$) and PDM ($B/f_s = 10$).

display the PAM/FM appears to be better than the PDM/FM. However, many people prefer the dimensional S/N parameter to the $S/K_1 f_S^{1/2}$, so the data of Fig. II-6-17 have been converted to S/N units as shown in Fig. 2.4.

It is interesting to compare the results of the Aeronutronics tests versus the theoretically derived curves. A look at Figs. 2.2 and 2.4 show that they are very close together, especially the PCM/FM curves. For rms errors below 1 percent, the PDM/FM and PAM/FM curves are close together. The difference between the theoretical curves and the experimental curves for PDM/FM and PAM/FM lies in the steepness of the slopes. The theoretical curves were derived on a purely linear basis, while the measured data curves show the effects of system nonlinearities.

However, there are enough similarities between the curves of Figs. 2.2 and 2.4 to warrant the acceptance of the curves of Fig. 2.4 as portraying valid experimental data. Therefore, the curves of Fig. 2.4 form a sound basis for comparing the performance of the three systems versus S/N ratio (Gaussian noise). It can be noted from Fig. 2.4 that the 3-bit PCM/FM is several

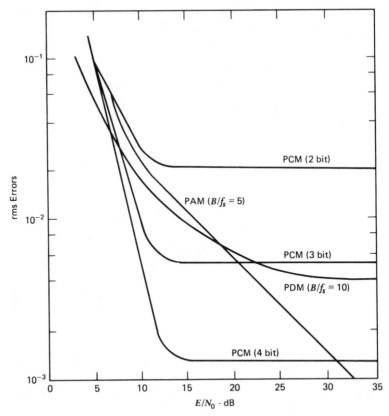

Figure 2.4. System performance comparison: PCM/FM, PDM/FM, and PAM/FM for PAM ($B/f_s = 10$) and PDM ($B/f_s = 5$).

decibels better in performance than are the PDM/FM and PAM/FM as regards to the S/N ratio in the rms error range of 0.5–1.5 percent, and that the 4-bit PCM/FM is always better than the other two systems for S/N ratios of 30 db or less. While these tests were run with Gaussian noise, other types of interference should be considered also. Since the carrier frequency for digital TV will likely be in the microwave region, atmospheric noise need not be considered. The other noise sources in the microwave regions are not high enough in amplitude to be more damaging than Gaussian noise. However, another source of interference to be considered is the effect of multipath transmissions.

However, the curves of Fig. 2.4 do indicate that a PCM/FM system is the preferred system, from the viewpoint of S/N performance.

chapter 3

PCM Encoding Principles

general

PCM is the term applied to a digital transmission system in which each analog sample is represented by a binary code group.

For example, if a 6-bit/sample system is used, the video signal would be coded in 64 discrete brightness levels. The digital bit rate would be 6 times the number of samples per second. If the highest video frequency was 4 MHz, and the sampling rate was 4 times this frequency, then the bit rate would be $4 \times 4 \times 6 = 96$ MHz. Aeronutronics has shown that the required bandwidth should equal the bit rate for optimum results.

Good image fidelity for PCM TV coding depends upon the following factors:

1   The pulse shape.
2   The number of bits per sample.
3   The sample spacing.
4   The shape of the reproducing aperture.
5   The overall tonal scale.
6   The placement of the quantization levels in the tone scale.
7   Noise and error rate.

3.1   THE PULSE SHAPE

The rise time of the sampling pulse determines the transmission bandwidth requirements. However, a bandwidth set equal to the bit rate has been proven to be adequate.

3.2   THE NUMBER OF BITS PER SAMPLE

The number of bits per sample sets the tone scale resolution. However, it takes at least 6 bits per sample level to avoid a phenomenon known as false contouring. In the low brightness gradient portion of the original picture, the quantized picture shows boundaries between areas differing in brightness by one quantum step. There are several methods of reducing the bits per sample below six, and still avoiding false contouring. These require special techniques, which will be described later.

3.3   THE SAMPLE SPACING

The sample spacing, or the number of picture elements per frame, sets the limiting resolution. Image sharpness is influenced by many system parameters, but resolution, which sets the ultimate limit of visibility of fine detail, depends only upon sample spacing.

3.4   THE SHAPE OF THE REPRODUCING APERTURE

The spatial spectrum of a matrix of samples being displayed consists of the spectrum of the picture centered about zero spatial frequency plus the same spectrums at reduced amplitude centered around the sampling frequency and the harmonics thereof. These extra spectra show up in the reproduced picture as a dot structure analogous to the line structure in a home TV set. To eliminate this visible dot structure without degrading the spectrum of the picture itself, the reproducing spot should act as an ideal low-pass filter.

3.5   THE TONAL SCALE

Tonal scale, although sometimes disregarded, is nevertheless important. Correct tonal scale is dependent upon the proper dc transfer characteristics of the amplifiers and transducers. Good quality cannot be achieved without optimizing these factors.

3.6   QUANTIZATION LEVELS

The best quality for a given number of brightness steps results when these steps are equally discernible, one from another, throughout the tone scale. A 5-bit picture with good quantization placement is as good as a 6-bit picture with poor placement.

3.7 NOISE

Noise in the analog video signal ahead of encoding is not greatly affected by PCM. False contouring is reduced by such noise. However, transmission errors produce an additional noise that differs from thermal noise in appearance because of its different amplitude probability distribution. If the channel error rate is R and B bits/sample are used, then the ratio of rms noise to peak signal after decoding is

$$\frac{N}{S} = RB\sqrt{\sum_{k=1}^{B} \frac{1}{4^k}} \tag{3.1}$$

provided that R is small enough to produce no more than one error per code group.

3.8 THE BINARY SYMMETRIC CHANNEL (BSC) AND PCM PICTURE QUALITY

Huang and Chikhaoui (1967) investigated the effect of the BSC noise on TV picture quality, when encoded in PCM.

It was found that for high SNR the white Gaussian noise is more objectionable to the observer than is BSC noise, while the reverse is true for low SNR. The crossover point lies approximately in the range 16–20 dB peak signal-to-rms noise ratio, and tends to occur at a higher SNR for the more detailed pictures.

For their tests BSC noise was simulated. For each signal bit, an independent random variate with uniform distribution over the unit interval [0, 1] was generated. The received bit was set to be the same as or different from the transmitted bit according to whether this random variate was greater than p, or it was less than or equal to p.

Let $s(x,y)$ denote the transmitted picture and $v(x,y)$ the received one. The noise is defined as

$$n(x,y) = v(x,y) - s(x,y) \tag{3.2}$$

It was shown by Young and Mott-Smith (1965) that, if the probability distribution of the brightness levels is flat or symmetrical with respect to the mean, the noise power is

$$\bar{n}^2 = \left(\frac{4^B - 1}{3} \right) p \tag{3.3}$$

where B is the number of bits per code word and p is the probability of error.

chapter 4

Bandwidth Reduction

General

It has been obvious for some time that the standard system of television is a very wasteful method of transmitting visual information. The picture is very redundant in nature. By removing redundancy bandwidth reductions of large ratios can be achieved. There are possible redundancy savings in each line, between lines and between frames. Many schemes have been successfully implemented to take advantage of these redundancies.

It is interesting to note that many TV patterns such as trees, grass, choppy seas, etc., have noise-like characteristics that tax ordinary transmission to the limit, and yet these are the redundant portions of the picture to the human observer who cares only for detail in such picture parts as faces, action scenes, and edge contours. As a rough estimate, something like 80 percent of the TV transmissions, the area of interest could be restricted to perhaps 10 foveae, that is, about 5 percent of the total picture.

There are fourteen general approaches that may be followed in order to reduce the total number of bits required to faithfully represent a frame of TV information:

1 Bandwidth reduction by picture degradation, including reduction of the number of lines, gray scale, and frame rate.
2 Statistical reduction by removing redundancies in the picture.
3 Direct digital processing techniques which permit the gray scale to be adequately represented by a smaller number of bits, without producing false contouring.
4 Psychophysical reduction that takes into account the human eye.
5 Systems that mask contouring due to quantization by the use of dither.
6 DPCM systems.
7 Delta modulation systems.
8 Synthetic highs and lows.
9 Bit-plane systems.

10 Intraframe encoding.

11 Conditional replenishment systems.

12 Significant bit selection.

13 Transform coding methods.

14 Channel-sharing systems.

4.1 REDUCTION BY PICTURE DEGRADATION

Bandwidth reduction can be obtained by reducing the spatial, temporal, and gray-scale resolution of the pictures. No matter how much picture degradation might be tolerable in the entertainment field, when the object of the transmission is reconnaissance or surveillance information, such degradation cannot be tolerated.

4.2 STATISTICAL REDUCTION

Each picture contains a considerable amount of redundant information because picture elements that are close in space or in time are more likely to be equal to each other than to be different. Successive picture elements, lines, and frames are all highly correlated, and the corresponding signal is very redundant.

Bandwidth reduction can be implemented by using PCM codes in such a manner as to remove these redundancies. One such concept takes advantage of the fact that short codes may be used for common events (no change in picture brightness level) and long codes for uncommon events such as the appearance of spatial or time differences in picture elements. Thus the average code length may be shortened.

Another way of achieving these results is to scan low gradient areas slowly and the high gradient areas swiftly. In this way the bandwidth may be set equal to the average information content of the picture, rather than the peak information content that occurs only in a relatively small portion of the picture.

However, these methods require a buffer storage, so that information generated at a varying rate in the picture is fed to the channel at a uniform rate. Redundancy removal also requires storage on a line-to-line basis or even on a frame-to-frame basis. Frame storage would require an even higher storage capacity capability.

4.3 REDUCTION BY DIRECT DIGITAL PROCESSING

Bandwidth reduction can be obtained by the use of direct digital processing techniques. The number of bits per sample needed to adequately represent a picture is set by the problems of false contouring rather than by proper gray-scale rendition. When coding is used to minimize false contouring, a small number of bits can adequately transmit picture information.

Delta modulation is a good example of this technique. This system, which has been investigated by Bell Laboratories, Colorado Research, and others, transmits a coarsely quantized (3 bits) version of the brightness difference between the picture and its reconstructed version. Delta modulation transmits only the changes in brightness level from element to element. Since most picture elements occur in high redundancy, a small number of bits can be used to represent this difference.

If the quantization pulses are tapered rather than uniform and the smallest step is made equivalent to 6 bit PCM, the problem of false contours can be avoided.

4.4   REDUCTION BY PSYCHOPHYSICAL EFFECTS

The human visual mechanism is a remarkable device. It is very sensitive to fine differences to such a degree that its performance is limited by the quantum of light arriving on the retina. Its spatial, temporal, and brightness resolution under correct conditions far exceeds that of the best television system ever devised. However, this visual acuity is a result of the ability of the visual mechanism to concentrate on the task at hand, or to a point position in space to the exclusion of all other points. That portion of the eye that has the highest spatial resolution is very insensitive to flicker. Large area changes in brightness are much more easily observed on the periphery of the retina. Very small differences in successive frames are easily detected, but in the presence of a small temporary changing stimulus, the spatial resolution of the eye decreases to a remarkable extent, as has been shown in repeated experiments. Slow gradients, either in space or time are largely ignored by the eye. The same effect is present for spatial resolution. For example, in the neighborhood of a sharp contour in the original scene, the eye is very insensitive to the exact contrast of the contour, while it is extremely sensitive to its location or to any difference in its location on successive scanning lines of the picture. On the other hand, this great sensitivity to gradients makes it imperative to avoid spurious effects such as false contouring as was mentioned previously.

4.5   BANDWIDTH REDUCTION BY DITHER MEANS

4.5.1   The Roberts' Method

Roberts (1962) proposed the masking of the input video signal by PRN, encoding at 3 bits/pixel, correlating it out at the receiver, to obtain good picture quality without contouring problems. Because the human eye is critical of contouring effects, but less critical of noise effects, the picture quality becomes quite usable.

Since 1962 others have proposed variations of the method presented by Roberts, and these are described later. However, the method described by Roberts of MIT is a very important one, and worth discussing in some detail.

In this method, noise produced by a digital PRN generator is added to the video signal before quantizing to 3 bits/pixel at the transmitter. In the receiver, this noise is removed after demodulation by a second pseudonoise (PN) generator synchronized to that at the transmitter.

The PN effectively randomizes the average video signal level and causes quantization at various levels rather than a single level. As a result the quantizing noise of the 3 bit A/D converter is removed and 3 bits can adequately represent the tonal scale. The Roberts' modulation technique is simpler than that of delta modulation and is far less sensitive to transmission errors.

This effect is the result of randomizing the video signal level before quantization, so the video signal becomes quantized at several levels instead of one. So, a slowly varying signal does not experience sharp quantum steps in the A/D conversion.

Instead of the sharp quantization steps ordinarily attributed to the use of a small number of bits, the transitions are smoothed out by the added PRN signal.

The implementation of a PRN generator can be very simple. It could consist of a shift register with the output of the final stage and that of an intermediate stage connected back into the shift register input through a modulo-2 adder. This circuit will pass through $2^n - 1$ possible states in a pseudorandom sequence when driven by a train of pulses. The quantity n is the number of stages used in the shift register.

The circuit will cycle through all possible states, except for one "forbidden" state, in a pseudorandom but predictable manner. The characteristic of this noise is near Gaussian in nature. Synchronization of two PRN generators can be accomplished by recognizing a particular state in the transmitter's PRN generator. When the output of all stages in the receiver PN generator, for example, are ones, they can be used in an "AND" gate to create a reset pulse to synchronize the receiver PN generator shift register. Synchronization is maintained by deriving the receiver's PRN driving signal from the received signal data rate, and utilizing the sync pulse when it arrives every $2^n - 1$ pulses.

Roberts' modulation has one great advantage over delta modulation. It is very insensitive to signal dropouts in transmission. Because of the differential nature of delta modulation, a signal dropout can cause resynchronization at a false level, which condition can persist for an extended length of time. In Roberts' modulation a signal dropout affects only that point in the picture where the error occurred.

Roberts' modulation is not bothered by sharp black-to-white transitions in the picture, while delta modulation is. Furthermore, false contouring is far less of a problem in Roberts' modulation than in delta modulation, even when 3 bits/sample are used.

Roberts' Modulation Errors

A generalized mathematical approach showing the effects of varying parameters on picture quality should prove useful. The mean-square error (MSE) is the usual point of reference. It will clearly indicate how the picture differs from what we want it to look like. It can be shown that the mean-square error for a digital memoryless channel has a minimum value dependent on channel capacity. For this reason, MSE will be divided into two components, which are suitable for separating channel processes and effectively predict the quality of the resultant pictures.

MSE

The output should be as similar to the input as possible. Correction for nonlinearity is introduced outside the channel. Therefore, the channel should have a linear transfer characteristic. Thus the MSE can be defined as

$$E = B \int_0^1 \int_0^1 P(y,x)(x-y)^2 \, dy \, dx. \tag{4.1}$$

A flat distribution for the input can be assumed, making $p(x)$ a constant. Thus we can write

$$E = A \int_0^1 \int_0^1 p(y|x)(x-y)^2 \, dy \, dx \tag{4.2}$$

where A is a normalization constant to be defined later. The signal during channel processing will be coded into 2^n samples, so there exists a lower limit on the MSE, which can be derived as follows. The input can be thought of as a large set of values X ranging from 1 to 0. At a time just before an input is presented, the channel must form a set of input points x_k. These points will be chosen to produce each code word K for all 2^n codes. Each code word will have a probability

$$P_k = \Sigma_{x_k} p(x). \tag{4.3}$$

When a code word is received, the most that can be known about the input is the set x_k. The best choices of an output y_k is one that will minimize the MSE.

$$E_k = \Sigma_{x_k} (x-y_k)^2. \tag{4.4}$$

The minimum error will be obtained when all the X in X_k are adjacent in a region x_k to $s_k + P_k$. The total range is P_k since all the inputs are equiprobable. The optimum value for y_k is at the center of the input range:

$$y_k = x_k + \tfrac{1}{2} P_k. \tag{4.5}$$

Thus going back to a continuous input, the error per code is

$$E_k \geqslant \int_{y_k}^{x_k + P_k} (x - y_k)^2 \, dx = \frac{P_k^3}{12}. \tag{4.6}$$

The error caused by each code is now known. The total error is the sum over these, corresponding to the integral over y in defining E:

$$E \geqslant A \Sigma_k P_k^3 / 12. \tag{4.7}$$

The probabilities of P_k must all sum to one, since the minimum mean-square error (MMSE) occurs when all code words are equiprobable. There are 2^n codes, so the probability is 2^{-n}.

Therefore, the MSE is

$$E \geqslant \sum_{k=1}^{2n} 2^{-3n} / 12 = \frac{A 2^{-2n}}{12} \tag{4.8}$$

and we will define A so as to make the MMSE 1:

$$A = 12(2^{2n}). \tag{4.9}$$

Variance

One of the important measures of picture quality is the amount of noise apparent that can be measured by looking at the variance in intensity in a portion of the picture that was of constant intensity at the input. The variance should be measured from the mean value of the output rather than from the input intensity. Thus a noise measure V is defined as the variance of the output averaged over all inputs:

$$V = A \int_0^1 \int_0^1 p(y - \bar{y}_x)^2 \, dy \, dx. \tag{4.10}$$

The normalizer A, as defined for the MSE, is used here to maintain the same units. The mean value of the output is defined as

$$\bar{y}_x = \int_0^1 p(y|x) y \, dy. \tag{4.11}$$

Deviation

The second measure of picture quality will be a measure of the deviation of the mean output $\bar{y}_x$ from the input. This quality is orthogonal to the variance V in form. It measures the tonal quality of the picture. We call the deviation D and define it so that

$$E = V + D. \tag{4.12}$$

In terms of the input and output, the deviation of the mean has the form

$$D = A \int_0^1 (x - \bar{y}_x)^2 \, dx. \tag{4.13}$$

Since there is very little noise in a straight PCM channel most of the error is caused by D.

There is always some minimum total error E so the main differences between channels are in the relative amounts of V and D.

The human observer is much more annoyed with tonal errors than with noise, because the eye averages out the noise. Thus in terms of the error functions developed, the best pictures result when D is reduced and V increased, without increasing E too much.

Channel Processes

A probability graph is useful for studying channel processes. A 2-bit channel is used for illustrative purposes. These graphs attempt to show $p(y|x)$ in the vertical dimension against the input x and the output y. A shaded area represents a constant probability over the area and a wall indicates that all or most of the probability is concentrated at a particular output.

The coding and decoding processes, as indicated by enclosing the signal to be quantized in brackets, is assumed to split the range between 0 and 1 into 2^n equal sections, assign a code to each section, transmit and receive the code, and then convert the code back to the average value of the section originating it. Any signal over 1 receives the code for 1, and any negative signal receives the code for 0. Thus the signal receives standard PCM transmission over a n-bit channel.

1. **Straight PCM.**

$$\text{Processing} \quad y = (x).$$

$$\text{Deviation} \quad D = 1.$$

$$\text{Variance} \quad V = 0.$$

$$\text{MSE} \quad E = 1. \tag{4.14}$$

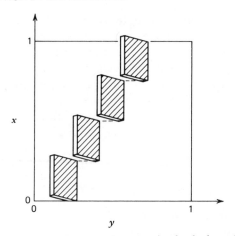

2. Receiver Noise. Perhaps the easiest method of changing the channel is to add random noise to the received signal, with an amplitude of plus or minus half a level. This only succeeds in making the channel worse.

Process $y = (x) + r. -2^{-n-1} \leqslant 1 < +2^{-n-1}$

Deviation $D = 1.$

Variance $V = 1.$ (4.15)

MSE $E = 2.$

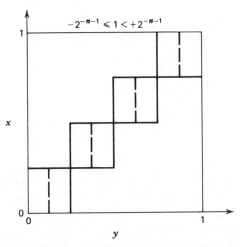

3. Transmitter Noise. The following processes create improved pictures by trading noise for tonal scale, or, in terms of error functions, reducing D while increasing V. As explained previously, noise tends to be more acceptable than loss of tonal scale. When random noise is added to the input before it is coded, a particular intensity level can result in different outputs at different times. In the graph shown in (4.16), the wedges pictured show the probability of one output, not a spread of outputs. Averaging in the y direction, the transfer curve becomes a diagonal except at the edges. For example, an input of $\frac{1}{2}$ would result in an output of $\frac{5}{8}$ half the time and $\frac{3}{8}$ the other half of the time, for a 2 bit channel. Thus the average intensity of an area in a picture would be reproduced correctly, but with more noise. Also, since no averaging has been done, the definition is not reduced. In the simple channel of this type, random noise is added with a peak-to-peak amplitude α between zero and one level, thus specifying a group of channels.

$$\text{Process} \quad Y=(x+r), \qquad -\alpha 2^{-n-1} \leqslant r < \alpha 2^{-n-1}.$$

$$\text{Deviation} \quad D=2^{-n}+(1-2^{-n})(1-\alpha)^2.$$

$$\text{Variance} \quad V=2\alpha(1-2^{-n}). \tag{4.16}$$

$$\text{MSE} \quad E=1+(1-2^{-n})\alpha^2.$$

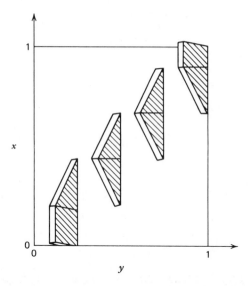

Without the addition of a slightly more complicated device at the receiver, to be described later, the best picture channel from experimental evidence seems to be one with about one level of random noise added to the signal before coding. The effect of doing this is to make the channel act more like an analog channel where the SNR determines the effective number of levels.

4. PRN. If one level of noise is added to a signal before coding, the tonal error can be reduced to zero, but at the cost of increasing the total MSE to twice its minimum value.

Minimum total error can be achieved by producing an output equal to the mean of the input set causing the code word at every time interval.

When noise is added at the transmitter, the input set is not known precisely from the code sent, unless the noise is also known. PRN is predictable and the receiver can generate the same noise used for coding and thus become coherent.

PN generators are easy to implement and synchronize. When noise is added to the input before coding and subtracted by coherent detection at the receiver, we have an output that is exactly the mean value of the inputs, which would have caused the code word.

As in the case of transmitter noise only, the amount of noise used will be varied from none to one level by the parameter D. Edge effects limit the reduction of D for low channel capacities, but when using a large number of bits per sample, the total error E is kept at its minimum value for all α. Thus with one level of noise all the tonal error can be eliminated in favor of an equal amount of noise error. The channel will now look as if it were completely analog. The pseudorandom channel has the following properties:

$$\text{Process} \quad Y=(x+r)-r, \qquad -\alpha2^{-n-1}\leqslant r<\alpha2^{-n-1}. \qquad (4.17)$$

$$\text{Deviation} \quad D=2^{-n}+(1-2^{-n})(1-\alpha)^{2}. \qquad (4.18)$$

$$\text{Variance} \quad V=2\alpha(1-2^{-n})-\alpha^{2}(1-2^{-n+1}). \qquad (4.19)$$

$$\text{MSE} \quad E=1+2^{-n}\alpha^{2}. \qquad (4.20)$$

Below are probability graphs for noise levels of $\alpha=\frac{1}{2}$ and $\alpha=1$.

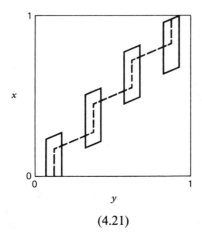

(4.21)

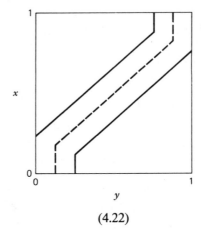

(4.22)

PCM/FM System with Roberts' Modulation

The theory behind the use of Roberts' modulation was just previously discussed. The video signal from the sensor is added linearly to the PRN, and then quantized in a 3-bit PCM encoder. The encoder output is used to FM modulate a transmitter. The receiver output is FM demodulated, and then the resulting PCM signal is decoded to its analog version. The decoder output contains the PN components, which are removed by a correlation detection process in which the synchronized PN generated in the receiver is used. The resultant analog video is then used in the display unit.

4.5.2 Dither Method by Thompson

Thompson (1971) described a method for using dither that was similar to that described by Roberts (1962). The main difference between the two systems is that Thompson suggested using pre-emphasis and de-emphasis of the video highs rather than companding.

Thompson describes a 3-bit/pixel quantizing system that employs one-dimensional correlated dither, and an empirical optimization of a complementary pair of pre-emphasis and de-emphasis networks.

To minimize the subjective visibility of the quantizing noise, the uniform amplitude distribution used by Roberts is optimum. But Thompson found that dither samples can be usefully autocorrelated so that the quantizing noise is concentrated in the higher frequencies of the video signal. However, the disadvantage is that the negative correlation achieved between quantizing noise samples is dependent upon the instantaneous value of the input signal, that is, the spectral distribution of the noise varies over the picture. However, this correlation between noise and picture was found to be barely perceptible in such cases. The advantage lies in a significant SNR improvement.

Dither is generated digitally and its amplitude distribution is defined only at a limited number of discrete values, so that a picture optimized to k levels and using an n-level dither appears to be quantized to nk levels.

In a 3-bit system an 8-level dither effects 64-level quantization with an unweighted peak-to-peak signal-to-quantizing noise ratio of 27.7 dB. The peak amplitude of the 8-level dither was set at $7q/8$, where q is the quantal step size, in order to achieve a uniform noise amplitude distribution without overlap between adjacent quantal steps.

The system can be mathematically analyzed by considering only the 8-level dither, sliced by a threshold detector of the encoder at a set-up level given by an assumed constant input signal amplitude. Quantization will generate one of seven distinct binary levels that can be designated as $Q_A(i, i+1)$, corresponding to the slicing dither between levels i and $i+1$, for $i = 1, 2, \ldots, 7$. Alternately, since the dither amplitude is only $7q/8$ it may be completely contained between two thresholds without causing threshold crossings. The resultant encoder output would be $Q_A(8, 1)$.

The functions $Q_A(i, i+1)$ therefore represent the eight possible quantizing noises that would be obtained if the dither were not subtracted at the decoder output.

Fig. 3 of Thompson (1971) is shown here as Fig. 4.1 and shows the pseudorandom dither generator used. The 8-level dither D is generated by adding three arbitrary binary $(+1, -1)$ signals x, y, and z with respective

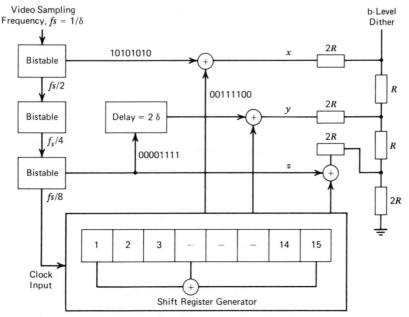

Figure 4.1. Optimized add-subtract pseudorandom dither generator.

weights $\frac{1}{2}$, $\frac{1}{4}$, and $\frac{1}{8}$. So that

$$D = \tfrac{1}{2} q \left(\tfrac{1}{2} x + \tfrac{1}{4} y + \tfrac{1}{8} z \right).$$ (4.23)

Each level is represented by

$$Q_A(1,2) = \frac{q(x+y+z-xz-xy-yz+xyz)}{8}$$ (4.24)

$$Q_A(2,3) = \frac{q(2x+2y-2xy)}{8}$$ (4.25)

$$Q_A(3,4) = \frac{q(3x+y+z+yz-xy-xz-xyz)}{8}$$ (4.26)

$$Q_A(4,5) = \frac{q(4x)}{8}$$ (4.27)

$$Q_A(5,6) = \frac{q(3x+y-2z-yz+xy+xz-xyz+z)}{8}$$ (4.28)

$$Q_A(6,7) = \frac{q(2y+2x+2xy)}{8}$$ (4.29)

$$Q_A(7,8) = \frac{q(z+x+y+yz+xz+xy+xyz)}{8}$$ (4.30)

$$Q_A(8,1) = \text{constant.}$$ (4.31)

Subtraction of the dither at the decoder yields

$$Q_{A-S}(i,i+1) = Q_A(i,i+1) - D \tag{4.32}$$

or

$$Q_{A-S}(1,2) = \frac{q\left(\frac{1}{2}z - x - yz - xy - xz + xyz\right)}{8} \tag{4.33}$$

$$Q_{A-S}(2,3) = \frac{q\left(y - \frac{1}{2}z - 2xy\right)}{8} \tag{4.34}$$

$$Q_{A-S}(3,4) = \frac{q\left(\frac{1}{2}z + x + yz - xy - xz - xyz\right)}{8} \tag{4.35}$$

$$Q_{A-S}(4,5) = \frac{q\left(2x - y - \frac{1}{2}z\right)}{8} \tag{4.36}$$

$$Q_{A-S}(5,6) = \frac{q\left(\frac{1}{2}z + x - yz + xy + xz - xyz\right)}{8} \tag{4.37}$$

$$Q_{A-S}(6,7) = \frac{q\left(y - \frac{1}{2}z + 2xy\right)}{8} \tag{4.38}$$

$$Q_{A-S}(7,8) = \frac{q\left(\frac{1}{2}z - x + yz + xy + xz + xyz\right)}{8} \tag{4.39}$$

$$Q_{A-S}(8,1) = \frac{q\left(-2x - y - \frac{1}{2}z\right)}{8} \tag{4.40}$$

Assuming the signal to have a uniform amplitude distribution, the eight types of noise are taken to occur in equal quantities over the entire picture. The visibility of each type can be evaluated by weighted integration of the Fourier transform of the autocorrelation function derived from the respective equation. The power spectral weighting function recommended for use is

$$W(\omega) = \frac{1}{1 + \omega^2 T^2} \tag{4.41}$$

where, according to CCIR recommendation, $T = 200$ nsec.

Proper Sequence Length

The proper sequence length of the pseudorandom sequence chosen by Thompson was chosen after considering relationships between the sampling and the line and field scanning frequencies. The sampling frequency is equal to $f_s = 1/\delta$, where δ is equal to the duration of the sampling pulse. A register of 16 stages or more produces PRN that is free of drifts or patterns for sampling frequencies below 13 MHz.

Thompson used a sample frequency of 12 MHz and 5.5 MHz system video bandwidth. It does not follow that the dither generated by the method shown in Fig. 4.1 is optimum for all sampling rates.

Pre- and De-Emphasis

Regarding quantizing noise as additive and random, the predominance of low-frequency energy in the television signal power spectrum has been exploited by complementary pre- and de-emphasis of the video high frequencies to increase the SNR. Unless the signal is noisy, however, the quantizing noise is highly correlated with the picture and pre-emphasis reduces contouring visibility by about 3 dB by smearing fast edges. Using a dither signal, and particularly one that confines randomized quantizing noise to high frequencies, emphasis is capable of producing a more significant improvement. However, there is some compromise between the amount of smearing and the attenuation of quantizing noise.

For a 12-MHz sampling rate subjective tests were made with various pre- and de-emphasis network pairs to determine optimization. It was found that a pair giving 6 dB of pre-emphasis gave the best subjective results. These networks were shown as Fig. 6 in Thompson, and here as Fig. 4.2.

This amount of pre- and de-emphasis gave about 3.5 dB improvement in SNR.

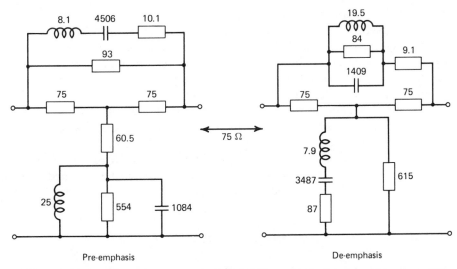

Pre-emphasis   De-emphasis

Figure 4.2. Component values of the preferred 6-dB emphasis networks (R in ohms, C in picofarads, L in microhenries).

It was found that the dithered system is considerably more sensitive to camera noise then DPCM system. However, it remains a simple and effective method of obtaining good quality pictures at 3 bits/pixel and is relatively insensitive to channel-noise-caused errors.

4.5.3 Nonrandom Dither

The types of dither discussed previously have been pseudorandom. Some authors have investigated the use of nonrandom dithers, constructed according to a fixed pattern. Some ordered-dither patterns have been constructed with a small square dither matrix repeated over the picture area to form a rectangular array of matrices.

It is claimed (Lippel and Kurland, 1971) that ordered dither gives improved performance when compared to pseudorandom dither. Not only is the picture quality better for a given quantizer, but it is not necessary to subtract the dither at the receiver, since the difference due to subtraction can be made almost imperceptible.

Ordered dither does more than just mask quantizing contouring effects. It is capable of restoring some of the information that the coarse quantizer without dither would remove. Spatial details are reinserted, as well as intermediate shades of gray.

In this system, the input picture is broken down into two components called background and the residue picture, and the output picture (without dither subtraction) is considered to consist of the same background plus an overlay. An overlay pattern or matrix is formed and added to the background to form the total picture.

In general it is found that the overlay tends to be a variable bandwidth approximation in which the more contrasty details of the picture are reproduced with higher spatial resolution than are the lower contrast details.

The background picture depends only on the quantizer. All effects that depend on the nature of the dither pattern manifest themselves only in the transformation from residue picture to overlay picture. The coarser the quantizer, the more information the residue contains.

The use of a uniform quantizer permits a fixed relation between dither amplitudes and step size. However, the mean quantizing error is set to be zero, in order to preserve picture luminance.

Lippel and Kurland have created optimum dither matrices, which give better picture quality for a 3-bit/pixel quantizer, than can be obtained with pseudorandom dither of the add-subtract type.

4.6 DIFFERENTIAL PULSE CODE MODULATION (DPCM)

Differential pulse code modulation (DPCM) and predictive quantizing are alternately used terms for a technique used to encode analog signals into digital pulses suitable for transmission over binary channels. O'Neal (1965) has described the basic system, with particular application to television signals. (Permission to use this material has been granted by American Telephone & Telegraph Co.)

A predictive communications system is one in which the difference between the actual signal and an estimate of the signal, based on its past, is transmitted. Both the transmitter and the receiver make an estimate or prediction of the signal's value based on the previously transmitted signal. The transmitter subtracts this prediction from the true value of the signal and transmits this difference. The receiver adds this prediction to the received difference signal yielding the true signal. Highly redundant systems, such as television, are well suited for predictive transmission, because of the accuracy possible in the predictions. When the signal is sampled, and if the difference signal is quantized and encoded into PCM, then we obtain a DPCM system.

DPCM systems are based primarily on an invention by Cutler (1952). He obtained a patent that used one or more integrators to perform the prediction function. His invention is based on transmitting the quantized difference between successive sample values rather than the sample values themselves. His invention is a special case of a predictive quantizing system.

Several studies have been made into the theory behind predictive quantizing. Weiner (1949) developed the theory of optimum linear prediction. By 1952, Oliver, Kretzmer and Harrison of the Bell Laboratories realized the importance of linear prediction in feedback communications systems and proposed to reduce redundancy by its use. Thus, the required power could be reduced in such highly periodic signals as television. Oliver (1952) explained how linear prediction could be used to reduce the bandwidth required to transmit redundant signals.

Kretzmer (1952) analyzed the statistical properties of television signals. Harrison (1952) implemented a linear prediction television system, wherein he reduced redundancy. Elias (1955) theoretically tied linear prediction to PCM to obtain DPCM.

Graham (1958) theorized that linear prediction theory could aptly describe Cutler's feedback system.

O'Neal (1966) studied nonadaptive systems using linear prediction in the feedback loop and a quantizer whose characteristics were independent of the instantaneous value of the input signal. In this study he considered a TV video sample (pixel) and prediction based on the prior pixel intensity value. He also studied the effect of prediction based on the previous-line feedback.

He used the performance criteria of SNR, where the noise was that due to quantization only. He did not include in his studies the effect of channel noise. Since noise in the sync pulses is seldom a limiting factor it was not considered. SNR is not the only criterion that could have been used, but it is a useful one.

In his studies the mean-square difference between the decoded output signal and the analog input signal was considered. All sampling was assumed to be at the Nyquist rate, and the resulting quantizing noise was considered to be in band.

In most DPCM systems the quantizing noise is highly correlated with the derivative of the signal.

O'Neal did not discuss nonlinear prediction; but only linear prediction. Permission to use this material was granted by American Telephone & Telegraph Co.

4.6.1 Linear Prediction Theory (O'Neal)

Let a stationary signal $S(t)$ with zero-mean and rms value σ be sampled at the times $t_1, t_2, \ldots, t_n$ and let the sample values be $S_1, S_2, \ldots, S_n$, respectively.

A linear estimate of the next sample value S_0 based on the previous n sample values $S_1, S_2, \ldots, S_n$ is defined to be

$$\hat{S}_0 = a_1 S_1 + a_2 S_2 + \cdots + a_n S_n. \tag{4.42}$$

For simplicity we can assume that a and S values are real numbers. A linear predictive encoder forms this estimate $\hat{S}_0$ and transmits the difference or error

$$e_0 = S_0 - \hat{S}_0.$$

Fig. 4.3 shows a typical block diagram of such a system. In this figure the D's represent delay elements.

The best estimate of S_0 is that value of $\hat{S}_0$ for which the expected value of the squared error is minimum. To find the values for the a's that satisfy these

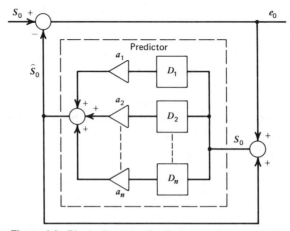

Figure 4.3. Block diagram of a linear predictive encoder.

conditions we take the partial derivatives of $E[(S_0 - \hat{S}_0)^2]$ with respect to each of the a's, where $E[\chi]$ denotes the expected value of χ.

$$\frac{E[(S_0 - \hat{S}_0)^2]}{\partial a_i} = \frac{\partial E\{[S_0 - (a_1 S_1 + a_2 S_2, + \cdots + a_n S_n)]^2\}}{\partial a_i}$$

$$= 2E\{[S_0 - (a_1 S_1 + a_2 S_2, + \cdots + a_n S_n)]S_i\}$$

$$\text{for } i = 1, 2, \ldots, n. \quad (4.44)$$

We set this equal to zero to obtain the minimum:

$$E\{[S_0 - (a_1 S_1 + a_2 S_2, + \cdots + a_n S_n)]S_i\} = 0$$

$$E[(S_0 - \hat{S}_0)S_i] = 0. \quad (4.45)$$

The covariance of S_i and S_j is given by

$$R_{ij} = E(S_i S_j) \quad (4.46)$$

and

$$R_{0i} = a_1 R_1 + a_2 R_2 + \cdots + a_n R_n \qquad \text{for } i = 1, 2, \ldots, n, \quad (4.47)$$

which defines a set of n simultaneous linear equations with the variable being a_i, $i = 1, 2, \ldots, n$.

The covariances are found from the autocovariances of the signal itself $\psi(\tau)$.

$$R_{ij} = \psi(t_i - t_j). \quad (4.48)$$

The variance of the differential signal σ_ε^2 is given by

$$\sigma_\varepsilon^2 = E\{[(S_0 - \hat{S}_0)]^2\} = E\{[(S_0 - \hat{S}_0)]S_0\} = R_{00}$$

$$- (a_1 R_{01} + a_2 R_{02}, + \cdots + a_n R_{0n}) \quad (4.49)$$

where R_{00} is the variance σ^2 of the original signal sequence $S_1, S_2, \ldots, S_n = S_i$. Since the television video signal is often classified with a Laplacian distribution, and sometimes as a Gaussian distribution, the value of σ^2 is taken to be unity.

The sequence of transmitted error samples is $e_0, e_1, \ldots = [e_i]$, where

$$e_i = S_i - \hat{S}_i \qquad \text{for } i = 0, 1, \ldots, \quad (4.50)$$

and

$$S_i = a_i S_{i+1} + a_2 S_{i+2} + \cdots + a_n S_{i+n}. \tag{4.51}$$

If the sequence of samples $S_0, S_1, \ldots = [S_i]$ is an rth order Markov sequence, then only r samples need be used in forming the best estimate of S_0 and the resulting sequence of error samples will be uncorrelated.

As an example of particular relevance to television, consider the first-order Markov sequence formed by sampling a signal whose autocorrelation is the exponential function $e^{-\alpha t}$. In this case, even if all the previous sample values are available, the estimate of S_0 that minimizes is

$$\hat{S}_0 = \frac{R_{01}}{\sigma^2} S_1, \tag{4.52}$$

where S_1 is the most recent sample value available. In this case the error sequence $[e_i]$ is completely uncorrelated:

$$E[e_i e_j] = 0 \quad (i \neq j) = \sigma_\varepsilon^2 (i = j). \tag{4.53}$$

The autocorrelation of one line of a TV signal is very similar to $e^{-\alpha t}$, so it can be argued that basing our estimate only on the previous sample value will be almost as good as using many sample values on the same line.

However, it has been proven that, if we base prediction on samples in an adjacent line and/or on the previous frame, the prediction will be improved. Experimentation has proved that line feedback in addition to prior sample feedback gives about 1.9 dB of improvement in SNR. In FCC standard monochrome entertainment TV the improvement due to line feedback would be considerably less than 1.9 dB.

DPCM provides more of an advantage for high resolution TV systems than for low resolution systems. For monochrome entertainment TV, prior-sample feedback DPCM can provide a signal-to-quantizing-noise-ratio approximately 15 dB higher than standard PCM. This improvement varies as much as 2 or 3 dB dependent on the pictorial scene complexity. A 2.8 dB improvement in SNR can be realized in standard PCM systems, if the sync pulses can be reconstructed by the decoder and are not transmitted. The net improvement would then be about 12 dB. These numbers include the effect of line feedback.

Since 6 dB quantizing noise is equivalent to 1 bit/sample (pixel), the advantage of DPCM over PCM can be expressed in bits/pixel. For a 12-dB improvement factor one could expect a reduction of 2 bits/pixel in DPCM over PCM or the equivalent of 1.5 times in bandwidth reduction. In other words, if DPCM were used rather than the standard PCM the bandwidth transmitted could be reduced by a factor 1.5 to 1.0.

It can be shown that, when only the prior sample is used for prediction

$$\sigma_\varepsilon^2 = \sigma^2 - \frac{R_{01}^2}{\sigma^2} \qquad (4.54)$$

and since $\sigma^2 = 1$

$$\sigma_e^2 = 1 - R_{01}^2. \qquad (4.55)$$

Typical values for R_{01} have been found to be 0.958 and 0.986.

When the two most recent previous samples are used in prediction

$$\sigma_e^2 = \left[1 - \frac{2R_{01}^2}{1 + R_{01}^{\sqrt{2}}} \right], \qquad (4.56)$$

which shows some advantage to using two rather than one prior samples.

For quantizing levels of 8 and greater, the approximate value for the variance of quantizing noise can be taken as

$$\sigma_q^2 = \frac{9}{2N^2} \sigma_e^2. \qquad (4.57)$$

4.6.2   The Effect of Entropy Encoding

As was mentioned previously DPCM can provide an advantage of about 2 bits/pixel over a standard PCM encoder. In addition, another bit can be saved by using entropy encoding at the DPCM output (Chow, 1971). By making certain assumptions about the statistics of television signals, it can be shown that DPCM plus entropy coding can attain a signal to quantizing noise ratio only 1.5 dB less than that predicted by the rate distortion function that bounds the performance of all source encoding systems (O'Neal, 1971).

An entropy-adding circuit was proposed by Agrawal and O'Neal (1973). They used an encoder that was a modification of one by Kaul (1970) and contained a uniform quantizer.

Subjective tests were performed on the DPCM coder when used with 12 different combinations of sampling frequency and a number of quantizing levels. The coder was operated at 9, 12, 18, and 24 Mbits/sec. Rating scale techniques were used to determine the optimum coder parameters at each of 4-bit rates.

Entropy measurements of the bit stream of each of the 12 different encoder configurations were made. From these entropy measurements, information rates were calculated and compared with subjective quality ratings. Surprisingly, at any bit rate the coder with the largest information rate had the lowest subjective quality rating.

A block diagram of the DPCM system used in these experiments is shown in Fig. 4.4. This system differs from that of Kaul by containing a digital multiplexer and the addition of a constant in the feedback loop. Both of these refinements improve the operation of the DPCM system and cause it to operate like a conventional analog system.

The analog TV signal is first sampled and converted into 8-bit PCM using a standard PCM encoder with equally spaced quantizing levels. Each PCM word is fed into logic circuits that perform the DPCM operation on the digital input signal.

The error signal is e_i and is equal to

$$e_i = s_i - \hat{s}_i. \tag{4.58}$$

The estimate $\hat{s}_i$ is, on the average, so close to s_i that the two most significant bits of e_i are almost equal to zero, and need not be transmitted. This means that the performance of an n-bit DPCM system is equivalent to that of an $n+2$-bit PCM system. In order to have the coder function as a n-bit DPCM system the $8-n$ most significant bits from the error signal e_i were dropped and not transmitted.

This procedure, which is equivalent to quantization in an analog DPCM system, introduces the slope overload noise characteristic of DPCM encoder.

The digital multiplier α in the feedback loop was chosen to be $0.96875 = 31/32$ because it could be easily implemented and caused very little delay in the feedback loop. Since $31/32$ can be written as $1 - 1/32$, multiplication of any binary number by $31/32$ can be accomplished by shifting the binary number 5 places to the right and subtracting the shifted number from the original number. Using a multiplier less than one improves the performance of the coder and prevents accumulation of digital errors introduced in the transmission path.

The constant $C(n)$ added to each sample entering the feedback loop serves to ensure that the quantizing noise has zero mean.

The DPCM encoder is a discrete source generating $N = 2^n$ different levels such that $P(i)$ is the probability of occurrence of the ith level and $P(i,j)$ the joint probability of occurrence of level i followed by level j. The first-order, second-order, and conditional entropies of the source are defined as

$$H(i) = \sum_{i=1}^{N} P(i)\log_2 P(i), \tag{4.59}$$

$$H(i,j) = \sum_{i=1}^{N} \sum_{j=1}^{N} P(i,j)\log_2 P(i,j), \tag{4.60}$$

$$H(j|i) = H(i,j) - H(i), \tag{4.61}$$

respectively. If the symbols are independent $H(j|i) = H(j)$ and the levels i

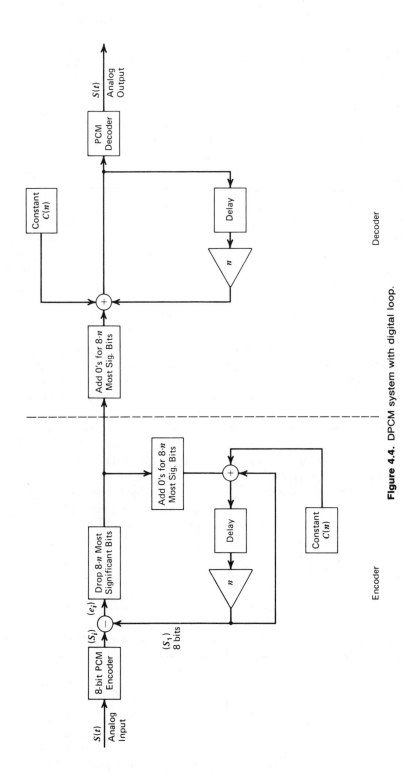

Figure 4.4. DPCM system with digital loop.

and j are uncorrelated. If the symbols are not independent, then $H(j|i) < H(i)$ and $H(i) - H(j|i)$ is a rough indication of the amount of redundancy that could be removed by using prediction based on the two previous samples instead of one.

The results of testing indicate that about 1 bit/pixel can be saved by using entropy encoding with uniform quantizers. When the number of bits n was held constant the entropy decreased as the sampling frequency f increased. This is true because the signal becomes more predictable as f increases, and the variance of the quantizer input signal decreases, causing a concommitant decrease in entropy. If f is held constant the entropy will increase by about 1 bit as n is increased by 1 bit. At a constant bit rate, if n is increased, f must decrease and both of these changes cause the entropy to increase.

However, when complicated picture scenes must be transmitted at low bit rates entropy coding is not likely to save 1 bit/pixel. In these cases, DPCM systems with tapered quantizers are probably preferable to systems with uniform quantizers and entropy coding.

4.6.3 DPCM and Channel Noise

Systems engineers deal with the total noise in the system. Prior to this only the effect of quantization noise has been considered. Actually total noise consists of quantization noise, sampling noise, and channel noise. MSE, used in the preceding text, was due to quantization noise produced errors. Now the total MSE is considered.

This case was analyzed by Essman and Wintz (1973). They state that there are three subsystems in DPCM systems that require careful design consideration; the predictor, the quantizer, and the reconstruction filter. The basic system they analyzed is shown here in Fig. 4.5.

Franks (1966) showed that a reasonable model for video data that has been scanned sequentially is the first-order Markov process. This model was further verified by Habibi and Wintz (1971). They derived an exponential autocorrelation function and a best estimate in the MSE sense of $f(t)$, based on the previous sample.

$$\hat{f}(kt) = A_f f(k-1)T \tag{4.62}$$

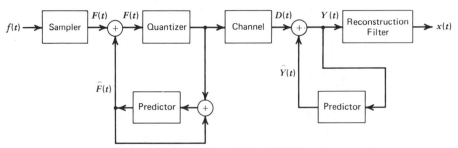

Figure 4.5. Block diagram fro a DPCM system.

where A_f is specified from linear prediction theory (Kuo, 1970) to be

$$A_f = \frac{\phi_{ff}(T)}{\phi_{ff}(0)} \tag{4.63}$$

where $\phi_{ff}(T)$ is the autocorrelation function of the process. The variance of the difference signal is given as

$$\sigma_e^2 = (1-\rho^2)\sigma_f^2 \tag{4.64}$$

where $\rho = A_f$. This agrees with the results of O'Neal as shown previously. O'Neal gave the value of $\sigma_f^2 = 1$.

Papoulis (1965) shows that the difference sequence $e(T), e(2T),\ldots$ is orthogonal so that

$$E\left[e(iT)e(jT) \right] = \begin{cases} 0 \, (i \neq j) \\ \sigma_e^2 \, (i=j) \end{cases}. \tag{4.65}$$

The MSE expected is (Essman and Wintz, 1973).

$$\hat{\varepsilon}(t) = \left(\overline{f(t) - x(t)} \right)^2. \tag{4.66}$$

The predicted version of the sampled signal is

$$\hat{F}(t) = A_f\left[Q(t-T) + \hat{F}(t-T) \right]. \tag{4.67}$$

At the receiver, the sampled output is

$$Y(t) = D(t) + A_f \hat{Y}(t-T). \tag{4.68}$$

The error between input and output is

$$\varepsilon(t) = f(t) - \sum_{k=-\infty}^{\infty} h(t-kT)\left[d(kT) + \hat{y}(kT) \right] \tag{4.69}$$

where $h(t)$ is the impulse response of the reconstruction filter.

The sampling error is incurred by sampling the signal and passing it through the reconstruction filter:

$$e_s(t) = f(t) - \sum_{k=-\infty}^{\infty} h(t-kT)f(kT). \tag{4.70}$$

The quantization error is

$$e_q(t) = \sum_{k=-\infty}^{\infty} h(t-kT)\left[e(kT) - q(kT) \right]. \tag{4.71}$$

The error due to channel noise is

$$e_c(t) = \sum_{k=-\infty}^{\infty} h(t-kT)\left\{ \sum_{n=0}^{\infty} A_f^n q[(k-n)T] - d[(k-n)T] \right\}. \quad (4.72)$$

The total error is

$$e(t) = e_s(t) + e_q(t) + e_c(t). \quad (4.73)$$

When uniform quantization is used and successive error samples are assumed to be uncorrelated and successive bits are assumed to be independent, it was shown (Essman and Wintz, 1973) that

$$\phi_{e_c e_c}(JT) = \sum_{n=0}^{\infty} A_f^{2n+|J|} 4V^2 P(E)(1-2^{-2B})/3 \quad (4.74)$$

where $\phi_{e_c e_c}(JT)$ is the autocorrelation function of channel error term, V is the quantization amplitude limit, B is the number of bits/per word in the DPCM code, and $P(E)$ is the probability of bit error.

For the sake of convenience define

$$\psi_{CH2} = \frac{4V^2(1-2^{-2B})P(E)}{3}, \quad (4.75)$$

which is the spectral density for standard PCM.

If nonuniform quantization is used for DPCM

$$\phi_{e_c e_c}(JT) = \frac{A_f^{|J|}}{1-A_f^2}\psi_{CH1}(B,V) \quad (4.76)$$

where

$$\psi_{CH1}(B,V) = \frac{63PE\sigma_e^2}{\sigma}. \quad (4.77)$$

Let us assume that the reconstruction filter takes the form of a zero-order-hold (ZOH) filter, and that the input signal is a sample function from a random process with a correlation function

$$\phi_{ff}(T) = \exp(-b|\tau|). \quad (4.78)$$

The MSE for DPCM is

$$\overline{e^2(t)} = 2\frac{1-[1-\exp(-b\tau)]}{b\tau} + \frac{9\sigma_e^2}{2^{2B+1}} + \frac{\psi_{CH1}(B,V)}{1-A_f^2}. \quad (4.79)$$

The MSE for PCM is

$$\overline{e^2(t)} = 2\frac{1-[1-\exp(-b\tau)]}{b\tau} + \frac{V^2}{(3)2^{2B}}$$

$$+ \psi_{CH2}(B,V) \tag{4.80}$$

for uniform quantization.

From p. 347 of Chang and Donaldson (1972) we find that when previous-sample plus previous-line prediction is used:

$$\frac{\sigma_e^2}{\sigma_x^2} = 1 + \alpha_1^2 + \alpha_N^2 + 2\alpha_1\alpha_N\rho_{N-1} - 2\alpha_1\rho_1 - Z\alpha_N\rho_N \tag{4.81}$$

where

$$\alpha_x^2 = 1 \tag{4.82}$$

$$\alpha_1 = \alpha_N = \rho\left(1+\rho^{\sqrt{2}}\right). \tag{4.83}$$

An optimum value for ρ is taken from Essman and Wintz (1973), $\rho = 0.83$.

$$\sigma_e^2 = 0.38, \tag{4.84}$$

which assumes that quantization and sampling errors are negligible.

Using this value for σ_e^2, and $A_f = \rho = 0.83$ we can find that

$$\overline{\varepsilon^2(t)} \text{ (DPCM)} = 0.0028 + 10.5PE. \tag{4.85}$$

When we let $V = 1$, we find that

$$\overline{\varepsilon^2(t)} \text{ (PCM)} = 0.006 + 1.3PE \tag{4.86}$$

A computer run was made on a Univac 1108 computer and the results are shown in Fig. 4.6. The abscissa is SNR versus MSEs. For low SNR PCM is better than DPCM, but for larger SNR DPCM is superior.

It is noted that the curve for PCM matches fairly close to the one from Viterbi shown in Fig. 2.2 for large SNR.

4.6.4 TV Visual Property and DPCM

The subjective quality of a TV picture can be used to remove redundancy in picture elements. A TV signal is highly correlated. Experiments have shown that the average power of prediction error signals $\bar{\varepsilon}_i^2$ is about one-tenth of the original signal power. In terms of average amplitude, it is about $\frac{1}{2}$ to $\frac{1}{4}$, and thereby a reduction of 1–2 bits/pixel is possible.

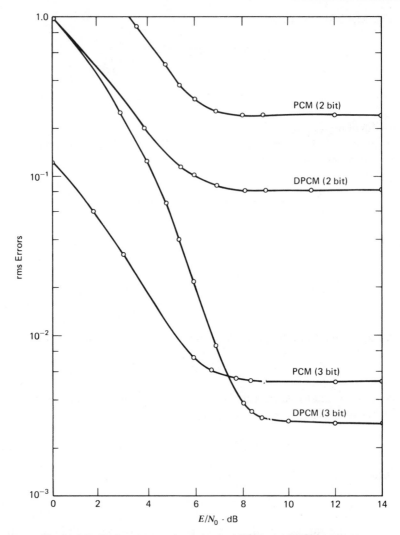

Figure 4.6. Performance comparison of PCM and DPCM systems.

There are several properties of human vision that can be used to obtain efficient coding. The property used in DPCM is called contrast sensitivity. When there is a small difference in brightness level between adjacent pixels, very small differences in brightness can be distinguished. On the other hand, when the differences in brightness level is large, the human eye sensitivity is reduced sharply. Since coarse quantizing is possible where there are large differences, it would be possible to obtain a reduction of 1–2 bits/pixels.

In DPCM the optimum prediction coefficient a is the ratio of the covariances of adjacent pixels.

Fukinuki (1974) has discussed the contrast sensitivity of vision. This is depicted in Fig. 4.7.

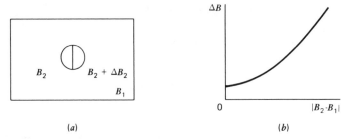

(a)   (b)

Figure 4.7. Contrast sensitivity of vision. (a) Test chart. (b) contrast sensitivity.

Let B_1 be the brightness of the background and B_2 and $B_2+\Delta B$ be those of the left and the right halves of the small circle, respectively. The difference in brightness ΔB, which is needed to distinguish the change from B_2 to $B_2+\Delta B$, varies in accord with the difference in brightness $|B_1 - B_2|$. Namely, when the difference is small, a slight amount of ΔB is sufficient to distinguish the change. When it becomes large, ΔB becomes large accordingly, as shown in Fig. 4.7(b).

However, this is over simplification. The size of the circle and the deviation of the images displayed make these contributions. Moreover, there are the effects of cathode ray tube brightness and room lighting. To keep the analysis simple Fukinuki (1974) assumed a threshold of ΔB.

In Fig. 4.8(a) brightness values as actually measured are shown (Schreiber, 1967). In cathode ray tubes, the brightness is approximately equal to the second power of the input signal. The graph shown in Fig. 4.8(b) is obtained by using this relationship ($B=C\chi^2, 1/\Delta\chi=2c^{1/2}/B^{1/2}\cdot B/\Delta B$, where c is a constant) and by plotting the signal χ on the horizontal axis with a linear scale and $1/\Delta\chi$ on the ordinate axis. As the figure shows brightness sensitivity is not simply a function of $|B_1 - B_2|$. This is shown in Fig. 4.8.

In order to effectively use this subjective information, it is necessary to optimize the quantization of the DPCM error signal $\varepsilon_i=(\chi_i-\chi_{i-1})$, taking

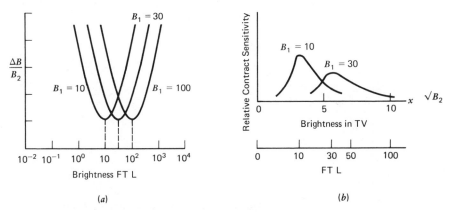

(a)   (b)

Figure 4.8. Experimental data of contrast sensitivity. (a) $\Delta B/B_2$. (b) $1/\Delta\chi$.

into consideration the contrast sensitivity $\delta_i = (\chi_i - \chi_{i-1})$ between adjacent pixels.

The average quantization noise power is

$$N_q = \int_{-\infty}^{\infty} \frac{\{\Delta/\xi\}^2}{12} p(\xi)\, d\xi \tag{4.87}$$

when the number of quantization levels is large.

In Fig. 4.9 is shown the compression characteristic of the companding process.

From this figure it can be seen that

$$y = F(\xi) \tag{4.88}$$

$$\frac{dy}{d\xi} = f(\xi). \tag{4.89}$$

The value $\Delta(\xi)$ can be approximated by

$$\Delta(\xi) \cong \frac{E_0}{f(\xi)} \tag{4.90}$$

since y is quantized linearly by the definite step E_0. From this we obtain

$$N_q = \frac{E_0^2}{12} \int_{-\infty}^{\infty} \frac{1}{[f(\xi)]^2} p(\xi)\, d\xi. \tag{4.91}$$

Fukinuki (1974) shows that

$$f(\xi) = \frac{2v}{\int_{-\infty}^{\infty} [p(\xi')]^{1/3}\, d\xi'} [p(\xi)]^{1/3} \tag{4.92}$$

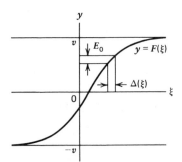

Figure 4.9. Instantaneous compression curve.

and

$$N_q = \frac{E_0^2}{48\,V^2} \left\{ \int_{-\infty}^{\infty} \left[p(\xi) \right]^{1/3} d\xi \right\}^3 \tag{4.93}$$

for linear compression.

He changed his variables as follows:

$$\nu(\delta) = \delta_i = \chi_i - \chi_{i-1} \tag{4.94}$$

$$\xi_i = \varepsilon \tag{4.95}$$

$$\delta_i = \delta. \tag{4.96}$$

When the error signal ε is compressed nonlinearly and quantized, the power of the quantization noise within a step is expressed as

$$\frac{1}{12} \left[\frac{E_0}{f(\varepsilon)} \right]^2. \tag{4.97}$$

When the quantization noise power is evaluated visually this can be expressed as

$$\frac{1}{12} \left[\frac{E_{0\nu(\delta)}}{f(\varepsilon)} \right]^2. \tag{4.98}$$

Therefore, the power of the visual quantization noise is given by

$$N_{qv} = \frac{E_0^2}{12} \int_{-\infty}^{\infty} \int_{-\infty}^{\infty} \frac{[\nu(\delta)]^2}{[f(\varepsilon)]^2} p(\varepsilon, \delta)\, d\delta\, d\varepsilon \tag{4.99}$$

where $p(\varepsilon, \delta)$ is the joint probability density function of ε and δ, or

$$N_{qv} = \frac{E_0^2}{12} \int_{-\infty}^{\infty} \frac{1}{[f(\varepsilon)]^2} P_v(\varepsilon)\, d\varepsilon \tag{4.100}$$

where

$$P_v(\varepsilon) = \int_{-\infty}^{\infty} \left[v(\delta) \right]^2 p(\varepsilon, \delta)\, d\delta. \tag{4.101}$$

Therefore, when nonlinear compression is used

$$f_V(\varepsilon) = \frac{2V}{\int_{-\infty}^{\infty} \left[P_v(\varepsilon') \right]^{1/3} d\varepsilon'} \left[P_v(\varepsilon) \right]^{1/3} \tag{4.102}$$

$$N_{qv} = \frac{E_0^2}{48\,V^2} \left\{ \int_{-\infty}^{\infty} \left[(P_v(\varepsilon) \right]^{1/3} d\varepsilon \right\}^3. \tag{4.103}$$

These results were obtained by optimizing both the prediction and the instantaneous companding simultaneously, with consideration of the statistical and visual properties.

Fukinuki (1974) has further shown that

$$p(\varepsilon,\delta) = \frac{1}{2\pi\sigma_\varepsilon\sigma_\delta(1-\rho_{\varepsilon\delta})^{1/2}}$$

$$\times \exp\left\{\frac{-1}{2(1-p_{\varepsilon\delta}^2)}\left[\frac{(\varepsilon-\bar{\varepsilon})^2}{\sigma_\varepsilon^2} - \frac{2p_{\varepsilon\delta}(\varepsilon-\bar{\varepsilon})\delta}{\sigma_\varepsilon\sigma_\delta} + \frac{\delta^2}{\sigma_\delta^2}\right]\right\} \quad (4.104)$$

where

$$\sigma_\varepsilon^2 = \overline{\varepsilon^2} = (1+a^2-2a\rho)\sigma^2 \quad (4.105)$$

$$\sigma_\delta^2 = \overline{\delta^2} = 2(1-\rho)\sigma^2 \quad (4.106)$$

$$\rho_{\varepsilon\delta}^2 = \frac{(\overline{\varepsilon\delta})^2}{\sigma_\varepsilon^2\sigma_\delta^2} = \frac{(1+a)^2(1-\rho)}{2(1+a^2-2a\rho)} \quad (4.107)$$

$$\rho = \frac{\phi(1)}{\phi(0)}. \quad (4.108)$$

By differentiating N_{qv} with respect to a and setting to zero, the optimum value for a may be found

$$a = \frac{\rho\sigma_v^2 + 2(1-\rho^2)\sigma^2}{\sigma_v^2 + 2(1-\rho^2)\sigma^2} \quad (4.109)$$

where

$$v(\delta) = \exp\left(-\frac{\delta^2}{2\sigma_v^2}\right). \quad (4.110)$$

For TV signals σ_v^2 has been found to vary in the range 0.3σ to 0.8σ, and $\sigma = 1$. By using nonlinear companding plus brightness of vision contrast 2–4 bits/pixel can be saved in bandwidth reduction.

4.6.5 DPCM and Color TV Systems

Much of the work done on monochrome TV DPCM systems is applicable, in an extended manner, to color TV DPCM systems.

There are two general categories for color TV systems; being associated with either the component luminance and color difference form (namely,

Y, U, V or Y, I, Q) or with the composite form of the signal [namely, phase alternation line (PAL)] or NTSC.

According to Thompson (1974) repeated decomposition of the color signal to its component form followed by recoding for composite analog transmission cumulatively degrades signal quality. A study is being made of data compression techniques that allow the signal to remain in composite form throughout such hybrid connections. This work is concerned mainly with encoding the (PAL) waveform.

Maintaining the signal in composite form gives additional advantages for the digital coders in avoiding problems of maintaining color balance. In the PAL system this allows many of the quantization errors to be alleviated by the alternate line phase reversal. However, it has been found that time division multiplexed encoding of the separate luminance and baseband chrominance signals are more efficient than encoding of the composite signal, since it allows savings through sub-Nyquist sampling (Golding, 1972). It is also possible to exploit correlation between the component images (Frei and Jaeger, 1973), and perceptual redundancy of the chrominance components (Rubinstein and Limb, 1973).

For data comparison of both component color signals (Pratt, 1971), and composite signals (Ohira, Hawakawa, Matsumoto, and Shibata, 1973) an alternative to DPCM is to operate in some transform domain such as the Walsh-Hadamard Transform coding of TV signals. They describe a practical system for encoding NTSC composite signals based on the transformation of an eight-element linear array of samples, whereby the signal spectrum is deemed to be specified as the energy content of eight discrete bands by parameters that can be later quantized for digital transmission according to their subjective importance.

Color TV transmitted by DPCM coding is readily implemented by sampling the signal exactly n/m times the color subcarrier frequency (when m and n are integers) and by arranging the delays in the prediction paths to be multiples of n samples. For example, when $m = 1$ and $n = 3$, a one-dimensional prediction of the third previous sample performs well in large uniform colored areas, while the trebled prediction distance is still relatively efficient in predicting changes in the low bandwidth chrominance signal. This works well for low bandwidth systems, but poorly on high-contrast wideband luminance transition systems, where excessive slope overload and edge business occurs in a 3-bits/pixel implementation.

A two-dimensional prediction can be realized by a technique termed "chrominance correction" that consists essentially of neutralizing the chrominance components of a prediction such as the previous sample, which is efficient for monochrome signals, and reinserting chrominance information derived by band-pass filtering of neighboring samples of suitable subcarrier phase.

Figs. 1 and 2 of Thompson (1974), shown here as Figs. 4.10 and 4.11, show that while $S2$ may be the nearest and best sample for wideband

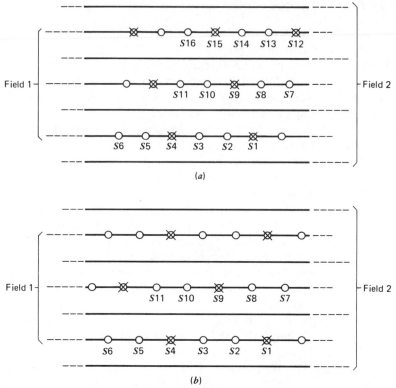

Figure 4.10. Diagram of picture points near the present sample ($S1$), which may be used for prediction. (*a*) PAL 625-line raster sampled at 13.3 MHz. (*b*) NTSC 525-line raster sampled at 10.7 MHz.

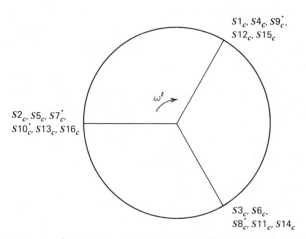

Figure 4.11. Vector diagram showing the relative subcarrier phases of the samples in Fig. 4.10 (relative phase is arbitrary). An asterisk represents the conjugate color signal in the PAL system. The subscript c denotes a band pass operation to separate the chrominance signal.

prediction of $S1$, its chrominance prediction is $2\pi/3$ radians out of phase. However, this can be corrected by subtraction of the chrominance of $S10$, that is, $S10_c^*$, and replacement by the chrominance of $S9$, $S9_c^*$.

The prediction $S1 = S2 - S10_c^* + S9_c^*$, where the subscript c denotes a band-pass operation to separate the chrominance signal.

This may be realized by using the block diagram of Fig. 3 of Thompson (1974), shown here as Fig. 4.12.

The conjugate chrominance signal is used from the previous line, since under normal operation of the PAL switch, $S10_c$ will only cancel $S2_c$ where there is no vertical component. The conjugate signal is derived by a simple modulation technique known as a PAL modifier (Bruch, 1964), of which the operation is illustrated in Fig. 4 of Thompson (1974) and shown here as Fig. 4.13.

The NTSC system can make use of Fig. 4.12 by omitting the modulator and the low-pass filter and substituting a chrominance bandpass filter. As shown in Fig. 4.12, the video input to the subtractor is sampled and held, and the quantizer output is clocked from bi-stables after the comparators have been allowed to settle, but the remainder of the encoder was constructed from analog devices.

Good broadcast color TV quality is obtainable with a chrominance correction system using 5 bits/pixel.

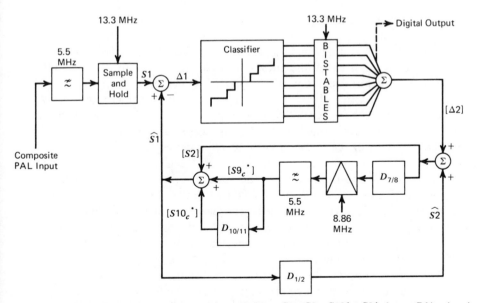

Figure 4.12. Chrominance-corrected prediction: $S1 = S2 - S10_c^* + S9_c^*$ for a PAL signal with 13.3 MHz sampling. Nomenclature: Δ_i denotes the prediction error on S_i, [ ] denotes quantized approximation, $D_{i,j}$ denotes the delay between samples S_i and S_j.

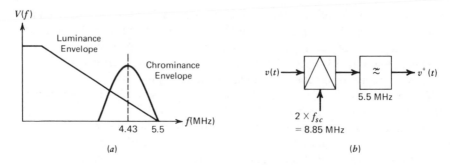

(a)

(b)

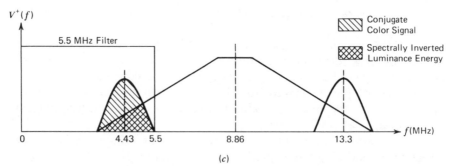

(c)

Figure 4.13. PAL modifier: Its Effects on the composite color signal. (a) Input signal spectrum. (b) PAL modifier configuration. (c) Output spectrum.

4.6.6 DPCM plus Dither

The technique of adding small-amplitude high-frequency waveforms to the input video signal, is called the adding of dither. This is equivalent to adding noise to the pictorial content of the signal. When the amplitude is small compared to the video signal level, it need not be removed at the receiver decoder. In this case, the visibility of contours due to PCM coding are greatly reduced. The penalty attached to adding dither is an increase in noise background level of the decoded video at the receiver.

The task in designing dither waveforms is to select that waveform which has the minimum visibility and is the least noticeable to the human eye. Dither can be used to advantage in a DPCM digital TV system.

When dither is used the number of quantizing levels needed to avoid contouring are reduced, with consequential bandwidth reduction.

When dither is removed at the receiver decoder still further bit per pixel reduction can be obtained.

Although pseudorandom waveforms are usually used, it has been found that certain deterministic waveforms give better results. For these waveforms it makes little difference whether or not the dither is removed at the receiver, since the picture quality is not much different in either case (Limb and

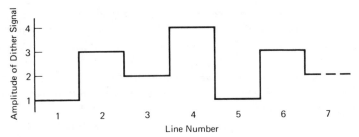

Figure 4.14. Four-step dither waveform for nonrandom dither.

Mounts, 1969). (This material is used with permission of American Telephone & Telegraph Co.)

A four-step deterministic waveform is shown in their Fig. 5; shown here as Fig. 4.14. The pattern sequence is $1,3,2,4$ applied to one dimension (vertical). This deterministic waveform has a zero mean and the levels are positioned uniformly within the quantizing interval. This waveform is optimum when subtraction is not used at the receiver. However, when subtraction is used at the receiver either the waveform shown in Fig. 4.14 or one having a pattern sequence of $1,2,4,3$ can be used. This $1,2,4,3$ sequence is not optimum for the case where subtraction is not used, because it becomes more noticeable to the human eye than the $1,3,2,4$ sequence.

Limb and Mounts (1969) have shown that deterministic one-dimensional dither reduces contour visibility by 7.1 dB. They also extend this concept to a two-dimensional 4×4 deterministic waveform dither pattern, and claim a 16.8 dB reduction in contour visibility.

When dither is used with DPCM the channel bandwidth needed can be reduced by 3 to 1 over that for straight PCM; or 2 bits/pixel.

4.6.7 DPCM TV Implementation

Kaul (1970) shows how to implement a digital TV system using DPCM coding. Fig. 2 of Kaul (1970) is reproduced here as Fig. 4.15. The system is all-digital in construction, and neither the coder nor the decoder are in the feedback loops. The propagation delay is less than 100 ns.

At the subtraction output the two most significant bits out of n bits will be 0's and there will be no need to transmit them. Only the $(n\text{-}2)$ least significant bits are transmitted. At the receiver the two most significant bits are assumed to be known, so the signal is reconstructed with n bits. In some cases, such as at sharp transitional points in the picture between white and black sections (the edges), the difference signal will exceed the range and cause an error to be transmitted. This is inherent in DPCM systems and correctional procedures consist of using some form of adaptive control or an image crispener.

To illustrate the subtraction system shown in Fig. 4.15 consider a 128-level encoder, and a peak-to-peak video input of 2 V. This is fed into the encoder and then to the subtractor. The encoder provides a straight binary code (i.e., for $+1.00$ V it puts out 1111111; for -1.00 V 0000000, and for $+0.00$ V 1000000). The subtractor output will have $\frac{1}{4}$ the dynamic range of the input signal. Hence only $N/4$ levels would need to be transmitted, assuming that the input signal is encoded into N levels. Assume that the subtractor has an input of two n-bit numbers, then the output of the subtractor should have only $(n-2)$ bits. Then the following conditions would be met.

1   The output is in straight binary code.

2   If $A_n > A_{n-1}$ by more than $N/4$ levels (where $N = 2^n$ and A_n is the nth sample), then all 1's are transmitted.

3   If A_{n-1}/A_n by more than $N/4$ levels, then all 0's are transmitted.

A subtraction algorithm was devised by Kaul (1970). The two 7-bit numbers A_n and A_{n-1} are placed in a 7-bit subtractor. If A_n is less than A_{n-1} there is an overflow. Thus the subtraction process automatically does the required comparison process. The overflow and the three most significant bits of the difference determine what to transmit on the following basis:

1   If there is no overflow and $D1, D2, D3$ (D denotes the difference) are not all 0's, then transmit $D3, D4, D5, D6,$ and $D7$ all as 1's.

2   If there is no overflow and $D1, D2,$ and $D3$ are all 0's, invert $D3$ and transmit $D4, D5, D6,$ and $D7$ as they are.

3   If overflow and $D1, D2,$ and $D3$ are all 1's, invert $D3$ and transmit $D4, D5, D6,$ and $D7$ as they are.

4   If overflow and $D1, D2,$ and $D3$ are not 1's, transmit $D3, D4, D5, D6,$ and $D7$ all as 0's.

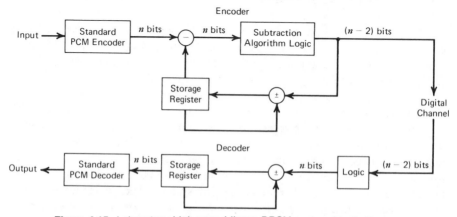

Figure 4.15. Laboratory high-speed linear DPCM system block diagram.

Thus a 5-bit code can be transmitted rather than the 7-bit code, with a saving of 2 bits.

4.6.8 Problem Areas in DPCM

One disadvantage of any DPCM system, however, is that channel noise will continue to propagate in the reconstruction process at the receiver. Catastrophic loss of sync can occur if some method of periodical correction is not used. The easiest correction maneuver is to reset the storage register at the receiver during every horizontal blanking interval. This can be done without too much extra circuitry for TV signals.

Unlike PCM systems which are amplitude limited, DPCM systems overload on the slope. If the sample-to-sample difference is greater than the largest quantizer step, the system is said to be in slope overload. A system exhibiting slope overload effectively reduces the horizontal resolution of the scene being processed.

In low-detail regions of the picture, the noise power is determined by the rather fine quantizing structure designed for small differences. This is termed the granular noise region. This situation is illustrated by Fig. 7 taken from Millard and Maunsell (1970), shown here as Fig. 4.16. (Permission to use this material has been granted by American Telephone & Telegraph Co.)

If the small steps are too large, the strong correlation between the quantizing noise and the signal can lead to visible patterns that are subjectively objectionable. In flat regions of the picture the DPCM loop tends to oscillate using alternately positive and negative small steps. The use of an integrator, in the feedback path, having resistive leakage is able to combat this defect. The time constant of the leakage path can be chosen for minimum oscillatory effects.

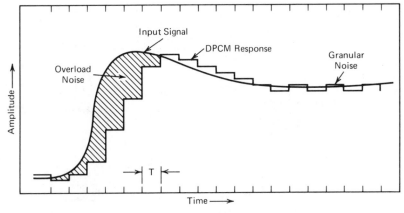

Figure 4.16. Illustrated video waveform and DPCM response showing a slope overload region and a granular noise region.

Edge busyness is also a problem, and is a function of noise, quantizer characteristic, threshold tolerances, and past history due to memory in the feedback loop. Synchronization acts to reduce these causes of edge busyness; but not completely.

Since edge busyness results from the variability in the selection of the set of steps to approximate the input ramp voltage, it follows that minimizing the horizontal differences between the two responses will minimize the edge busyness in a mean-square sense. The horizontal differences are minimized by making the differences between adjacent steps as small as possible. However, the existence of other impairments with their conflicting requirements makes the use of a uniform quantizer impractical.

Millard and Maunsell (1970) show a set of DPCM TV block diagrams. Their experimental transmitter is shown in their Fig. 15, shown here as Fig. 4.17.

The coder is shown in their Fig. 16, shown here as Fig. 4.18.

At each sampling period of time the input is compared directly against the decoded output signal last transmitted. The output difference amplifier is connected in parallel to the inputs of seven threshold detectors set to 0, ±2 percent, ±5 percent, and ±11 percent of the nominal 1.5 V peak-to-peak video input signal. The cyclic coded outputs of these detectors are the output of the coder. The nonlinear companding is achieved by making the current sources appropriate for output voltage changes, $e_q(nT)$ of magnitude ±1 percent, ±3 percent, ±7 percent, and ±15 percent of the peak-to-peak quantizer input video signal.

After conversion to a serial bit stream, the coder output bit stream passes through a digital store, with a capacity of four 3-bit words each of which occupies a fixed position within the memory.

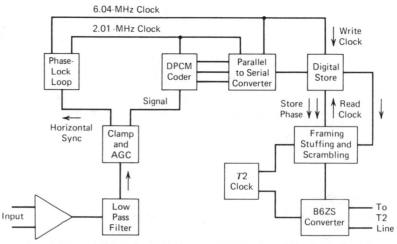

Figure 4.17. Block diagram of a DPCM transmitting circuit.

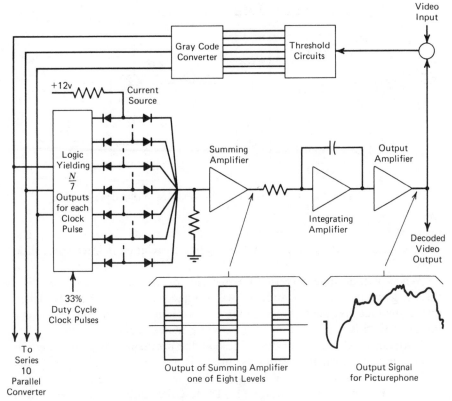

Figure 4.18. DPCM coder diagram.

In their digital TV receiver, which is shown here, the unipolar signal is recovered from the B6ZS input signal and is descrambled to yield the 108-bit code words. The descrambler contains a 1-bit shift register, the input and output of which are applied as inputs to an exclusive OR logic circuit to derive the output data stream. The bit stream is passed through a 12-cell store using a write clock that is obtained by analysis of the 108-bit words and which clock only DPCM information bits into the store. The signaling bits are recovered and the digital store framing information is used to ensure that the four 3-bit words are placed in the decoder signal store in the same position that were taken out of the coder digital store.

After passing through the digital store, the bit stream is processed in a serial-to-parallel converter that provides the required inputs for the DPCM decoder.

Millard and Maunsell's receiver is shown in their Fig. 17 and here as Fig. 4.19.

The implementation of DPCM TV systems using line-to-line prediction is discussed by Connor, Pease, and Scholes (1971). (Permission to use this

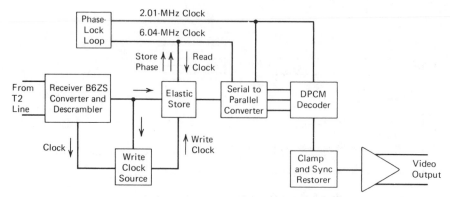

Figure 4.19. Block diagram for a DPCM receiving circuit.

material has been granted by American Telephone & Telegraph Co.) They discuss various methods of utilizing line-to-line correlation in interlace scanned pictures.

When interlaced scan is used the vertical correlation of intensity within a field is less than for correlation between adjacent horizontal pixels. This is due to the fact that the distance between a pixel on line 1 and the next line of the same field is twice that existing between adjacent pixels on line 1.

Vertical correlation is needed to resolve the problem of edge busyness. By making use of information from the previous line, it is possible to anticipate the vertical edges and most of the diagonal ones.

Connor, Pease, and Scholes (1971) describe three classes of vertical prediction. They describe these classes while referring to their Fig. 1, which is shown here as Fig. 4.20.

In Class 1, the prediction is the average of the value of the previous pixel A and the value of the vertically adjacent pixels B, C, and D. This is termed "averaged prediction."

In Class 2, the prediction is the sum of A and the scaled horizontal element difference between the pixels on the previous line, e.g., $C - B$ or $\frac{1}{2}(D - B)$. This is termed "planar prediction." The two classes use linear prediction.

In Class 3, nonlinear prediction is used. The prediction can be either A or some linear combination of A, B, C, and/or D. The choice is dependent on the magnitude of intensity differences between pixels in the previously coded neighborhood of the current pixel X. This is termed "optional prediction" (Graham, 1958).

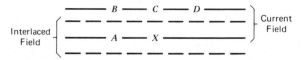

Figure 4.20. Diagram of picture elements (pixels) near the current pixel.

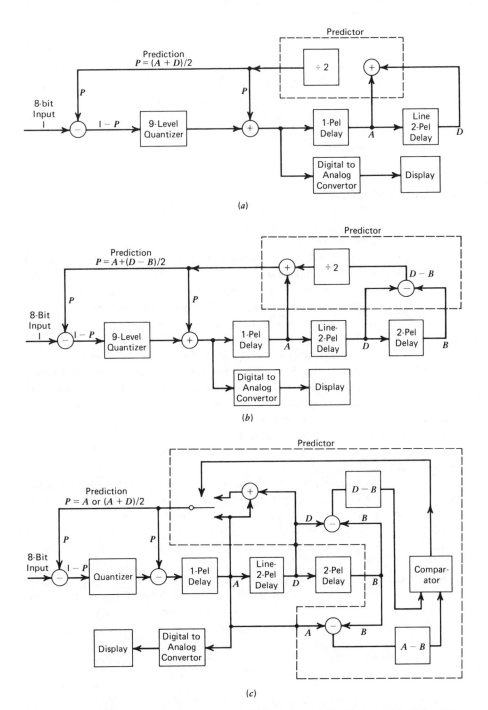

Figure 4.21.(a) Diagram of apparatus used to evaluate coder (*i*) in which the value $(A+D)/2$ is used as a prediction of the current picture element. The displayed picture is equivalent to the received picture providing there are no channel errors. (*b*) Apparatus to evaluate coder (*v*) in which $A+(D-B)/2$ is used as a prediction of the current pel (pixel). The position of pels A, B, and D is shown in Fig. 4.20. (*c*) Apparatus used to test the optional prediction encoder. If $|A-B|>|D-D|$, A is used as the prediction. If $|A-B|<|D-B|$, $(A+D)/2$ is used. The position of A, B, and D are shown in Fig. 4.20.

All three types of vertical prediction coders were tested. The test circuits used are shown in Fig. 4 of Connor, Pease, and Scholes (1971), and are also shown here as Fig. 4.21. The average linear prediction scheme and the optional prediction scheme performed better than did the linear planar prediction method. However, all three systems appeared to be quite vulnerable to channel type noise.

4.7   DELTA MODULATION

A simplified 1-bit delta modulation system consists of a 1-bit delay and a summer. This forms a digital integrator that, subtracted from the input analog signal, causes a signal difference to be detected and transmitted. Since in most picture areas the tonal change between successive picture elements is small, a differential system, such as the delta, that transmits only picture differences would appear to offer a method of bandwidth reduction.

Bell Laboratories have investigated delta modulation techniques for use with digital TV. In their investigation they varied the number of bits per sample. The scheme used by Bell made use of higher bit rates than one. In the n-bit TV system there is an additional important parameter, and that is the quantizer characteristic. Bell Laboratories found that a "tapered" quantizer gives the best results. Tapering makes the quantization error more nearly proportional to the signal amplitude.

Bell Laboratories concluded that 3 bits/sample with a suitably tapered quantizer, in pictures of 100 lines/frame, gave little lower quality than for 6-bit PCM.

The main problem with delta modulation occurs at sharp picture edges (sudden change from black to white, or vice versa), which condition demands a large number of bits for adequate representation. A small number of bits cannot adequately represent a sharp brightness jump. Therefore, tapering the quantized code is a necessity to partially overcome this difficulty and 3 bits/sample is the practical lower limit. An additional problem with delta modulation is its relatively high error rate sensitivity due to the differential nature of the transmission. That is, if a bit drops out at the receiver, the receiver will reconstruct the picture from an incorrect difference and this error will continue for quite some time. This is considered to be very undesirable for use with surveillance systems that must have a high degree of reliability. However, this condition may be overcome by resetting the receiver storage register during the horizontal blanking intervals.

4.8   SYNTHETIC HIGHS AND LOWS

In this technique (Graham, 1967) the video from the sensor is separated into two channels; a low frequency band and a high frequency band. The low frequency band is quantized at a low rate, the high band at a high rate. In this system there must be some sort of nonlinear operation, such as thresholding, to determine if an edge point is important. The location and character of each edge point must be coded for transmission. This system requires very good synchronization upon recombination at the receiver of the highs and lows. It also requires the use of storage registers. If the received signal is stored in a recorder, the recombining process is smoothed out very well. However, sharp picture edges are not continuous in nature, but are statistically arranged. Therefore, maximum bandwidth reduction is not achievable when the pictures are transmitted in real time. However, bandwidth reductions of 4:1 can be achieved for real time TV.

4.9 BIT-PLANE ENCODING

Bandwidth reduction in digital TV can be obtained by using statistical techniques such as run-length encoding and line-to-line encoding. It has been found (Spencer and Huang, 1969) that the choice of quantization means can cause better performance at lower bandwidths, when added to either of the two above-mentioned statistical techniques.

The technique of bit-plane encoding transforms a picture of 2^n possible shades of gray into n pictures, each of which has only two shades; nominally black or white. The 2^n shades are represented by an n-bit binary word at the output of the intensity quantizer. A picture is represented by the two-dimensional matrix of n-bit words, one per element.

The jth bit plane may be defined as the picture that results when each n-bit word $b_1, b_2, b_3, \ldots, b_n$ is replaced by its jth bit b_j.

The complete process of bit-plane encoding consists of the generation of these n-bit planes from the original picture. This means that the two-dimensional matrix of n-bit words may be equivalently represented as n-two-dimensional matrices of 1-bit words. Each matrix is called a bit plane, and is either black or white. The transformation is reversible; the original picture being accurately reconstructed from the n-bit planes.

Spencer and Huang (1969) analyzed the performance of both run-length coding and line-to-line correlation by assuming three types of codes: binary, gray, and DPCM.

It was found that the gray code consistently out-performed the binary code, and that line-to-line correlation was better than run-length coding. However, when the gray code was used with an optimum combination of run-length coding and line-to-line correlation, that superior performance was attained. Picture quality was good at 1 bit/pixel with this optimum combination.

Spencer and Huang show in their Fig. 9, a comparison in performance between bit-plane encoding and DPCM coding. The optimum bit-plane method used less entropy per bit for a crowd scene than did the DPCM

		Continuous Tone		Gray Code Bit-Plane Encoding	
		DPCM	RL	Corr	RL-Corr
1-Bit	Girl's Face	0.3953	0.4216	0.4262	0.3742
	Cameraman	0.5050	0.5212	0.6279	0.5111
	Crowd Scene	0.6660	0.7297	0.9027	0.7017
6-Bit	Girl's Face	0.1611	0.1593	0.0812	0.0812
	Cameraman	0.1377	0.1115	0.0807	0.0807
	Crowd Scene	0.4931	0.5036	0.4464	0.4464

Figure 4.22. Bit-plane encoding versus DPCM: Entropy per bit.

system for a 1-bit/pixel system, but was slightly inferior for a 6-bits/pixel system. Comparison of performance when the picture was the face of a girl showed that bit-plane encoding is superior to DPCM when 6 bits/pixel and 1 bit/pixel are used. Spencer and Huang's Fig. 9 is shown here as Fig. 4.22.

The use of a four-point spatial filter gave added performance improvements.

4.10 INTRAFRAME REDUNDANCY REMOVAL BY USE OF CONTOUR INTERPOLATION

The redundancy existing in a TV frame can be removed by several means, in order to achieve bandwidth reduction. A method called "contour interpolation" was proposed by Gabor and Hill (1961). They made use of the correlation existing between fields and frames.

An interlaced system causes an increase in bandwidth. In order to reduce flicker, 50–60 fields or half-frames are used instead of 25–30 full noninterlaced frames. Therefore, a 2:1 Bandwidth increase is used to eliminate flicker. This sacrifices almost one-half of the vertical definition. An interlaced field, therefore, contributes nothing to the information content of a picture and increases the bandwidth.

Methods, other than interlace, can be used to reduce flicker. Repetition in TV could be achieved by attaching to the receiver tube a memory device storing a full frame, by using a storage tube for the display in which the old picture is erased just prior to recording a new one. A some-what better result could be obtained by merging one frame continuously into the next, by superposing the new and the old frame in continuously changing proportions. This method is called "linear interpolation."

A method called "contour interpolation" transmitts only one field out of two and constructs from memory the missing interlaced field. This method is based on the physiological fact that the eye focusses on contours.

This principle of "contour interpolation" can best be explained by use of Gabor and Hill's (1961) Fig. 1; shown here as Fig. 4.23.

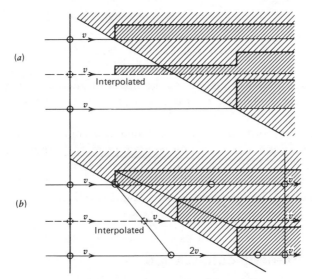

Figure 4.23. Linear and contour interpolation.

The first and third lines, of the two diagrams shown, portray intensity distributions along two consecutive lines of a field at the receiver. The middle lines belong to the interlaced field; the one not transmitted. The missing interlaced portion of the picture is reconstructed at the receiver by use of interpolation. Both examples relate to a slanting contour that separates dark and bright areas.

Assume that the two received lines have been stored in some sort of storage device. The upper diagram shows linear interpolation. Further assume that in the storage system the two lines can be scanned by two separate spots, both running at the same speed v, and the intensity I in the missing line is approximated by the arithmetical mean of the two intensities I_1 and I_2:

$$I(X) = \tfrac{1}{2}[I_1(X) + I_2(X)]. \tag{4.111}$$

Linear interpolation works well for gently shaded areas and for vertical edges, but for a slanting contour a stepped amplitude profile arises that causes noticeable picture distortion.

In contour interpolation the two moving spots run with equal speeds v until one of them reaches a slanting contour edge. Here it stops, while the other spot, which has not yet reached the contour, continues with a speed $2V$, so that the interpolated spot midway between the two, moves with a velocity v. This continues until the second spot has reached the contour and the retarded spot gradually catches up, and they again move at the same speed v. The intensity rate is such that the interpolated intensity I is formed as the arithmetical mean of the two intensities, but weighted with the velocities

$$I\left[\tfrac{1}{2}(X_1 + X_2)\right] = \frac{v_1 I_1(X) + v_2 I_2(X)}{v_1 + v_2}. \tag{4.112}$$

As is shown in Fig. 4.23 just the right amount of interpolation is given to the interlaced field; and at the right timing required to give a slanted contour the proper fidelity.

When one frame out of two is transmitted and contour interpolation is used, flicker is eliminated with no required increase in bandwidth. This amounts to a $2:1$ bandwith reduction, when compared to the commonly used method of interlace. Gabor and Hill (1961) state that a $4:1$ compression might be achieved without appreciable amounts of flicker, but this is the practical limit.

4.11 FRAME-TO-FRAME CORRELATION

Frame-to-frame correlation, to reduce bandwidth, has been proposed by many authors. Mounts (1969) proposed a method for utilizing this technique. He called it "conditional replenishment." (Permission to use this material has been granted by American Telephone & Telegraph Co.) In this method only those elements that change between frames are transmitted. A memory buffer is used to store a reference picture.

At the receiver this information is used to update a similar stored reference picture (frame) in order to track the one stored at the transmitter. In order for the receiver to correctly update the picture elements, two pieces of information must be conveyed to the receiver—the new value and the position of the element to be replenished. Because this information occurs at a random rate, buffers are used to redistribute and present the information to the transmission channel at a uniform bit rate. In order to regulate the average replenishment rate to match the channel capacity, the threshold used to determine which changes are significant, is varied as a function of the amount of information stored in the buffer.

This method of encoding requires that only the information pertaining to the active region of the picture be transmitted. The receiver reinserts the horizontal and vertical blanking interval within the reconstructed video information. A bandwidth reduced to 1 bit/pixel is claimed for this system.

Mounts (1969) shows how to implement this system. His block diagram for the transmitter is shown in his Fig. 1 and is reproduced here as Fig. 4.24.

The video elements are encoded at first by utilizing 8-bit PCM. A selector switch directs the signal to either the reference frame memory or to recirculate the information stored in the frame memory. If the new information is significantly different from prior information the information in the reference frame memory is updated; if not the information is recirculated but not used in transmission.

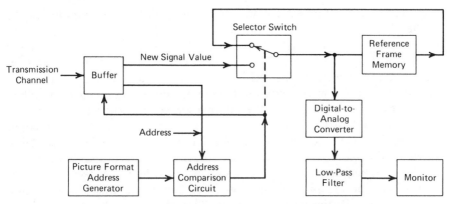

Figure 4.24. Conditional replenishment transmitter terminal.

The frame memory consisted of delay lines in the Mounts' version, but could take on any of the new modern memory forms. It has enough capacity to store a complete TV frame, when encoded in 8-bit PCM.

The subtractor circuit is used to compare new information from the camera with that in the reference frame memory. It yields the absolute difference between the new sample and the reference value corresponding to a given picture element. During each sample period the control logic makes a decision, depending on the magnitude of the difference signal, as to whether a difference exists. If the difference is significant the output from the control logic operates the selector switch in order to strobe the new information into the reference frame memory. The control logic also causes the new signal value to be stored in the buffer, along with its address. The buffer store matches the varying data rate to the constant bit rate of the transmission channel. This information is read out at a constant rate, in the order received.

The video is encoded with 8-bit PCM with an additional 7 bits being used to identify the position information. A total of 15 bits thus comprises one word. Ambiguity in the vertical sense is avoided by always sending the first active sample of each line whether it changes or not.

In order to force the average replenishment rate to match the channel capacity, the significant change threshold is varied as a function of the amount of information stored in the buffer. However, it is desirable to keep some minimum amount of data in the buffer at all times, so that data are always available for transmission, which could happen during vertical blanking periods when data leaves the buffer, but none enters. To guard against loss of buffer storage, the significant change threshold is lowered to zero whenever the buffer count falls below a certain threshold. On the other hand, when the buffer is full, replenishment is stopped, which can cause momentary picture breakup.

At the receiver a buffer is used to store a TV frame until it can be strobed. Mounts (1969) shows the receiver block diagram in his Fig. 3, which is reproduced here as Fig. 4.25. A transfer of new information from the buffer to the frame memory occurs whenever the output of the picture format address generator agrees with the address information of the picture element to be read from the buffer. This agreement is determined by the address comparison circuit that operates the selector switch to enable the new amplitude information to flow from the buffer into the frame memory. When the addresses do not coincide, the information stored in the frame memory recirculates and readout cell of the buffer is held fixed. The information stored in the frame memory, when decoded, provides the video information for visual display.

During experimentation with this system it was found that as the motion of a subject in the picture becomes more rapid, the number of picture elements stored in the buffer increases. This causes the significant change threshold to be increased so that small changes in the picture are not reproduced. As a result picture fidelity suffers. The result then resembles a picture as viewed through a dirty window.

When the subject size is large and it is in rapid motion the buffer is caused to overflow, and picture breakup occurs. This could be a serious defect of the system. However, it would appear likely that modern techniques could be used to overcome this defect.

It was also demonstrated by Mounts that it is more efficient to gradually update all picture elements according to predetermined pattern, rather than to lower the threshold to permit noise to cause replenishment.

A discussion of the conditional replenishment method is found in Haskell, Mounts, and Candy (1972). They refer to the impractically large buffer needed to obtain very much saving in transmission capacity because peaks of high activity often last for long periods of time. Smoothing these peaks is

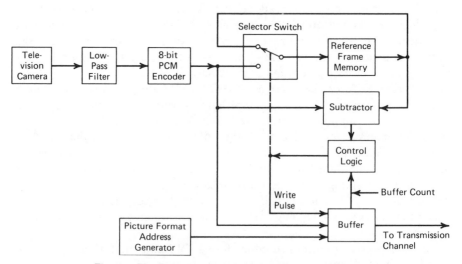

Figure 4.25. Conditional replenishment receiver terminal.

impractical for two reasons: First, the required buffer size would be too large to be economical. Second, the delay introduced would become intolerable in face-to-face amplications such as videophones. More than one-third of a second delay between talker and viewer has been found to be distracting.

Rather than smoothing large data peaks, it is more practical to transmit changing areas with reduced amplitude and spatial resolution during active periods. A useful threshold for measuring such activity would be the fullness of the buffer. Thus when the number of bits in the transmitter buffer exceeds a prescribed amount, indicative of a too high data rate, resolution reduction could be used.

Significant saving of transmission capacity by buffering alone does not occur until the buffer is large enough to smooth the data between one surge of activity and the next. Such large buffers are impractical when the requirement for storage reaches the 1 to 2-Mbit region.

The biggest problem to overcome with the conditional replenishment frame-to-frame correlation systems, is that of degraded performance when very rapid motion occurs in a scene. Several methods have been suggested to overcome this problem.

4.11.1 Picture Segmentation

A method for overcoming the rapid motion defect in frame-to-frame correlation systems has been proposed by Rocca and Zanoletti (1970). They devised a method in which the image is divided into a mosaic of small zones and for each zone a replacement vector is determined, such that there is a maximal correlation between the samples of the displaced zones. The displacement vector is the same for all zones in the case of certain camera movements, and differs from zone to zone in the case of panning, or of rapid movements of the object televised.

The displacement vectors are transmitted together with the variations of the corresponding picture elements. An optimum zone size exists and was found to consist of 70–80 pixels per zone.

This system would appear to require a complete bookkeeping arrangement.

4.11.2 Dot Interlace Systems

A method of raster scan called dot interlace was proposed (Mounts, 1967) consisting 4:1 dot interlacing. Only one-fourth of the picture elements in a frame are transmitted every $\frac{1}{60}$ of a second. At the receiver each element is displayed for $\frac{1}{15}$ of a second. A number of selective replenishment patterns yielding 4:1 dot interlace are possible. One such sampling is shown in Fig. 1 (Haskell, Mounts, and Candy, 1972). It is reproduced here as Fig. 4.26.

When this system is used picture quality is good during periods of little or no movement. When motion becomes more rapid, however, edges of the

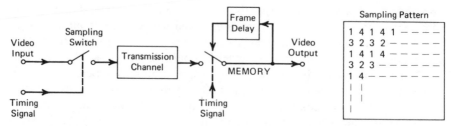

Figure 4.26. Four-to-one dot-interlaced transmission of video signals. Elements labeled 1 are sent in the first frame, elements labeled 2 in the second frame, etc. At the display, the untransmitted elements in each frame are obtained from the previous frame. A delay of one frame is required.

moving images in the picture tend to be degraded in such a way that the sampling pattern is visible.

An improvement over the simple dot interlace system can be obtained by filtering the frames in the temporal direction, that is, for each picture element a weighted average of the brightness values in several successive frames is formed. Then during each frame period a subset of these average values is selected for transmission. Also, rather than displaying each transmitted element several times at the receiver, frame-to-frame interpolation between the received values is used to reduce the distortion.

4.11.3 Frame Differential Coding

That portion of a TV frame that is affected by motion can be segregated from the rest of the picture. This is one form of segmentation. Basic to this scheme is a good method for detecting motion in a picture. It has been found that picture element changes of 1.5 percent or less of the maximum signal amplitude can be regarded as unchanged. If the change is 3 percent or greater in noise face pictures, motion is present. When the threshold is set at 1.5 percent the changing area can be detected, while good picture quality is maintained.

A frame-to-frame differential coder can be used, such as is shown in Fig. 2 of Haskell, Mounts, and Candy. It is shown here as Fig. 4.27.

The signal is delayed by one frame with the output used as a reference, which is subtracted from the input signal. The resulting differential signal is quantized, transmitted, and added to the reference frame, both at the transmitter and at the receiver, to obtain a representation of the new input frame. This representation is then fed to the delay to be used as the reference when coding the next frame.

The levels of the quantizer are companded so that small changes are reproduced more accurately than larger ones. Thus good picture quality is maintained in nonmotion areas of the picture. Frame differences due to motion are quantized more coarsely, since the human eye is tolerant of a blurred motion image.

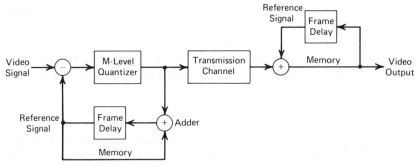

Figure 4.27. Frame-to-frame differential coder. 64-level quantization provides near perfect reproduction. 16-level quantization provides good reproduction except when an object containing large brightness transitions moves rapidly.

A 64-level quantizer provides nearly perfect picture quality. A 16-level quantizer gives good picture quality except when rapid motion occurs. Moving edges of this type appear in the display as a sequence of brightness steps, each step differing from its neighbor by an amount equal to the largest quantizer level. This effect is analogous to slope overload in DPCM. But it is caused here by motion and has been called "temporal overload."

When pixel-to-pixel correlation is used in DPCM systems nonmotion parts of the picture tend to be distorted; whereas in frame-to-frame differential correlation the motion parts of the scene are distorted. As with DPCM systems channel noise is hard to handle, as in frame-to-frame differential coding. Some form of error correction is needed for such systems.

Frame differential coding takes advantage of the observer's higher tolerance for distortion in moving images by reducing amplitude resolution for motional picture segments. Rather than reduced amplitude resolution, spatial resolution can be reduced to obtain the desired effect.

The simplest and most direct method of reducing spatial resolution is obtained by subsampling. In this scheme information is transmitted for only a fraction of the picture elements in the areas of motion. The rest of the elements are set to an average value. Excellent picture quality has been obtained by transmitting only every other element along a line in rapidly moving segments. This is 2:1 horizontal subsampling. The values of the intervening elements are set equal to the average of their neighbors along the scan line. Thus horizontal resolution is halved in areas of motion. This is a simplified scheme. More elaborate schemes have been used by applying dot interlace techniques and interpolation.

4.11.4 Combined Frame Differential Coding and Dot Interlace

Frame differential coding and dot interlace can be combined into a useful system. Frame-to-frame difference values are transmitted every other picture element along a line, while values for the intervening elements are obtained by repeating their values from the previous frame. Thus 2 : 1 dot interlace is used in stationary areas. In motion areas, every other element is transmitted as in the stationary areas; but intervening elements are not replaced by the previous frame values, but rather by an average of their neighbors along the line, in the same frame. A 2 : 1 reduction of the horizontal spatial resolution is thus used in the motion areas of the picture.

Since the unsampled elements are treated differently in stationary and changing areas, the receiver must be told where these areas lie. Addressing information is transmitted, and since this information occurs at an irregular rate, a small amount of buffering is required.

Using 5-bit frame differential coding to transmit amplitudes, a channel rate of 2.5 bits/pixel is needed in this system. Picture quality is excellent in stationary areas and in rapidly moving areas. However, objects containing sharp edges moving with moderate speeds are visibly degraded.

4.11.5   Combined DPCM and Frame Repeating

In this system, entire fields are classified as either containing motion areas or not, depending on the number of large frame differences per field.

In the stationary mode of operation, pictures are displayed with reduced temporal resolution. Every other line in each field is transmitted with element-to-element DPCM, while the intervening lines are obtained by a linear interpolation between the two adjacent frames. The sampling pattern is inverted every two fields to maintain full spatial resolution (see Figs. 5 and 6 of Haskell, Mounts, and Candy, 1972).

When a field is deemed to contain significant movement, temporal resolution in increased, but vertical spatial resolution in the whole field is reduced. Every other field is transmitted in its entirety using element-differential PCM, while the intervening fields are obtained by interpolation between the two adjacent fields.

Using 4-bit DPCM to transmit the lines a channel rate of 2 bits/pixel is required. Also, enough buffering must be provided to spread the transmission of each sampled field over two field intervals.

This system was nicknamed CRIDEC and tested in videophone experiments. It was found that in the motion segments of a picture there was adequate picture quality, except in large areas of fast motion where some jerkiness was noted. In regions of high vertical detail some aliasing was noted.

A big advantage in CRIDEC lies in the fact that lines are transmitted intact by DPCM. This avoids sending frame differences in areas of motion that require 5-bit quantization for adequate picture quality. (Permission to use this material was granted by the American Telephone & Telegraph Co.)

4.11.6   Buffer and Channel Sharing

If, instead of one, several coders are used and their outputs are combined in a conditional replenishment system, a 1-bit/pixel system results. A large amount of smoothing can be obtained without adding much delay or requiring large buffers (Haskell, 1972). Thus more buffer space and more channel capacity can be allocated to those coders that are producing the most data, at the expense of those which are not.

With this type of multiplexed transmission it is the average data rather than the peak that determine the performance. Thus resolution reduction techniques should be employed during all periods of active motion. The buffer is shared by all of the coders and cannot be used to control the resolution of any one source if resolution reduction is achieved by using the number of changed elements per field as a threshold. When 12 to 15 conditional-replenishment coders are used, the required channel rate per coder can be halved.

As in all redundancy removal schemes for digital TV, bandwidth reduction is obtained at the expense of more susceptibility to channel noise. One should, therefore, seriously consider the use of error-correcting coders.

4.11.7 Combined Intraframe and Frame-to-Frame Coding

In this coding system for digital TV, coding within the frame is combined with frame-to-frame coding. Changes from frame-to-frame are detected, but are encoded on an intraframe basis. The best features of both methods are retained in the combined system.

In interframe coding the element in the previous frame corresponding to the element being encoded is a good prediction when an object is moving slowly, whereas a spatially adjacent element in the same frame is a better prediction of the current element when the object is moving fast. A combined system can handle both fast and slow rates of object movement optimally.

In an ideal situation, it is easy to determine the changeover point at which the element difference is smaller than the frame difference. Above a certain movement rate threshold frame difference is greater than element difference. Therefore, fast motion can be determined by using frame differences, but the encoding is done on an intraframe basis.

The encoding of moving areas by intraframe techniques becomes more efficient with faster movement. This is in contrast to most other frame-to-frame coding schemes in which efficiency decreases with the speed of movement. There are other advantages to using intraframe coding such as the minimization of requantizing effects and less storage space required.

When these techniques are combined with conditional replenishment a great amount of redundancy can be removed from the TV picture.

This system has been implemented experimentally by Limb, Pease, and Walsh (1974). It was named by them the CR/IR conditional replenishment intraframe coding system. They describe the system in great detail. (Permission to use this material has been granted by American Telephone & Telegraph Co.)

The difference between the stored frame and the current frame is both spatially and temporally filtered, and applied to a varying threshold that is under control of a modified element-difference signal. This compensates for the larger errors introduced in high detail areas by the differential quantizer. They also use a technique called "blocking in," an operation that produces a more contiguous moving area and rejects small isolated changes.

As the speed of the moving object increases, the resolution of the resulting image in the direction of the movement decreases, because of the light-integrating action of the camera target. For horizontal movement, this in turn reduces the amplitude of the element-to-element differences. Although the eye can apparently detect smearing in the picture because of camera target integration, the usual observer is quite tolerant of this type of defect. These are first-order effects in entropy reduction. However, there remains much redundancy in a fast-moving object when the signal is essentially oversampled.

One method for removing this redundancy is to switch to a submultiple of the sampling rate, but this degrades picture quality quite noticeably. Another method is to sample every other element on a scan line, using interpolation for the intervening pixels. When the threshold is low, nearly all points are encoded normally. As the threshold increases, more and more conditional points are extrapolated until, if the threshold is high enough, the signal becomes effectively subsampled. In order to obtain a bit-rate advantage with horizontal conditional subsampling, a variable-length code is required, because otherwise the data are transmitted about all points, including the conditional points, unless the signal is fully subsampled.

As mentioned previously this system was implemented experimentally. This was done as a development project for Picturephone® systems. It was found that during violent motion of arms and hands movements, the system performed acceptably. However, there was some distortion from noise due to sampling. During fast panning of the camera, the system performed poorly.

4.11.8 Combined Interframe Coding and Fourier Transforms

Haskell (1974) has proposed the use of frame-to-frame encoding, combined with two-dimensional Fourier transforms and conditional replenishment. Changes due to the motion of an object in translation are represented by brightness differences between two pixels in the moving object itself. However, on the leading and trailing edges of such objects the comparison is between moving and stationary areas of the picture. But, if the distance traveled by the moving object in one frame period is small compared with the object size, the edge pixels are small in number compared to the number of pixels in the whole movement area.

By this reasoning, if the velocity of moving objects does not change much between frames, then the fame differences in the changed area of one frame should be about the same as they were in the previous frame except for a linear displacement in the direction of motion. Around the edges this is not so, but the number of elements involved are small.

The Fourier transform of a time-shifted version of a signal differs from the Fourier transform of the original signal only by a linear phase shift $\omega\Delta\tau$. The frame difference signal during one frame should be nearly a time shifted version of the frame difference signal from the previous frame. Thus the two-dimensional Fourier transform of the frame difference corresponding to one frame should be similar to the Fourier transform of the frame-difference signal corresponding to the previous frame, except for a linear phase shift $\omega_x\Delta x + \omega_y\Delta y$. The phase shift represents the frame-to-frame displacement undergone by the frame difference signal. Thus it is only necessary to transmit Δx and Δy plus information needed to describe those parts of the frame difference that are not moving in pure translation such as edges of the area of movement.

The use of a Fourier transform simplifies the frame-to-frame prediction since translations are measured simply by averaging phase shifts, and movement compensation is done by shifting a phase.

When the previously described method is combined with conditional replenishment additional redundancy is removed. Predictive coding of the frame-difference signal in the changed area of the picture can be used. This method is realizable by taking the frame-difference Fourier transforms corresponding to the previous frame and the present frame, estimating the linear phase shift and applying it to the frame-difference Fourier transform from the prior frame; then taking the inverse Fourier transform to get a prediction of the frame difference during the present frame. This is then encoded and transmitted along with the required addressing information.

Since the velocity of movement in a picture will not be constant over the whole changed area, it is beneficial to segmentize the picture frame into many subframes, taking Fourier transforms and estimating phase shifts for each separate segment. If the frame-to-frame displacement is small with respect to the size of the subblocks, then all previous observations hold.

However, the advantages gained by subdivision may be outweighed by the amount of bookkeeping memory required to keep track of each subblock.

In regions where frame differences are small due to slow movement or low detail, defining the changed areas consistently frame after frame is difficult. In a practical system the Fourier transform of the changed area pixel brightness may change considerably from frame to frame due to variations in changed area definition. The use of a Fourier transform minimizes the effects of these variations, since the frame differences are small in those regions of the picture where definition of the changed area is small.

4.12 SIGNIFICANT BIT SELECTION

There are various methods available for assigning the number of bits transmitted per pixel. Bisignani, Richards, and Whelan (1966) devised two such coding schemes.

The first of the two methods they called "improved gray scale." This scheme consists of digitizing the original signal to 6 bits, transmitting the three most significant bits, and retaining the three least significant bits to be modulo 8 added to the intensity of the next picture element.

The cumulative effect of the retained three least significant bits is to change the most significant bits in such a way that the displayed levels average out to about 6-bit accuracy representation of the true intensity.

So, in spite of the fact that 3 bits/pixel are transmitted, the picture quality looks as if 6 bits/pixel had been transmitted. This makes for a two-to-one reduction in bandwidth.

In this system the three most significant bits represent coarse data and the three least significant bits fine data. Provision is made for overflow resulting from the most recent addition into the coarse data of the present word. When an overflow occurs, the carry is added to the three most significant bits of the present word. This addition is inhibited if the three most significant bits are all in the "one" state. In this case the three most significant bits are transmitted unmodified except when a carry modifies the coarse data.

The second method is called "coarse–fine PCM." In this system the original signal is digitalized to 6-bit PCM, but only 3 bits are transmitted. If the information is coarse, the three most significant bits are transmitted; if fine, the three least significant bits. When the original signal varies extremely rapidly over a range greater than one eighth of full scale, such that there is a change in the three most significant bits at each picture element, then these bits are transmitted as absolute levels. However, if the signal falls below this threshold, the three least significant bits are transmitted.

The results at the receiver display are as though 6 bits/pixel had been transmitted.

A code has been established for determining whether to send coarse or fine bits, and is transmitted to inform the receiver which of the two groups to expect. A three word memory is required at each end, and three successive words are compared at all times.

Of coarse it is not quite that simple to encode all the situations encountered. The code must be made intricate enough to handle seven separate situational possibilities. These situations and their solutions are discussed in detail by Bisignani, Richards, and Whelan (1966).

4.13   TRANSFORM CODING

Transform coding of the video signals for digital TV usage is a subclass of the more general class of digital image processing.

Jet Propulsion Laboratory (JPL) made use of digital image processing techniques in the processing of space televised images. Most of this was done in nonreal time and took months of work before the pictures were in the final form. However, real time TV was also used for obtaining pictures from space.

Transform coding in some of its forms is very complex in implementation, requiring computer software. This is usually done in nonreal time. Real time versions of transform coding are less complex to implement. Both nonreal time systems and real time systems are considered here.

Necessity is not the sole reason for the growth of transform coding of images. Images can also be processed by optical means, the laser being one of the tools in these processes.

Digital image processing is growing because of (Hunt, 1975):

1 Hardware advances: The continued development of solid-state integrated-circuit technology.
2 Software advances: The discovery of fast convolutional algorithms.
3 Flexibility: The possession of much greater flexibility by digital systems to carry out nonlinear processing algorithms, iterative processes, and processes requiring tests and decision making, when compared to optical processors.

The great quantity of information in an image, and inherent nonlinearities combine to make digital image processing potentially one of the most complex branches of the field of digital signal processing.

Every visual scene is an image to the human eye. A mathematical basis can be derived for image formation processes.

4.13.1 Basic Mathematical Models for Image Processing

Consider a point in the plane of an image, point (x,y). The image at point (x,y) is a function of contributions in a neighborhood of x_1, y_1. Represent the image radiant energy distribution as $g(x,y)$ and the object radiant energy distribution as $f(x,y)$. The most general description of g is

$$g(x,y) = \int_{-\infty}^{\infty} \int_{-\infty}^{\infty} h\left[x, y, x_1, y_1, f(x_1, y_1) \right] dx_1 \, dy_1. \qquad (4.113)$$

The function h involving the object distribution as an argument allows for nonlinear processes in the formation of the image, since we have assumed nothing about linearity. The function h is referred to as the point-spread function.

This point-spread function can be simplified by assuming that it merely weights the object distribution as a scaler multiplier. From this assumption we get

$$h\left[x, y, x_1, y_1, f(x_1, y_1) \right] = h(x, y, x_1, y_1) f(x_1, y_1). \qquad (4.114)$$

The resulting equation is now linear

$$g(x, y) = \int_{-\infty}^{\infty} \int_{-\infty}^{\infty} h(x, y, x_1, y_1) f(x_1, y_1) dx_1 \, dy_1. \qquad (4.115)$$

Next, we make the point-spread function invariant as a function of position. Thus

$$h(x, y, x_1, y_1) = h(x - x_1, y - y_1) \tag{4.116}$$

which leads to a two-dimensional convolution

$$g(x, y) = \int_{-\infty}^{\infty} \int_{-\infty}^{\infty} h(x - x_1, y - y_1) f(x_1, y_1) dx_1 dy_1. \tag{4.117}$$

We can also assume that the point-spread function is separable (Nathan, 1968):

$$h(x, y, x_1, y_1) = h_1(x, x_1) h_2(y, y_1) \tag{4.118}$$

$$h(x - x_1, y - y_1) = h_1(x - x_1) h_2(y - y_2). \tag{4.119}$$

This leads to 2 one-dimensional operations on the object distribution

$$g(x, y) = \int_{-\infty}^{\infty} h(x, x_1) \left[\int_{-\infty}^{\infty} h(y, y_1) f(x_1, y_1) dy_1 \right] dx_1 \tag{4.120}$$

$$g(x, y) = \int_{-\infty}^{\infty} h(x - x_1) \left[\int_{-\infty}^{\infty} h(y - y_1) f(x_1, y_1) dy_1 \right] dx_1 \tag{4.121}$$

and the order of integration is arbitrary.

4.13.2   Image Recording Processes

Some means of recording the image is necessary in digital image processing. Current computing technology is not capable of performing digital image processing operations of interest on pictures even of moderate quality of commercial TV, at the present bandwidth of commercial TV. Bandwidth reduction is a necessity for digital TV image processing. Thus, digital image processing does not take place in real time, as many one-dimensional digital signal processing operations do. However, one-dimensional digital TV processing is practical in the area of transform coding.

4.13.3   Coding Processes

Coding as used in digital TV takes into account the statistical nature of the visual scene. When a scene is scanned in TV each intensity spot on a scan line carries a sample of the video signal. Such dots of light intensity have been labeled pixels. There is a great amount of statistical predictability between adjacent pixels. A way to describe this is to use the image covariance matrix:

$$R_g = E\left\{ [g - E(g)][g - E(g)]^T \right\} \tag{4.122}$$

where g is an N^2 by one-column vector formed by lexicographically ordering the rows of the image sample matrix, that is, row 1 of $g(j,k)$ occupies elements 1 through N of g, row 2 is in elements $N+1$ through $2N$, etc. E is the ensemble operator.

Cole (1973) has shown that all pictures look basically the same in a covariance or power spectrum sense. Habibi (1971) used a Markov process model to develop image covariance properties, which has been used successfully. Covariance statistics are not adequate to describe image structure, and the ensemble average is probably meaningful over ensembles of pictures, that look basically the same, but differ in minor random details.

The matrix R_g describes the degree of statistical dependence between pixels in the digital image. Another representation of R_g is

$$R_g = \Phi \wedge \Phi^T = \sum_{i=1}^{N} \lambda_i \phi_i \phi_i^T \tag{4.123}$$

where Φ is the N^2 by N^2 collection of orthogonal column eigenvectors and $\wedge$ is the diagonal matrix of (real) eigenvalues. The equation R_g describes a rotation of the data vector g into a space with uncorrelated coordinates. Furthermore, it has been shown that in the rotated (eigenvalues) space a great amount of statistical variability in the image data can be accounted for with a small number of coordinates in the transformed space and with minimum-mean-square error (MMSE) (Chen, 1973; Pearl, Andrews, and Pratt, 1972; Pratt and Andrews, 1970). Such behavior is a general property of R_g, for the discrete Karhunen-Loeve expansion (Fukanaga, 1973).

Besides the examination of statistical correlations between pixels, the occupancy of quantization levels must also be considered. In either the original or the transformed space, the assignment of an equal number of R bits to each pixel leads to inefficiency, since not all of the 2^R quantization levels will be present with equal probability. Therefore a number of schemes have been devised to make optimal use of bits assigned to quantization of data, including uniform and nonuniform quantization based on objective and subjective criteria.

The human eye is the final judge of pictorial quality. It has known limitations and some partially known idiosynchrasies. The use of these properties is referred to as psychovisual coding. The human eye is more sensitive to image spatial frequencies lying in the mid-range rather than in the low or high frequencies. Coding schemes can take advantage of these factors. Error criteria have been developed using weighting to account for these phenomena. The human eye also possesses behavior equivalent to logarithmic transformation. The incorporation of the logarithmic eye response into an image coding model results in image coding schemes that are less sensitive to quantization errors and more immune to channel noise (Stockham, 1972).

Another property of the human eye is neural inhibition in the retina, which leads to high-pass filter-like behavior (Hartline, 1964). Such behavior results in making high-frequency edges in images of more importance, and is utilized in a coding scheme known as synthetic highs (Graham, 1967).

The discovery of fast transform techniques and their applications have been of great importance to digital signal processing. In image coding, fast transforms have been used to approximate the coding efficiency of the Karhunen-Loeve expansions, while avoiding the computational requirements of Karhunen-Loeve.

Given the eigenvalues and eigenvectors from R_g we order the eigenvalues and their associated eigenvector

$$\lambda_1 \geqslant \lambda_2 \geqslant \lambda_3 \geqslant, \ldots, \lambda_N \qquad (4.124)$$

Then the coded image is derived by the transformation

$$g_1 = \Phi g. \qquad (4.125)$$

This means that the components of g_1 represent variations in descending order. By selecting and saving components of g_1, one can generate a vector of length $M < N^2$. Saving the largest components of g_1 is one approach that is used. The saved components are the only ones transmitted, along with bookkeeping information to indicate components actually saved. Thus, a reduction in bandwidth is achieved. At the receiver, the vector of length M is made into a vector g_2 of length N^2 by inserting zeros for the components not transmitted, and the image is reconstructed by the transformation

$$g_3 = \Phi^T g_2 \qquad (4.126)$$

the MSE is thus

$$\| g_3 - g \|^2 = \varepsilon \qquad (4.127)$$

and is minimal over all possible transformations.

Karhunen-Loeve expansion transformations require on the order of N^4 computations, for a typical $N = 500$. This is excessive. Experimental studies have shown that the covariance between two pixels rapidly approaches zero as the distance between them increases. Consequently, it is possible to code an image by breaking it up into subblocks of size P by P. Covariance and coding computations are performed at the subblock level where the computing time is proportional to $Q^2 P^4$, where $Q = N/P$. An appreciable reduction in computing time is thus achieved.

In transform coding the eigenvector transformation of g_1 is replaced by one of the known fast transform algorithms. The fast Fourier, Hadamard, Haar, and slant transforms have all been tried with success.

The purpose of the transform step is to create a domain in which the data are uncorrelated and signal energy is compacted into a small number of components. The previously mentioned fast transforms are noted for their energy compaction properties. The fast Fourier transform (FFT) algorithms can generate uncorrelated components as the value of N goes to infinity (Davenport and Root, 1958).

The gain in computation by the use of fast transforms is great. An entire $N \times N$ image can be transformed in $N^2 \log_2 N^2$ operations. Encoding a picture in P by P subblocks results in a still further reduction in computations operations.

PCM coding requires 6 bits/pixel, whereas fast transform coding requires only 1 bit/pixel. This amounts to a 6:1 bandwidth reduction. The picture quality resulting from the use of fast transform coding is about equal to that of Karhunen-Loeve coding (Chen, 1973; Pratt and Andrews, 1970).

Phase is also of importance in image coding. Tescher (1973) has shown that by use of the fast Fourier transform and the use of phase synchronization coding requirements are reduced to $\frac{1}{3}$ to $\frac{1}{2}$ bit/pixel. These are bandwidth reductions of 16 to 24 to 1.

4.13.4 Image Reconstruction

The fast transforms are used to reconstruct the original image with as much fidelity as possible. There are various degrading influences, however, in the transmission of the signal and errors creep in. An optimum solution to overcoming these degrading factors can take many forms, tailored to a given error-causing environment.

The density image restoration can be expressed as

$$g = H_T f + n \tag{4.128}$$

The MMSE is

$$\min E(f - \hat{f}) \tag{4.129}$$

where $\hat{f}$ denotes the restored image estimate. Since H_T is block Toeplitz, the block circulant approximation may be used. We can assume stationary image and noise covariances, and decoy to zero in a finite-time interval. This makes the resulting covariance matrices of f and n also block Toeplitz. The resulting algorithm is a linear digital spatial filter. It can be written in the frequency domain as

$$H_w(m,n) = \frac{\overline{H(m,n)}}{|H(m,n)|^2 + \dfrac{\Phi_n(m,n)}{\Phi_f(m,n)}} \tag{4.130}$$

The Wiener estimate can be used and is described by

$$\hat{F}(m,n) = H_w(m,n)\, G(m,n). \tag{4.131}$$

These are two-dimensional discrete Fourier transforms having lower case quantities for arguments. The Φ's are power spectra of noise and image, and the overbar denotes complex conjugate. The point-spread function (PSF) must be known, so that its Fourier transform may be computed for the filter. The application of Weiner filtering has been used successfully in several problem areas. (Harris, 1968; Helstrom, 1967; Horner, 1970; McGlamery, 1967; Pratt, 1972; Rushforth and Harris, 1968).

However, Weiner filtering has the undesired need for extensive *a priori* information, namely the PSF and detailed knowledge of image and noise covariance functions. Constrained least squares estimation can be used to overcome the requirement for covariance information (Hunt, 1973).

The constrained least squares estimate is obtained by solving the minimization problem

$$\text{minimize:} \, f^T C^T C f$$

$$\text{subject to:} \, \left[H_T f - g \right]^T \left[H_T f - g \right] = e \qquad (4.132)$$

where C is a constraint matrix, e is proportional to the noise variance, and T superscript denotes matrix transpose. Given the previous assumption on H_T and assuming that C is expressible in Toeplitz form, the discrete Fourier transform can be used to obtain a solution, and the frequency domain filter is (Hunt, 1973)

$$H_C(m, n) = \frac{\overline{H(m, n)}}{|H(m, n)|^2 + \lambda |C(m, n)|^2} \qquad (4.133)$$

where λ is a parameter determined by iteration.

In some works the point spread has been assumed to be unknown and then estimated from the degraded image by taking averages of image segments in the log spectral domain (Cannon, 1974; Cole, 1973; Stockham, 1972). These are homomorphic techniques describable in the frequency domain as

$$H_H(m,n) = \sqrt{\frac{1}{|H(m,n)|^2 + \dfrac{\Phi_n(m,n)}{\Phi_f(m,n)}}} \, . \qquad (4.134)$$

When a zero phase point-spread function is assumed, the homomorphic filter is the geometric mean between the Weiner filter and the inverse filter where

$$H_I(m,n) = \frac{1}{H(m,n)} \, . \qquad (4.135)$$

4.13.5 The FFT

The Cooley-Tukey FFT algorithm has found wide usage in the transform coding of images. The FFT is useful in obtaining bandwidth compression in image transmission. A practical scheme utilizing the FFT has been presented by Anderson and Huang (1971). They treated pictures adaptively on a piecewise basis rather than on an entire frame basis. They expanded picture subsections in Fourier series and discarded relatively low-energy Fourier coefficients to derive sets of numbers that are smaller in size for good picture quality reconstruction.

Consider a two-dimensional array of picture values $\beta_{m,n}$ obtained by sampling a picture at points of rectangular lattice, where $m=0,1,\dots,M-1$ and $n=0,1,\dots,N-1$. Each $\beta_{m,n}$ is proportional to brightness of the picture sample position labeled by m and n. The discrete Fourier transform of the discrete function $\beta_{m,n}$ is given as

$$F_{k,i} = \frac{1}{\sqrt{MN}} \sum_{m=0}^{M-1} \sum_{n=0}^{N-1} \beta_{m,n} \exp\left(-j2\pi\left(\frac{km}{M}+\frac{in}{N}\right)\right) \qquad (4.136)$$

where $k=0,1,\dots,M-1$ and $i=0,1,\dots,N-1$. The inversion is

$$\beta_{m,n} = \frac{1}{\sqrt{MN}} \sum_{m=0}^{M-1} \sum_{n=0}^{N-1} F_{k,i} \exp\left(+j2\pi\left(\frac{km}{M}+\frac{in}{N}\right)\right). \qquad (4.137)$$

In order to represent a matrix of picture elements, with a subset of size $L \leqslant MN$, of the complex exponentials of the discrete Fourier transform, we choose L exponentials with the largest coefficient magnitudes. This minimizes the MSE.

The method used here divided the picture into rectangular subblocks. Each subblock was expanded in a discrete Fourier transform, and the coefficients were subjected to adaptive thresholding. A quantity L was determined for each subblock that was linearly proportional to the integer part of

$$\sigma = \left[\frac{1}{MN} \sum_{m=0}^{M-1} \sum_{n=0}^{N-1} (\beta_{m,n} - \alpha)^2\right]^{1/2} \qquad (4.138)$$

where σ is the standard deviation of the subblock picture elements. The parameter α is the average value of the picture points in the subblock:

$$\alpha = \frac{1}{MN} \sum_{m=0}^{M-1} \sum_{n=0}^{N-1} \beta_{m,n}. \qquad (4.139)$$

The frequencies and complex amplitudes of the L Fourier coefficients of largest value are transmitted.

Linear adaptive quantization of the Fourier coefficients are used prior to transmission. The maximum Fourier coefficient, aside from $F_{0,0}$, determines the value of the largest quantization level within each subblock. The phases of the complex Fourier coefficients are linearly quantized on a scale from 0 to 2π. The number of bits used to quantize the Fourier coefficients are determined by σ_1 assuming that it is a measure of importance within each subblock. Either 2, 3, 4, or 5 bits are used to quantize all the phases and magnitude samples within a subblock. MN real numbers are needed to specify the discrete Fourier transform. $F_{0,0}$ is also finely quantized prior to transmission, independent of the value for σ. Coefficients not transmitted are set to zero in the receiver decoder.

The characteristics of the human eye are also taken into consideration in this system. The Weber fraction $\Delta B/B$, which relates the just noticeable difference in brightness ΔB to the background brightness B, is nearly constant over a wide range of values of B at about 2 percent. This contrast sensitivity of the eye causes errors of given brightness magnitude to be more noticeable in the darker areas of the picture. If a picture is subjected to an error environment, then the subjective quality of the resulting picture normally improves if the effects of the errors can be more evenly spread throughout the picture, rather than being more pronounced in the darker areas. Advantage of this is taken by using the logarithm of picture brightness before processing at the transmitter, and exponentiating after processing at the receiver.

There are several things to consider in choosing a subblock size. The value of σ usually increases with increase in subblock area, because correlation between picture points usually decrease as the distance between pixels becomes greater. An increase in error of the reconstructed picture is seen if the total number of Fourier coefficients is kept constant. The larger the subblock, the more likely that areas of little variation in brightness are included with areas of large variation. Areas that vary little are likely to be lost, because of the presence in the same subblock of the areas of large variation. This results from the use of a threshold that is adaptive. Smaller subblock sizes result in simple Fourier spectrum structure.

This simple spectrum structure can result in a good picture quality representation at the receiver, with the need for fewer Fourier coefficients. Subblock size can become too small and severely hamper data compression. The subblock size chosen for hardware implementation is 16×16.

In the implementation of this system all of the magnitude and phase samples within a subblock are linearly quantized to either 2, 3, 4, or 5 bits depending on the value of σ within that particular subblock. When σ is beneath the threshold level, 2 bits are used to quantize. Depending on how many of the three quantization thresholds are beneath σ, the number of quantization bits is increased to 3, 4, or 5 bits per magnitude or phase sample to transmit $F_{0,0}$ required 7 bits.

Run-length coding is used for position data based on an examination of the threshold of the Fourier spectra. Different sets of codewords are used to code the run lengths in different rows of the discrete Fourier transform. The Huffman (1952) coding procedure is used to select the codes for position data. Run-length coding is usually implemented by transmitting the intensity of the first pixel in a horizontal scan line of imagery data, followed by a run-length code representing the number of following pixels before a change in intensity occurs. When a change does occur, the intensity of the sample is again transmitted and the process repeated. Run-length encoding takes advantage of the statistics of the run lengths in a typical imagery and utilizes a coding scheme matched to the statistics to permit a reduction in average bits per pixel required for good quality picture reconstruction.

The method of encoding previously described is very complex to implement for TV. To divide a frame into 16×16 subblocks requires a large amount of bookkeeping as well as computer software. A special and very complex scan technique is required. This can be done for nonreal time TV, but is hardly practical for real time TV usage.

4.13.6 The Haar Transform

Haar (1955) proposed a simple transform composed of ones and minus ones, and zero elements. A 4×4 Haar matrix is as follows:

$$H_4 = \frac{1}{\sqrt{4}} \begin{vmatrix} 1 & 1 & 1 & 1 \\ 1 & 1 & -1 & -1 \\ \sqrt{2} & -\sqrt{2} & 0 & 0 \\ 0 & 0 & \sqrt{2} & -\sqrt{2} \end{vmatrix}. \qquad (4.140a)$$

When we extend this to an 8×8 matrix we obtain

$$H_8 = \frac{1}{\sqrt{8}} \begin{vmatrix} 1 & 1 & 1 & 1 & 1 & 1 & 1 & 1 \\ 1 & 1 & 1 & 1 & -1 & -1 & -1 & -1 \\ \sqrt{2} & \sqrt{2} & -\sqrt{2} & -\sqrt{2} & 0 & 0 & 0 & 0 \\ 0 & 0 & 0 & 0 & \sqrt{2} & \sqrt{2} & -\sqrt{2} & -\sqrt{2} \\ 2 & -2 & 0 & 0 & 0 & 0 & 0 & 0 \\ 0 & 0 & 2 & -2 & 0 & 0 & 0 & 0 \\ 0 & 0 & 0 & 0 & 2 & -2 & 0 & 0 \\ 0 & 0 & 0 & 0 & 0 & 0 & 2 & -2 \end{vmatrix}.$$

Extensions to higher order matrices are obvious.

The Haar transform has been likened to a sampling process in which rows of the transform matrix sample the input data sequence with finer and finer resolution increasing in powers of two. When it is used for imaging applications, it provides a transform domain in which a type of differential energy is concentrated in localized regions.

However, the picture quality resulting from the use of the Haar transform is poorer than that of other transforms.

4.13.7 The Walsh-Hadamard Transform

Whereas the FFT deals with complex numbers, the fast Walsh-Hadamard transform does not. The discrete Walsh-Hadamard function matrix is as shown in the following example for $n=4$:

$$
\begin{array}{c}
\phantom{\text{WAL}(k,j)} \\
j \\
\text{WAL}(k,j)
\end{array}
\begin{array}{cccc}
0 & 1 & 2 & 3 \\
\end{array}
\left[
\begin{array}{cccc}
1 & 1 & 1 & 1 \\
1 & 1 & -1 & -1 \\
1 & -1 & -1 & -1 \\
1 & -1 & -1 & -1 \\
\end{array}
\right]
\begin{array}{c}
k=0 \\
1 \\
2 \\
3 \\
\end{array}
\qquad (4.140)
$$

The matrix contains no complex number, but only ± 1's. This greatly reduces the number of computations needed to $N\log_2 N$ additions versus N^2 operations needed for Karhunen-Loeve. The FFT requires $N\log_2 N$ additions; but also requires $N\log_2 N$ multiplications.

The fast Walsh-Hadamard transform can be implemented in a parallel implementation or in a cascade implementation for both the one- and two-dimensional transforms. The two-dimensional fast Walsh-Hadamard transform can be implemented with proper choice of internal delay storage and timing as two-cascaded one-dimensional transforms. This eliminates the need for a buffer array storage and matrix rotation between the transforms. The two-dimensional version operates in real time on the input data stream (fed in row by row) with no input buffer and with only a fixed delay of $2(N-1)$ word times required.

The fast Walsh-Hadamard transformation matrix can be described in the following:

$$
[A_w] = (-1)\sum_{i=0}^{n-1} u_i x_i \qquad (4.141)
$$

where u_i and x_i are the binary bits representing the row and column indexes.

For symmetric Walsh-Hadamard matrices of the order $N=2^n$, the two-dimensional Walsh-Hadamard transform may be written in series form as (Andrews and Pratt, 1970):

$$
F(u,v) = \frac{1}{N}\sum_{x=0}^{N-1}\sum_{y=0}^{N-1} f(x,y)(-1)p(x,y,u,v) \qquad (4.142)
$$

where

$$
p(x,y,u,v) = \sum_{j=0}^{N-1} (u_j x_j + v_j y_j) \qquad (4.148)
$$

the terms u_j, x_j, v_j, y_j are binary representations of u, x, v, and y, respectively.

In the implementation of Walsh-Hadamard transform image processors, subblocks are used. By dividing the main block into subblocks and applying a transformation to each subblock rather than the entire image, the reduction in hardware is on the order of $\log_2 N / N_S$ for $N_S \times N_S$ subblocks. Mathematically speaking subblock performance can be indicated by noting that the sum of the squares of the cross correlations between different subblocks is invariant under the orthogonal subblock transformations. Thus if the subblocks are chosen large enough so that the tail of the cross-correlation function is relatively small, then the subblock performance should be comparable to the full transformation. Sakrison (1971) has analytic results based on rate distortion considerations, which show that for a fixed distortion no more than a factor of 2 or 3 can be saved in the number of bits required for optimal two-dimensional as opposed to line-by-line processing. Sakrison further points out that savings on the order of 10 or more are possible if the MSE distortion criteria is relaxed to a weighted-mean-squared-error criteria. This is equivalent to applying the MSE criteria to a spatially filtered version of the image. According to Sakrison, a good spatial filter should emphasize spatial differences on a local scale and de-emphasize large distance variations. We note that subframe processing has an inherent tendency to do this. Experimentally it has been shown that very little picture quality is sacrificed using subblocks versus full frame processing.

In some experimental results (Claire, Faber, and Green, 1971), the picture quality was improved in the 16×16 subblock version versus the full frame version.

In the implementation of Walsh-Hadamard bandwidth reduction schemes, the dynamic range of the transform domain coefficients and the general complexity of the processor increase with transform size, making the processing of successively small subblocks the desirable approach. For two-dimensional cascade processors, the complexity increases as $2\log_2 N$ and for parallel processors as N^2 as N increases.

After the image has been transformed either in one- or two-dimensional processors, the redundancy reduction is accomplished by selecting and efficiently quantizing the high energy coefficients. There are two preferred types of coefficient selection schemes, zonal and threshold sampling. Zonal sampling is the simpler of the two to implement, and assumes that the highest energy coefficients will usually lie within a certain zone of the transform domain. Threshold sampling is more difficult to implement, but permits the highest energy coefficients to be selected wherever they are. For digital TV usage threshold sampling is optimal for minimizing error for a given amount of redundancy reduction. However, threshold sampling requires position bits to identify the particular coefficients selected and the resulting bandwidth reduction is less because of the required bookkeeping, for a distortion level.

Fortunately, experimental data have shown that the bulk of the transform domain energy for a two-dimensional Walsh-Hadamard transform lies in a hyperbolic zone about the origin and extending along both axes (Pratt and Andrews, 1970). This permits the zonal sampling techniques (which requires no bookkeeping bits) to be used. However, there is increased distortion above that for threshold sampling; for a given fraction of transmitted coefficients.

It has been established that uniform block quantization for the selected coefficients is the most efficient and practical method (Wintz and Kurtenbach, 1968). In this method the quantization error is minimized by dividing M total bits available for n^2 selected samples, with variances σ_{ij}^2, $i,j = 1, 2, \ldots, n$, by assigning m_{ij} bits to the samples in proportion to the logarithm of their variance. Thus the optimum number of bits per sample is

$$m_{ij} = \text{integer}\left[\frac{M}{n^2} + 2\log \sigma_{ij}^2 - \frac{1}{n^2} \sum_i^n \sum_j^n \log \sigma_{ij}^2 \right] \qquad (4.144)$$

with the constraint

$$M = \sum_{i=1, j=1}^{n} m_{ij}. \qquad (4.145)$$

The most practical nonadaptive transform image encoding scheme would use the two-dimensional fast Walsh-Hadamard transform on subblocks of the full frame, with zonal sampling and uniform block quantization. However, experimental results indicate that the goal of good picture quality and a 1 bit/pixel average transmitted cannot be achieved with this system for pictures with moderate high detail. The use of an adaptive scheme is necessary to achieve such a goal.

However, threshold sampling requires considerable hardware complexity, because a large memory to store multiple subblock transforms and threshold circuitry are required.

At the present writing, the fast Walsh-Hadamard transform is the best one to use when combined performance and hardware feasibility are considered. It is assumed that subblock techniques are necessary for dynamic range and transform size implementation reasons.

A promising fixed-rate transform scheme, which uses an adaptive two-dimensional subblock fast Walsh-Hadamard transform, has been proposed (Claire, 1970). This scheme proposes a combined zonal and threshold sampling technique. Claire says:

Suppose that an estimate of your anticipated typical imagery spatial correlation function is available such that you can specify the best subframe size. All literature references thus far said that an $N \times N$ subframe with $N \leqslant 16$ is entirely sufficient for highly detailed reconnaissance data. For this example say $N = 8$ is selected.

After performing the two-dimensional transform on the 8×8 array (subframe) of 6 bit pixels, the 64 resulting transform domain coefficients must be sampled and quantized to provide the desired 1 bit per pixel average. A suggested scheme would be to use zonal sampling set up to always select the same 9 transform domain coefficients. Then, block quantize these using the optimum uniform block quantization rule of Wintz and Kurtenbach, to say $M=40$ bits with the bit assignment based on their variances as discussed in the preceding section. An additional four high sequence coefficients would be selected using threshold sampling and quantized to 2 bits, two with a 4 bit ID and two with a 3 bit ID, to identify which position in particular blocks of 16 and 8 coefficients that they were chosen from. Then, assuming an average of 2 bits per subframe for sync (16 bit frame-sync word every 8 subframes) the total is 64 bits per subframe or an average of 1 bit per pixel.

4.13.8   Slant Transform Coding

A new image transform, called the slant transform and specifically designed for image coding, has been developed (Pratt, Chen, and Welch, 1974). The transformation possesses a discrete sawtoothlike basis vector that efficiently represents brightness variations along an image line. A fast computational algorithm is used with it to give superior reproduction of picture quality.

The slant transform has performance characteristics that are a little less optimum than for the Karhunen-Loeve transform. It is less costly to implement then Karhunen-Loeve but more costly than fast Fourier.

The slant transform has the following properties: 1) orthonormal set of basis vectors; 2) one constant basis vector; 3) one slant basis vector; 4) sequency property; 5) variable size transformation; fast computational algorithm, and 6) high energy compaction.

The slant transform is a two-dimensional transform. Let $[F]$ be an $N \times N$ matrix of picture element brightness values of an image and let $[f_i]$ be an $N \times 1$ vector representing the ith column of $[F]$. The one-dimensional transform of the ith image line is then

$$[f_i] = [S][f_i]$$ (4.146)

where $[S]$ is the $N \times N$ unitary slant matrix. A two-dimensional slant transform is performed by sequential row and column transformations on $[F]$, yielding

$$[\mathscr{F}] = [S][F][S]^T.$$ (4.147)

The inverse transformation $[F]$ from the transform components $[\mathscr{F}]$ is given by

$$[F] = [S]^T[\mathscr{F}][S].$$ (4.148)

A series representation can also be shown:

$$\mathscr{F}(u,v) = \sum_{j=1}^{N} \sum_{k=1}^{N} F(j,k) S(u,j) S(k,v)$$ (4.149)

and

$$F(j,k)= \sum_{u=1}^{N} \sum_{v=1}^{N} \mathscr{F}(u,v)S(j,u)S(v,k). \qquad (4.150)$$

The slant transform matrix of order 2 is given by

$$S_2 = \frac{1}{\sqrt{2}}\begin{bmatrix} 1 & 1 \\ 1 & -1 \end{bmatrix}. \qquad (4.151)$$

The forth-order matrix is obtained by

$$S_4 = \frac{1}{\sqrt{2}}\begin{bmatrix} 1 & 0 & 1 & 0 \\ a_4 & b_4 & -a_4 & b_4 \\ 0 & 1 & 0 & -1 \\ -b_4 & a_4 & b_4 & a_4 \end{bmatrix}\begin{bmatrix} S_2 & 0 \\ 0 & S_2 \end{bmatrix}. \qquad (4.152)$$

or

$$S_4 = \frac{1}{\sqrt{2}}\begin{bmatrix} 1 & 1 & 1 & 1 \\ (a_4+b_4) & (a_4-b_4) & (-a_4+b_4) & (-a_4-b_4) \\ 1 & -1 & -1 & 1 \\ (a_4-b_4) & (-a_4-b_4) & (a_4+b_4) & (-a_4+b_4) \end{bmatrix} \qquad (4.153)$$

where a_4 and b_4 are scaling constants. The first matrix row is made up of positive constants. The second row is linear with a negative slope. The step sizes between adjacent elements of the slant vector are $2b_4$, $2a_4-2b_4$, and $2b_4$. By setting these step sizes equal, one finds that

$$a_4 = 2b_4. \qquad (4.154)$$

The orthonormality condition $[S_4][S_4]^T = [I]$ leads to

$$b_4 = \frac{1}{\sqrt{5}}. \qquad (4.155)$$

When the values for a_4 and b_4 are substituted into the matrix we get

$$S_4 = \frac{1}{2}\begin{bmatrix} 1 & 1 & 1 & 1 \\ \dfrac{3}{\sqrt{5}} & \dfrac{1}{\sqrt{5}} & \dfrac{-1}{\sqrt{5}} & \dfrac{-3}{\sqrt{5}} \\ 1 & -1 & -1 & 1 \\ \dfrac{1}{\sqrt{5}} & \dfrac{-3}{\sqrt{5}} & \dfrac{3}{\sqrt{5}} & \dfrac{-1}{\sqrt{5}} \end{bmatrix}. \qquad (4.156)$$

The rows of S_4 form an orthonormal set. Furthermore, S_4 possesses the sequency property; each row has a distinct number of sign reversals and each integer from 0 to 3 is the number of sign reversals of some row.

A general matrix form has been derived of the order of $N(n=2^n, n=3,\ldots)$. An order $N/2$ form is as follows:

$$
S_N = \frac{1}{\sqrt{2}}
\left[
\begin{array}{cccc|cccc}
1 & & 0 & & & 1 & 0 & \\
 & 0 & & & & & & 0 \\
a_N & b_N & & & & -a_N & b_N & \\
\hline
 & 0 & & I_{(N/2)-2} & & 0 & & I_{(N/2)-2} \\
0 & & 1 & & & 0 & -1 & \\
 & 0 & & & & & & 0 \\
-b_N & a_N & & & & b_N & a_N & \\
\hline
 & 0 & & I_{(N/2)-2} & & 0 & & -I_{(N/2)-2}
\end{array}
\right]
\left[
\begin{array}{c|c}
S_{N/2} & 0 \\
\hline
0 & S_{N/2}
\end{array}
\right]
$$

(4.157)

where $I_{(N/2)-2}$ is the identity matrix of dimension $(N/2)-2$ and the various partition blocks are determined by the same considerations as described previously for 4. The constants a_N and b_N may be computed from

$$a_2 = 1 \tag{4.158}$$

$$b_N = \left[1 + 4(a_{N/2})^2\right]^{-1/2} \tag{4.159}$$

$$a_N = 2b_N a_{N/2} \tag{4.160}$$

$$a_{2N} = \left(\frac{3N^2}{4N^2-1}\right)^{1/2} \tag{4.161}$$

$$b_{2N} = \left(\frac{N^2-1}{4N^2-1}\right)^{1/2}. \tag{4.162}$$

The basic waveforms for a slant transform are shown in Fig. 4.28.

This slant transform requires a total of $N\log_2 N + (N/2) - 2$ additions and subtractions plus $2N-4$ multiplications for an N-dimensional data vector. This can be compared to the computations required for the Walsh-Hadamard, which is $N\log_2 N$. When N is equal to 16 the slant transform requires 53 percent more computations than for the Walsh-Hadamard transform.

For statistical analysis it is assumed that the image array $F(j,k)$ is a sample of the stochastic process with mean

$$E[F(j,k)] \equiv \overline{F(j,k)} \tag{4.163}$$

Wave
Number
u

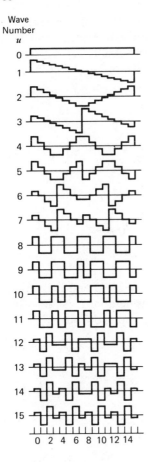

0 2 4 6 8 10 12 14   **Figure 4.28.** Basic slant transform waveforms.

and covariance of

$$E\left\{\left[\,F(j_1,k_1)-\overline{F(j_1,k_1)}\,\right]\left[\,F(j_2,k_2)-\overline{F(j_2,k_1)}\,\right]\right\}\equiv C_F(j_1,k_1,j_2,k_2).\quad (4.164)$$

A convenient, and reasonably accurate, covariance model for an image is the first-order Markov process model for which

$$C_F(j_1,k_1,j_2,k_2)=\sigma_R^2\sigma_C^2 pR^{|j_1-j_2|}pC^{|k_1-k_2|}\qquad (4.165)$$

where σ_R^2 and σ_C^2 are the row and column variances and pR and pC are the adjacent pixel correlation factors.

The mean and variance of the slant transform samples can be expressed as

$$E\left[\,\mathcal{F}(u,v)\,\right]\equiv\overline{\mathcal{F}(u,u)}\;=\sum_{j}^{N}\sum_{k}^{N}\overline{F(j,k)}\,S(u,j)S(k,v)\qquad (4.166)$$

and

$$C_{\mathcal{F}}(u_1,u_2,v_1,v_2) = \sum_{j_1}^{N} \sum_{j_2}^{N} \sum_{k_1}^{N} \sum_{k_2}^{N}$$

$$C_F(j_1,j_2,k_1,k_2) \cdot S(u_1,j_1)$$

$$S(k_1,v_1)S(u_2,j_2)S(k_2,v_2). \tag{4.167}$$

If the image array is stationary, then as a result of the orthogonality of the kernel

$$\overline{\mathcal{F}(0,0)} = \overline{NF(j,k)} \tag{4.168}$$

$$\overline{\mathcal{F}(u,v)} = 0 u, v \neq 0. \tag{4.169}$$

No closed form expression has been found for the covariance of the slant transform samples, but the covariance may be computed as the two-dimensional slant transform of the function $C\{j_1 - j_2, k_1 - k_2\}$ if the original image field is stationary.

The transform domain sample $\mathcal{F}(0,0)$ is a nonnegative weighted sum of pixel values. Its histogram will generally follow the histogram of pixel values, which is often modeled by a Rayleigh density

$$p_{\mathcal{F}(0,0)}(x) = \left[2\pi\sigma^2(u,v)\right]^{-1/2} \exp\left[-\frac{x^2}{2\sigma^2(u,v)}\right] \quad \text{for } x \geqslant 0; \ (u,v) \neq (0,0)$$

$$\tag{4.170}$$

where $\sigma^2(u,v) = C_{\mathcal{F}}(u,u,v,v)$ is the variance of the transform samples.

Either zonal or threshold sampling can be used with the slant transform. There are a number of zones that could be used for zonal sampling. However, analysis and experimental studies have indicated that the optimum zone is the so-called maximum variance zone. The MSE can be written as

$$\overline{\varepsilon^2}(t) = 1/N^2 \sum_{u}^{N} \sum_{v}^{N} E\left\{\mathcal{F}(u,v)\left[1 - T(u,v)\right]^2\right\} \tag{4.171}$$

where $T(u,v)$ is a transform domain sampling function, which is valued at one for transmitted samples and zero for those samples not transmitted. For the maximum variance zone $T(u,v)$ is chosen to be unity for those samples having the largest variance values (Pratt, Chen, and Welch, 1974). Their Fig. 6 shows a plot of MSE versus block size for the Karhunen-Loeve transform, the slant transform, the Fourier transform, the Hadamard transform, and the Haar transform, for zonal sampling conditions. Of course, the Karhunen-Loeve transform gives the best results. However, the slant transform is not degraded very much from the results for the Karhunen-Loeve transform. For large block sizes, the Fourier transform is progressively better, than are the Hadamard and Haar transforms. This is shown here as Fig. 4.29.

When zonal sampling is used with the slant transform, bandwidth compression of 1.5 bits/pixel can be obtained for good quality pictures.

Pratt, Chen, and Welch (1974) combined threshold sampling with runlength coding to obtain somewhat better results than with zonal sampling. They also applied slant transform coding to color pictures and obtained 2 : 1 bandwidth compression ratios.

The version of slant transform image coding described by Pratt, Chen, and Welch requires a considerable amount of bookkeeping to keep track of the various subblock samplings. In this form it requires a prohibitive amount of hardware to implement it. This could have an application in nonreal time digital RV. It is not likely that commercial TV could use it for real time digital TV.

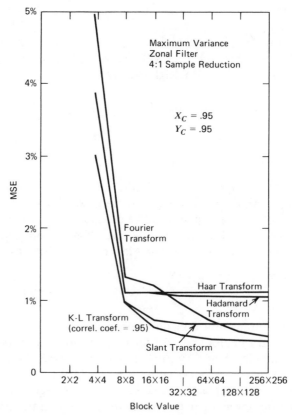

Figure 4.29. Zonal sampling MSE performance of various image transforms as a function of block size.

However, another version of the slant transform has been experimentally tried in Japan by Enomoto and Shibata (1970). In their report, it was used to implement digital TV. In this system a bandwidth compression of 2.5 bits/pixel was obtained.

They claim that use of the slant transformation for TV images is most advantageous because linear variation of brightness occurs in many portions of TV pictures.

An output component of the orthogonal transformation of images should be the dc components, and slant of brightness should be expressed as an output of the orthogonal transformation.

They made use of two scanning approaches: (1) a one-line method in which eight sample values were taken in sequence; and (2) a two-line method in which four sample values (total eight) were taken from two adjacent scan lines. This was used rather than a subblock system that requires much bookkeeping.

In this system each of the eight samples has its associated slant transform. These eight outputs are fed in parallel to the encoder. A simplified block diagram is shown in Fig. 4.30.

The coding system used was PCM. A more detailed block diagram of the system is shown in Enomoto and Shibata's Fig. 17 and shown here as Fig. 4.31. This depicts the one-line version, which is simpler than a two-line version, while the performance of the two systems is about equal. The system works as follows. The video input signal is delayed 0.1 μsec by 0.1 μsec delay circuits and 8 parallel video signals are obtained. From the 8 video signals, x_i, 8 orthogonal parallel transform outputs, and h_j were obtained through the orthogonal transform circuit consisting of a resistor matrix network. These 8 outputs were simultaneously sampled at 1.25 MHz ($\frac{1}{8}$ of 10 MHz), they were then encoded by PCM.

The inverse orthogonal transform equipment decodes the transmitted PCM signal to make PAM signals. A parallel to serial transformation is used and the output video is obtained. This system is simple to implement and has real time TV applications.

4.13.9 Cosine Transform Coding

The discrete cosine transform (DCT) can be implemented with a fast computational algorithm, and has a performance that is close to that of

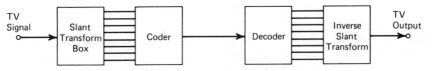

Figure 4.30. Block diagram of a slant transform model.

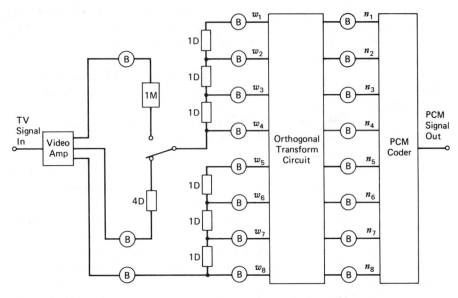

(a) Block Diagram of Equipment of Orthogonal Transformation

WHERE B: BUFFER AMPL.
nD: n PICTURE ELEMENT INTERNAL DELAY CIRCUIT
1M: HORIZONTAL SWEEP INTERNAL DELAY CIRCUIT

(b) Block Diagram of Equipment of Inverse-orthogonal Transformation

Figure 4.31. Block diagram of an orthogonal transform PCM system. (a) Equipment of orthogonal transformation. (b) Equipment of inverse-orthogonal transformation. B=buffer amp; nD=n picture element interval delay circuit (n =1–4, D=0.1 msec); 1H=horizontal sweep interval delay circuit (H=635 msec).

Karhunen-Loeve transform. Although the Karhunen-Loeve transform is considered to be the optimal transform, there is no general algorithm that enables its fast computation.

The DCT of a data sequence $X(m)$, $m = 0, 1, 2, \ldots, (M-1)$ is defined by Ahmed, Natarajan, and Rao (1974) as

$$G_x(0) = \frac{\sqrt{2}}{M \sum\limits_{m=0}^{M-1} X(m)} \tag{4.172}$$

$$G_x(k) = \frac{2}{M \sum\limits_{m=0}^{M-1} X(m)} \cos \frac{(2m+1)k\pi}{2M}, \qquad k = 1, 2, \ldots (M-1) \tag{4.173}$$

where $G_x(k)$ is the kth DCT coefficient. It is worthwhile noting that the set of basis vectors $\{1/\sqrt{2}, \cos[(2m+1)k\pi]/2M\}$ is actually a class of discrete Chebyshev polynomials. This can be seen by recalling that Chebyshev polynomials can be defined as

$$\hat{T}_0(\xi_p) = \frac{1}{\sqrt{2}} \tag{4.174}$$

$$\hat{T}_k(\xi_p) = \cos(k \cos^{-1}\xi_p), \qquad k, p = 1, 2, \ldots, M \tag{4.175}$$

where $\hat{T}_k(\xi_p)$ is the kth Chebyshev polynomial.

The expression ξ_p is chosen to be the pth zero of $\hat{T}_M(\xi)$, which is denoted by

$$\xi_p = \cos \frac{(2p-1)\pi}{2M}, \qquad p = 1, 2, \ldots, M. \tag{4.176}$$

When expression (4.176) is substituted into expression (4.173), we obtain a set of Chebyshev polynomials

$$\hat{T}_0(p) = \frac{1}{\sqrt{2}} \tag{4.177}$$

$$\hat{T}_k(p) = \cos \frac{(2p-1)k\pi}{2M}, \qquad k, p = 1, 2, \ldots, M. \tag{4.178}$$

It follows that $\hat{T}_k(p)$ can be equivalently defined as

$$T_0(m) = \frac{1}{\sqrt{2}} \tag{4.179}$$

$$T_k(m) = \cos \frac{(2m+1)k\pi}{2M}, \qquad k = 1, 2, \ldots, (m-1) \, m = 0, 1, \ldots, (m-1). \tag{4.180}$$

The inverse DCT can be written as

$$X(m) = \frac{1}{\sqrt{2}} G_x(0) + \sum_{k=1}^{M-1} G_x(k)\cos\frac{(2m+1)k\pi}{2M} \qquad \text{for } m = 0, 1, \ldots, (M-1).$$

(4.181)

The two-dimensional form of the inverse DCT can be written as

$$X(m, n) = \frac{2}{M} \sum_{k=0}^{M-1} \sum_{j=0}^{M-1} G_x(k, j)\cos\frac{(2m+1)k\pi}{2M}\cos\frac{(2n+1)j\pi}{2M},$$

(4.182)

which of course gives a forward DCT of

$$G_x(k, j) = \frac{2}{M} \sum_{k=0}^{M-1} \sum_{j=0}^{M-1} X(m, n)\cos\frac{(2m+1)k\pi}{2M}\cos\frac{(2n+1)j\pi}{2M},$$

(4.183)

which can also be written as

$$G_x(k, j) = \frac{2}{M} R_e\left\{ \exp\frac{i\pi m}{2M} \sum_{k=0}^{M-1} \sum_{j=0}^{M-1} X(m, n)\exp\left[-\frac{\pi i}{M}(mk + nj) \right] \right\}.$$

(4.184)

It is also of interest to note that all of the M DCT coefficients can be computed using a $2M$-point FFT. The inverse can, likewise, be computed using the FFT.

4.13.10 Bandwidth Compression Application of the DCT

As was mentioned above in Section 4.13.8, the DCT gives picture performance, which is very close to that of the Karhunen-Loeve transform. Martinson (1978) has made practical application of the DCT in a bandwidth reduction television system, using a two-dimensional FFT implementation. The DCT can be computed with the FFT algorithm by either using a data set twice as long with zeros inserted in the second half, or by using a symmetric FFT aperture created by folding the input data sequence. The latter approach is used in this bandwidth compression model, which processes two 32-point real symmetric image data channels in the complex FFT operation to obtain two 16-point real output transforms.

The application of bandwidth reduction techniques to wideband image transmission requires real time two-dimensional transformations of image samples at rates of about 10 megasamples/sec. Martinson has made use of radar matched filtering techniques, which use pipelined FFT processors. Excellent image quality has been achieved with a 4:1 bandwidth reduction ratio.

In this system, the input video, which has been sampled by a scanner and an A/D converter to form a time series of the image, is first organized into subpictures. The spatial domain subpictures then undergo a two-dimensional Fourier transform, which provides a frequency domain representation. The usual 16×16 pixel subpicture is chosen.

After this transformation an algorithm is applied, which assigns a specified number of bits to each frequency coefficient. The algorithm retains the low-frequency terms by varying degrees, while eliminating or reducing the quantization levels of the high-frequency terms.

A functional block diagram is shown in Martinson's Figure 3. It is reproduced here as Fig. 4.32. The input data are stored in an input buffer, which outputs 2048 replications of each subpicture on two output parallel data word lines to the two-dimensional FFT processor. The data are outputted from the buffer in line pairs that make up the in-phase (I) and quadrature (Q) words of the FFT processor. The data rate at this point is 5 MHz for each parallel word. The input data next passes through a selector switch, which alternately selects input data and first-dimensional transform output from the two-dimensional corner turning buffer. The output data rate from the switch is 10 MHz.

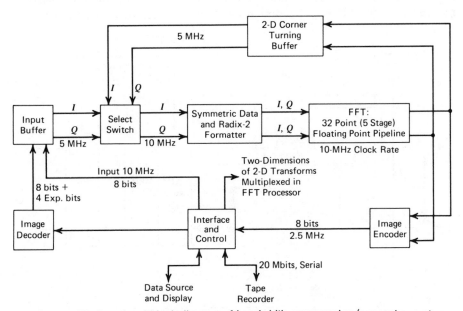

Figure 4.32. Functional block diagram of bandwidth compression/expansion system.

Next, formatting the data to provide a 32-sample symmetric signal and radix-2 operation is accomplished. Essentially, four parallel data lines are fed from the data formatter to the 32-point pipeline FFT. Given the single input word rate of 10 MHz, the processing rate effectively becomes 40 MHz, a factor of 2 increase for the 32-point symmetry and another factor of 2 for the second dimension of the two-dimensional transform. The four parallel lines processed by the FFT, $I, Q(2)$, and radix-2(2) compensate and give a processing clock rate of 10 MHz for the input data word rate of 10 MHz.

Transform data are fed from the FFT to the image encoder where the bandwidth compression algorithm is applied. The algorithm used organizes the data in diagonal band groupings, which are normalized relative to the maximum magnitude of the group. All data points other than the dc term pass through a linear-log quantizer. The final output bit sequence from a 16×16 subpicture consists of 512 bits assigned as follows: synchronization—16 bits; dc term—8 bits; diagonal band magnitudes—24 bits; 173 nonzero image coefficients—464 bits.

The final function in the decode mode is the conversion of the quantized data to a continuous serial bit stream with no gaps between subpictures.

The decode functions are essentially the inverse of the encoding functions.

Of course, the key functional element in this bandwidth reduction system is the pipeline FFT subsystem. The functional flow of the radix-2 pipeline FFT is shown in Martinson's Fig. 4, and is reproduced here as Fig. 4.33. Each stage in the FFT consists of a complex multiplier, an adder, a subtractor, phase reference, a memory unit, and a switching and control system. Computer simulation studies of radar matched filters have shown that performance with a fixed-point FFT design was limited to a narrow dynamic range.

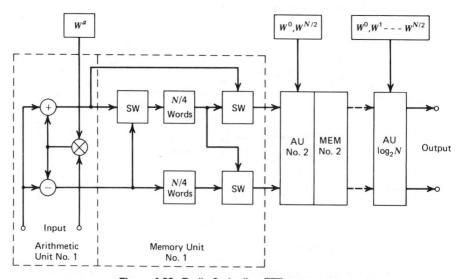

Figure 4.33. Radix-2 pipeline FFT processor.

As the signal is processed in an FFT its magnitude grows, resulting in a "bit growth." The bit growth can be most efficiently handled by using a floating-point approach. The floating-point approach uses a complex word consisting of two fixed-word-size mantissas and a single-word-size exponent representing powers of 2.

The computation time of the system should be mentioned here. The two-dimensional FFT consisting of 32 cosine transforms of 16 sample points must be computed in 25.6×10^{-6} sec. A 32-point real symmetric sequence requires $(32/2) \log_2 32 = 80$ computational steps for radix-2 operation. If two real transforms are computed in the complex FFT, a total of $16 \times 80 = 1280$ complex arithmetic operations must be performed per subpicture. If a single arithmetic unit were used for the FFT computation, its speed would have to be $25.6 \times 10^{-6}/1280 = 20$ nsec per computation. A pipeline FFT of 5 cascaded stages increases the required computation time to 100 nsec (10 MHz rate).

4.13.11 Sine Transform Coding

The sine transform was introduced by Jain (1974) as a fast algorithmic substitute for the Karhunen-Loeve transform of a Markov process. Jain and Angel (1974) make use of the tridiagonal matrix. They suppose Q to be an arbitrary symmetric tridiagonal matrix of order n

$$Q = \begin{bmatrix} \alpha & \beta & 0 & \cdot & \cdot & \cdot & \cdot & 0 \\ \beta & \alpha & \beta & \cdot & \cdot & \cdot & \cdot & \cdot \\ 0 & \beta & \alpha & \beta & \cdot & \cdot & \cdot & \cdot \\ 0 & \cdot & \cdot & \cdot & \beta & \alpha & \beta & 0 \\ 0 & \cdot & \cdot & \cdot & \cdot & \beta & \alpha & \beta \\ 0 & \cdot & \cdot & \cdot & \cdot & \cdot & \beta & \alpha \end{bmatrix}. \tag{4.185}$$

Q can be written as

$$Q = \alpha I + \beta P \tag{4.186}$$

where

$$P = \begin{bmatrix} 0 & 1 & 0 & \cdot & \cdot & \cdot & \cdot & 0 \\ 1 & 0 & 1 & 0 & \cdot & \cdot & \cdot & \cdot \\ 0 & 1 & 0 & 1 & 0 & \cdot & \cdot & \cdot \\ 0 & \cdot & \cdot & \cdot & \cdot & \cdot & \cdot & \cdot \\ 0 & \cdot & \cdot & \cdot & \cdot & \cdot & \cdot & \cdot \\ 0 & \cdot & \cdot & \cdot & 1 & 0 & 1 & 0 \\ 0 & \cdot & \cdot & \cdot & \cdot & 1 & 0 & 1 \\ 0 & \cdot & \cdot & \cdot & \cdot & 0 & 1 & 0 \end{bmatrix}. \tag{4.187}$$

If

$$P = [M][\Lambda][M]^T \tag{4.188}$$

where $[\Lambda]$ is a diagonal matrix, $[M]$ is orthogonal, and $[M]^T$ is the transpose of $[M]$, then

$$[Q] = [M][D][M]^T \tag{4.189}$$

where the diagonal matrix

$$[D] = \alpha I + \beta \Lambda. \tag{4.190}$$

When a vector h is defined as

$$h = [h_i] = \left[\sin \frac{ki\pi}{n+1}\right] \tag{4.191}$$

for k an arbitrary integer. Then, using some trigonometric identities, it is easy to show that the vector d is given by

$$d = Ph \tag{4.192}$$

and can be written as

$$d = [d_i] = \left[2\sin \frac{ki\pi}{n+1} \cos \frac{k\pi}{n+1}\right]. \tag{4.193}$$

Thus

$$Ph = \lambda h, \tag{4.194}$$

if and only if

$$\lambda = 2\cos \frac{k\pi}{n+1}. \tag{4.195}$$

Since k is arbitrary, a complete set of eigenvalues is given by

$$\lambda_j = 2\cos \frac{j\pi}{n+1} \qquad j = 1, 2, \ldots, n \tag{4.196}$$

and the corresponding eigenvectors are given by

$$h_i = [h(i,j)] = \left[\sin \frac{ij\pi}{n+1}\right]. \tag{4.197}$$

$[M]$ is then constructed from $\{h_j\}$ by normalizing each vector

$$[M] = \sqrt{\frac{2}{n+1}} \; [h_1, h_2, \ldots, h_n], \qquad (4.198)$$

which can also be written as

$$[M(i,j)] = \sqrt{\frac{2}{n+1}} \; \sin\frac{ij\pi}{n+1} \qquad \text{for } i,j = 0, 1, 2, \ldots, (n-1), \quad (4.199)$$

which is the basic one-dimensional sine transform.

The two-dimensional sine transform may be written as

$$G_x(k,j) = \frac{2}{M} \sum_{k=0}^{M-1} \sum_{j=0}^{M-1} X(m,n) \sin\frac{(k+1)(m+1)\pi}{N+1} \sin\frac{(j+1)(n+1)\pi}{N+1}.$$

$$(4.200)$$

Discussion on Fast Computation

Jain and Angel (1974) give the following means of achieving fast computation. Suppose we have a two-dimensional monochromatic image, with $N \times N$ pixels. Let $x_{n,m}$ represent the intensity at the spatial coordinate (n,m). We assume the image belongs to a class of images whose autocorrelation function can be represented by

$$E[x_{n,m} x_{n+1,m+1}] = R_{ij} = \sigma^2 \exp(-\gamma_i|i| - \gamma_2|j|). \qquad (4.201)$$

The model for image representation is based on the assumption that the random variable $x_{i,j}$ is correlated directly with its nearest neighbors as is shown in Fig. 4.34.

From Fig. 4.34 we may write

$$x_{i,j} = \alpha_1 x_{i,j+1} + \alpha_2 x_{i+1,j} + \alpha_3 x_{i,j-1} + \alpha_i x_{i-1,j} + \beta u_{i,j}, \qquad (4.202)$$

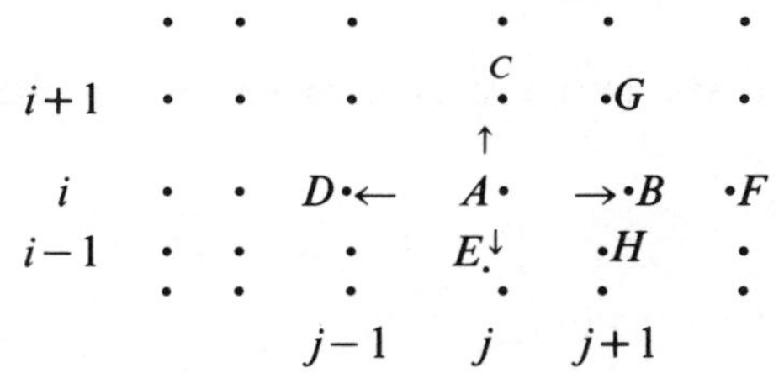

Figure 4.34. Nearest neighbors model for two-dimensional images.

where $u_{i,j}$ are zero-mean uncorrelated random variables. It should be noted that, in this assumption, each point is correlated with every other point through its nearest neighbors. Thus while point A in Fig. 4.32 is directly correlated with only B, C, D, and E, it is, for example, indirectly correlated with F, G, and H since B is directly correlated with A, F, G, and H. This model is in striking contrast with other models. This model lies in the realm of two-point boundary value problems since every point is correlated with every other point. Thus the model requires more sophisticated numerical algorithms, which fortunately turn out to be computationally feasible.

First, let us find values for α_1, α_2, α_3, and α_4, so that the model has an autocorrelation matrix close to that given by Expression (4.201). Writing equations for the regression coefficients α_1, α_2, α_3, and α_4 by taking the correlation of $x_{i,j}$ with its nearest neighbors, it is found that

$$R_{01} = \alpha_1 R_{00} + \alpha_2 R_{11} + \alpha_3 R_{02} + \alpha_4 R_{11} \tag{4.203}$$

$$R_{10} = \alpha_1 R_{11} + \alpha_2 R_{00} + \alpha_3 R_{11} + \alpha_4 R_{20} \tag{4.204}$$

$$R_{01} = \alpha_1 R_{01} + \alpha_2 R_{11} + \alpha_3 R_{00} + \alpha_4 R_{11} \tag{4.205}$$

$$R_{10} = \alpha_1 R_{11} + \alpha_2 R_{20} + \alpha_3 R_{11} + \alpha_4 R_{00}. \tag{4.206}$$

Solving for α_1, α_2, α_3, and α_4 we get

$$\alpha_1 = \alpha_3 = \frac{(1 + r_{02})r_{10} - 2r_{11}r_{01}}{(1 + r_{02})(1 + r_{20}) - 4r_{11}^2} \tag{4.207}$$

$$\alpha_2 = \alpha_4 = \frac{(1 + r_{20})r_{01} - 2r_{11}r_{10}}{(1 + r_{02})(1 + r_{20}) - 4r_{11}^2} \tag{4.208}$$

where

$$r_{ij} = \frac{R_{ij}}{R_{00}}, \qquad r_{00} = \sigma^2 \tag{4.209}$$

and

$$\beta^2 = |1 - 2(\alpha_1 r_{10} + \alpha_2 r_{01})|, \tag{4.210}$$

where each u_{ij} is assumed to be a zero-mean random variable with variance σ^2.

Next, we need to set up a two-dimensional vector filter. We define N vectors of order N, x_i, as

$$x_i = [x_{ij}], \qquad ij = 1, 2, \ldots, N, \tag{4.211}$$

and write as the vector equation

$$x_{i+1} = Qx_i - x_{i-1} + bu_i, \tag{4.212}$$

with

$$u_i = [u_{ij}], \qquad E[u_i] = 0, \qquad E[u_i u_j^T] = \sigma^2 I \delta_{ij}, \qquad (4.213)$$

$$b = -\frac{\beta}{\alpha_2} \qquad (4.214)$$

where δ_{ij} is the Kronecker delta and Q is the tridiagonal matrix. When we assume that the observations y_{ij} are corrupted with white Gaussian noise, we have

$$y_i = [y_{ij}] = x_i + \eta_i \qquad (4.215)$$

where η_i is zero mean with covariance

$$E(\eta_i \eta_j^T) = \frac{\sigma_n^2}{\delta_{ij}} \qquad (4.216)$$

and η_i and η_j are uncorrelated.

The recursive filtering problem is that of finding an estimate $\hat{x}_i$ of x_i, such that $E[(\hat{x}_i - x_i)^T(\hat{x}_i - x_i)|y_i, y_{i-1}, \ldots, y_i]$ is minimal. The optimal interpolation problem is that of minimizing $E[(\hat{x}_i - x_i)^T(\hat{x}_i - x_i)|y_N, y_{N-1}, \ldots, y_i]$. While the two problems are somewhat similar in form and possess a common theoretical basis, the interpolation problem is computationally more difficult since all the observations are used to make an estimate at any point, thus leading to large storage requirements. By converting (4.212) to a first-order equation, it is easy to show that the dual problem is to minimize the functional

$$J = \frac{1}{\sigma_n^2} \sum_{i=0}^{N} (y_i - x_i)^T (y_i - x_i) + \frac{1}{\sigma^2} [x_0^T x_0 + x_1^T x_1] + \frac{1}{\sigma^2} \sum_{i=1}^{N} f_i^T f_i \quad (4.217)$$

over $\{x_i\}$ and $\{f_i\}$ subject to x_i and f_i being related by

$$x_{i+1} = Q x_i - x_{i-1} + b f_i. \qquad (4.218)$$

Introducing Lagrange multipliers λ_i there are a number of methods of obtaining necessary conditions for minimizing. If v_i and w_i denote N-dimensional vectors of multipliers, the necessary conditions are

$$x_{i+1} = Q x_i - x_{i-1} + \sigma^2 b^2 v_{i+1} \qquad (4.219)$$

$$v_i = Q v_{i+1} + w_{i+1} + \frac{1}{\sigma^2}(y_i - x_i) \qquad (4.220)$$

$$w_i = -v_{i+1} \qquad (4.221)$$

subject to

$$v_{N+1} = w_{N+1} = 0 \tag{4.222}$$

$$x_i = \sigma^2 v_i. \tag{4.223}$$

We find that

$$x_0 = \frac{\sigma^2}{1 + (\sigma^2/\sigma_n^2)} \left(w_1 + \frac{1}{\sigma_n^2} y_0 \right). \tag{4.224}$$

This linear two-point boundary value problem can be converted to an initial-value problem by looking for a relation in the form of

$$x_i = R_i v_i + T_i w_i + s_i \tag{4.225}$$

where R_i, T_i, and s_i satisfy initial value problems independent of x_i, v_i, and w_i. The desired initial value problems can be found by first assuming x_i, can also be written by

$$x_i = D_i v_{i+1} + E_i w_{i+1} + g_i \tag{4.226}$$

where use of (4.219)–(4.224) leads to

$$D_i = F_i (R_i Q - T_i) \tag{4.227}$$

$$E_i = \sigma_n^2 (I - F_i) \tag{4.228}$$

$$F_i = \left(I + \frac{1}{\sigma^2} R_i \right)^{-1} \tag{4.229}$$

$$g_i = F_i \left(\frac{1}{\sigma_n^2} + s_i \right). \tag{4.230}$$

After some further manipulation the following relations are derived

$$R_{i+1} = Q D_i - D_{i-1} F_i \left(Q + \frac{1}{\sigma^2} T_i \right) + \sigma^2 b^2 I + E_{i-1}, \tag{4.231}$$

$$T_{i+1} = Q E_i - D_{i-1} F_i \tag{4.232}$$

$$s_{i+1} = Q g_i - \frac{1}{\sigma^2} D_{i-1} F_i (y_i - s_i) - g_{i-1}. \tag{4.233}$$

To satisfy the initial conditions of (4.223) we must have

$$R_i = \sigma^2 I, \qquad T_i = 0, \qquad s_i = 0 \tag{4.234}$$

and to also satisfy (4.224) it is found that

$$E_0 = \frac{\sigma^2}{1+\left(\sigma^2/\sigma_n^2\right)} I, \qquad D_0 = 0, \qquad g_0 = \frac{\sigma^2/\sigma_n^2}{1+\left(\sigma^2/\sigma_n^2\right)} y_0. \qquad (4.235)$$

These equations determine both the optimal estimator and optimal inter-polator. We reason as follows. For an on-line value of x_j, assume the process of (4.217) has j rather than N stages. Hence, the appropriate boundary conditions are

$$w_{j+1} = v_{j+1} = 0 \qquad (4.236)$$

and by (4.226)

$$x_j = g_i, \qquad (4.237)$$

which is the on-line estimate. If the process has N stages, then x_j is the interpolated value. Thus to find the optimal estimate one needs to solve (4.227)–(4.233). However, to find the optimal interpolated value, one must solve (4.227)–(4.233), store the results, and then work backwards using relations based on the original dynamics, (4.219)–(4.221), to obtain the desired x_i. The storage requirements for this problem can be very large but, in general, the interpolated values are much better than the estimated values. To get around this problem we will use a "one-step" interpolator that can be implemented on-line, adding only a fixed delay. Such an interpolator gives x_{N-1} as

$$x_{N-1} = \sigma_n^2 D_N F_N (y_N - s_N) + g_N. \qquad (4.238)$$

This expression needs observations only one stage past the desired stage. Thus the "one-step" interpolator

$$x_i = \sigma_n^2 D_{i+1} F_{i+1} (y_{i+1} - s_{i+1}) + g_{i+1} \qquad (4.239)$$

necessitates a fixed delay on one observation so that y_{i+1} can be received and D_{i+1}, F_{i+1}, s_{i+1}, and g_{i+1} computed.

Now, let us return to (4.197) and manipulate it to get further reduction in dimensionality. Let $[M]$ denote the matrix of orthonormal vectors obtained from $[H] = [h_{i,j}]$. Hence,

$$[M][M]^T = I \qquad (4.240)$$

and, if Q is given by (4.186), then

$$[\Lambda] = [M]^T [Q][M] \qquad (4.241)$$

is the diagonal matrix. Suppose we transform the vectors x_i, v_i, w_i, and $y_i vy [M]^T$ and define the new vectors as

$$\hat{x}_i = [M]^T x_i \tag{4.242}$$

$$\hat{v}_i = [M]^T v_i \tag{4.243}$$

$$\hat{w}_i = [M]^T w_i \tag{4.244}$$

$$\hat{y}_i = [M]^T y_i. \tag{4.245}$$

Then, using (4.219)–(4.224), and (4.241), we find that these quantities are determined by the equations

$$\hat{x}_{i+1} = \Lambda \hat{x}_i - \hat{x}_{i+1} + \sigma^2 b^2 v_{i+1} \tag{4.246}$$

$$\hat{v}_i = \Lambda \hat{v}_{i+1} + \hat{w}_{i+1} + \frac{1}{\sigma^2}(\hat{y}_i - \hat{x}_i) \tag{4.247}$$

$$\hat{w}_i = -\hat{v}_{i+1} \tag{4.248}$$

subject to

$$\hat{v}_{N+1} = \hat{w}_{N+1} = 0 \tag{4.249}$$

$$\hat{x}_1 = \sigma^2 \hat{v}_1 \tag{4.250}$$

$$\hat{x}_0 = \frac{\sigma^2}{1 + (\sigma^2/\sigma_n^2)}\left(\hat{w}_i + \frac{1}{\sigma_n^2}\hat{y}_0\right). \tag{4.251}$$

Since Λ is diagonal, each of the vector equations is actually N decoupled scaler equations. Thus, if λ_j is the jth diagonal element of Λ, then (4.246)–(4.251) can be rewritten as the N sets of coupled scaler equations

$$\hat{x}_{j,i+1} = \lambda_j \hat{x}_{j,i} - \hat{x}_{j,i-1} + \sigma^2 b^2 \hat{v}_{j,i+1} \tag{4.252}$$

$$\hat{v}_{ji} = \lambda_j \hat{v}_{j,i+1} + \hat{w}_{j,i+1} + \frac{1}{\sigma_n^2}(\hat{y}_{ji} - \hat{x}_{ji}), \tag{4.253}$$

$$\hat{w}_{ji} = -v_{j,i+1}, \qquad j = 1, 2, \ldots, N, \tag{4.254}$$

subject to

$$\hat{v}_{j,N+1} = \hat{w}_{j,N+1} = 0 \qquad (4.255)$$

$$\hat{x}_{j,1} = \sigma^2 \hat{v}_{j1} \qquad (4.256)$$

$$\hat{x}_{j,0} = \frac{\sigma^2}{1 + (\sigma^2/\sigma_n^2)} \left(\hat{w}_{ji} + \frac{1}{\sigma_n^2} \hat{y}_{j0} \right). \qquad (4.257)$$

Therefore, at the expense of transforming the observations to obtain $\{y\}$ and transforming the solution to obtain x_i from $\hat{x}_i$, we have eliminated all matrix computations. A count of the number of operations required in the computations show that the transformation of the observations and the transformation of the solution require $0(N^3)$ operations, while all the other computations can be done in $0(N^2)$ operations. This is a considerable savings over the solution, which did not transform the observations and did not transform the solution $0(N^4)$. Fortunately, even further reductions are possible. The elements of the transformation matrix H (or M) contains terms

$$h_{ij} = \sin \frac{ij\pi}{N+1}. \qquad (4.258)$$

Therefore, the elements h_{ij} are related to the Fourier transform and hence a FFT algorithm can be employed to perform all the necessary transformations. Thus the $0(N^3)$ operations are reduced to $0(N^2 \log_2 N)$ operations, which is a tremendous reduction from the original $0(N^4)$.

The difference between the one-step interpolator and the optimal interpolator is less than 1 dB. Thus a one-step interpolator gives an estimate very close to that of the optimal interpolator. Even closer estimates can be obtained by implementing a two-step interpolator. The computation time on an IBM 360/44 computer for a 32×32 image was found to be reduced by a factor of 30 without the use of the FFT algorithm. Further reduction in computation can of course be achieved by using the FFT algorithm.

4.13.12 Singular Value Decomposition (SVD) Transform Coding

Andrews and Patterson (1975) describe a technique called singular value decomposition (SVD). It is a form of transform coding that makes use of numerical analysis techniques. The use of SVD when combined with pseudo-inverse techniques represents an attempt to apply results from the field of numerical analysis to linear imaging system models.

Any matrix $[G]$ can be represented in a space defined by orthonormal matrices $[U]$ and $[V]$ as

$$[G] = [U][\Lambda]^{1/2}[V]^T \qquad (4.259)$$

where $[\Lambda]$ is a diagonal matrix composed of singular values of $[G]$, $[U]$ is the row eigenvector system of $[G]$, and $[V]$ is the column eigenvector system of $[G]$. Thus

$$[G][G]^T = [U][\Lambda][U]^T \tag{4.260}$$

$$[G]^T[G] = [V][\Lambda][V]^T \tag{4.261}$$

and $[G]$ need not be square, but for notational simplicity we assume that $[G]$ is $N \times N$. $[U]$ and $[V]$ are orthonormal in nature, so that

$$[U][U]^T = 1 \tag{4.262}$$

$$[V]^T[V] = 1. \tag{4.263}$$

If the quantities $[U]$ and $[V]$ are written in the form

$$[U] = [u_1 u_2 \cdots u_N], \tag{4.264}$$

$$[V] = [v_1 v_2 \cdots v_N], \tag{4.265}$$

respectively, then

$$[G] = [u_1 u_2 \cdots u_N][\Lambda]^{1/2} \begin{matrix} v_1^T \\ v_2^T \\ \vdots \\ v_N^T \end{matrix}. \tag{4.266}$$

If the matrix $[\Lambda]^{1/2}$ is written as a sum of the form

$$[\Lambda]^{1/2} = \begin{bmatrix} \lambda_1^{1/2} & 0 & 0 & \cdots & 0 & 0 & 0 & 0\cdots \\ 0 & 0 & 0 & \cdots & 0 & \lambda_2^{1/2} & 0 \\ & \vdots & & + & \vdots & + & \cdots \\ 0 & \cdots & & 0 & 0 & \cdots & 0\cdots \end{bmatrix}, \tag{4.267}$$

it follows that the expansion of the matrix $[G]$ can be represented in vector product notation as

$$[G] = \sum_{i=1}^{R} \lambda_i^{1/2} u_i v_i^T, \tag{4.268}$$

where the u_i and v_i are the column vectors of $[U]$ and $[V]$; λ_i the diagonal terms of $[\Lambda]$. R is the limit of the summation and represents rank (number of nonzero singular values) of $[G]$. However, if $[G]$ is nonsingular, R is equal to N. If only partial sums are provided, and if we order the singular values in decreasing monotonic order, we have

$$[G]_K = \sum_{i=1}^{R} \lambda_i^{1/2} u_i v_i^T. \tag{4.269}$$

This process is called truncation and yields a truncation error of

$$\|[G] - [G_K]\|^2 = \sum_{i=K}^{R} \lambda_i \tag{4.270}$$

where matrix least square error is the Euclidean measure

$$\text{tr}[G]^T[G] = \|[G]\|^2. \tag{4.271}$$

Andrews and Patterson (1976) call the SVD a transform in the deterministic sense, and that it is better for image energy compression than any other deterministic orthonormal transform. The Karhunen-Loeve transform is optimal only in a statistical or MSE sense, while the SVD is optimal in a least square sense.

The SVD transform is a two-dimensional unitary transform based on the singular value decomposition of image matrices. The forward transform is written as

$$[G] = [A][\alpha][B]^T, \tag{4.272}$$

the inverse of which is

$$[\alpha] = [A]^T[G][B]. \tag{4.273}$$

The $[\alpha]$ matrix is the transform of the image matrix, where $[A]^T$ transforms the columns of the image and $[B]$ transforms the rows of the image.

In terms of potentially independent samples or degrees of freedom of the picture, there exist $N^2 = N \times N$ brightness values in either $[G]$ or $[\alpha]$. If $[A]$ and $[B]$ are Fourier matrices, the $[\alpha]$ is the matrix of coefficients associated with the two-dimensional Fourier transform of the image. However, when $[A]$ and $[B]$ are deterministic, (i.e., not computed or adapted from $[G]$, such as Fourier, cosine, Walsh, etc.) then the N^2 potential degrees of freedom of the image map into N^2 coefficients in the transform domain $[\alpha]$.

SVD Image Restoration Processes

As mentioned previously, it is possible to use the algebra of matrices and outer products as a tool in image restoration. One method of accomplishing image restoration is to use the PSF matrix $[H]$. When the picture degradation is nonseparable from the image, the point-spread function is a useful tool. When the pictorial object is imaged through the nonseparable space-invariant point-spread function (SIPSF), the Fredholm integral of the first kind is useful;

$$g(x, y) = \int_{-\infty}^{\infty} \int f(\zeta, \eta) h(x, y, \zeta, \eta) d\zeta d\eta \qquad (4.274)$$

where the object $f(\xi, \eta)$ is imaged through the space-variant point-spread function (SVPSF) $H(x, y, \xi, \eta)$. A one-dimensional solution is the Fredholm integral, using the SVD method, which was investigated by Hanson (1971) and Varah (1973), to obtain a pseudoinverse matrix. Sondhi (1972) investigated the use of the SVD as a tool for image processing restoration of spatially invariant degradations, by using a scanning or stacking operator. They simplified the Fredholm integral to

$$g = [H] f \qquad (4.275)$$

where g (the image) is N^2 by 1, f (the object) is N^2 by 1, and $[H]$ (the PSF) is $N^2 \times N^2$. Treitel and Shanks (1971) have provided planar filters using the SVD approach. A digital computer was used to simulate a similar method. The results of the simulation showed that there is considerable smooth structure in the point-spread definition. Since the image representation of the PSF is just a matrix $[G]$ (for each column of $[H]$), and since the entries in $[G]$ are smooth, correlated, and well-behaved and go to zero rapidly, it is expected that a considerable savings may be achieved by using an SVD representation of $[G]$.

Experimental verification of this simulation has been achieved. Extremely rapid convergence was achieved. It was found that only 5 out of 101 eigenimages need be saved. This would be a considerable saving in computer memory. If we let $[\beta_{\xi, \eta}]$ represent the truncated object matrix, then badly aberrated systems will contain many off diagonal entries. For well-behaved systems, $[\beta_{\xi, \eta}]$ will be close to diagonal. Computer storage savings are provided by saving Ku vectors Kv vectors, $K\lambda$'s, for the on axis column of $[H]$, and then saving only a few β scalers for each new column (point source position). Thus $[G_{\zeta, \eta}]$ is approximated by a few diagonal and off diagonal terms in an orthogonal expansion in the basis system $[U_{0,0}]$, $[V_{0,0}]^T$, where

$$[G_{\zeta, \eta}] = [U_{0,0}][\beta_{\xi, \eta}][V_{0,0}]^T \qquad (4.276)$$

with $[G_{\zeta, \eta}]$ centered at (ζ, η). This is an expansion of off axis point-spread function in terms of on axis basis functions.

Pseudoinversion of Separable SVPSF. When the object is imaged through a separable SVPSF matrix $[H]$, pseudoinversion is needed for image restoration. Restoration is performed in the eigenspace of the distortion. For

SIPSF's, the eigenspace of the distortion is provided by the Fourier transform, because $[H]$ takes on the form of a circulant matrix that is well known to be diagonalized by $[\ \ ] = [\exp\ (2\pi i/N)km]$, $k, m = 0, 1, \ldots, N-1$. However, for SVPSF the Fourier decomposition is not valid, SVD must be resorted to for restoration processing. In either case the restoration filter is scaler (i.e., a diagonal weighting matrix) in the appropriate space, because proper decoupling exists in such a space for simple scaler filter compensation.

When the SVPSF is nonseparable, as was discussed previously, that is, when $h(x, y, \zeta, \eta)$ is not factorable, the discrete representation of $[H]$ requires $N^2 \times N^2$ locations. When the SVPSF can be represented in separable form

$$h(x, y, \zeta, \eta) = a(x, \zeta)b(\eta, y),$$

the discrete matrix becomes

$$[H] = [A] \circledast [B] \tag{4.277}$$

where $[H]$ is $N^2 \times N^2$, $[A]$, and $[B]$ are both $N \times N$, and $\circledast$ is the Kronecker or direct product operation. The advantage of representing the SVPSF as a separable phenomenon is one of computational simplicity. In matrix notation the image $[G]$ becomes

$$[G] = [A][F][B] \tag{4.278}$$

where the object $[F]$ is operated upon by the column operator $[A]$ and the row operator $[B]$ in a separable manner. This model assumes a noiseless environment and the rank of the image $[G]$ is determined by the minimum of the ranks of $[A]$, $[F]$, and $[B]$. Space-variant blur is introduced by $[A]$ and $[B]$, and it is desirable to obtain a minimum norm estimate of the object $[\hat{F}]$ by pseudoinversion of the blur matrices to recover up to the ranks of the individual blurs. The pseudoinverse matrix can be represented by

$$[\hat{F}] = [A]^+[G][B]^+, \tag{4.279}$$

where $[A]^+$ and $[B]^+$ are the pseudoinverses of $[A]$ and $[B]$, respectively. When the pseudoinverse matrix is put in its SVD form where

$$[A] = [U_a][\Lambda_a]^{1/2}[V_a]^T \tag{4.280}$$

and

$$[B] = [U_b][\Lambda_b]^{1/2}[V_b]^T, \tag{4.281}$$

then

$$[A]^+ = [V_a][\Lambda_a]^{-1/2}[U_a]^T \tag{4.282}$$

and

$$[B]^+ = [B_b][\Lambda_b]^{-1/2}[U_b]^T. \tag{4.283}$$

Here the $[\Lambda_a]^{-1/2}$ and $[\Lambda_b]^{-1/2}$ matrices are diagonal with nonzero entries. Vector notation produces

$$[A] = \sum_{i=1}^{R_a} \lambda_{ai}^{1/2} u_{ai} v_{ai}^T \tag{4.284}$$

$$[B] = \sum_{i=1}^{R_b} \lambda_{bi}^{1/2} u_{bi} v_{bi}^T \tag{4.285}$$

and

$$[A]^+ = \sum_{i=1}^{R_a} \lambda_{ai}^{-1/2} v_{ai} u_{ai}^T \tag{4.286}$$

$$[B]^+ = \sum_{i=1}^{R_b} \lambda_{bi}^{-1/2} v_{bi} u_{bi}^T \tag{4.287}$$

and R_a and R_b are the respective ranks of the blur matrices $[A]$ and $[B]$. Using the SVD expansions and pseudoinverses as described here, an estimate of the object can be obtained as

$$[\hat{F}] = [V_a][\Lambda_a]^{-1/2}[U_a]^T[G][V_b][\Lambda_b]^{-1/2}[U_b]^T$$

$$= [V_a][S][U_b]^T \tag{4.288}$$

or

$$= \sum_{i=1}^{R_a} \sum_{j=1}^{R_b} s_{ij} v_{ai} u_{bj}^T, \tag{4.289}$$

where the matrix $[S]$ is given by

$$[S] = [\Lambda_a]^{-1/2}[U_a]^T[G][V_b][\Lambda_b]^{-1/2} \tag{4.290}$$

and the ijth entry of $[S]$ is

$$s_{ij} = \lambda_{ai}^{-1/2} \lambda_{bj}^{-1/2} u_{ai}^T[G] u_{bj}. \tag{4.291}$$

Computationally, it is difficult to determine R_a and R_b because of roundoff noise. Thus a truncated estimate of the object is helpful, where

$$[\hat{F}_K] = [V_a][S_K][U_b]^T \qquad (4.292)$$

$$= \sum_{i=1}^{K} \sum_{j=1}^{K} s_{ij} u_{ai} v_{bj}^T. \qquad (4.293)$$

This is a two-dimensional orthogonal basis system defined by the column and row orthogonal eigenvectors of the respective blur matrices. Since the pseudo-inverse sought is an optimal minimal norm mean-square estimate of the original object, the expansion defined by the $[V_a]$ and $[U_b]^T$ spaces are the most efficient for such inversion.

The previous analysis was described for a noise-free case in which the pseudoinverse results in the optimal mean-square estimate filter. Generalization to the additive noise case and separable SVPSF results in a pseudoin-verse Wiener filter, where the noise power spectrum would be represented in the $[V_a]$, $[U_b]^T$ space, rather than the two-dimensional Fourier space.

The previous theory was tested experimentally with matrices of the size $N = 128$. For computational simplicity both horizontal and vertical blur functions were made equal. Thus $[A] = [B]$. Two test cases were developed, referred to as "moderate" and "severe" distortion. It was found that small singular values gave rise to computational instability, but the singular vectors remained orthogonal. To overcome this problem filtering was tried. It was assumed that the computational error in the calculation of the singular values was uncorrelated and independent of a given index. Scaler Wiener filtering theory was applied to de-emphasize these incorrect singular values. From a filtering viewpoint the traditional pseudoinverse filter multiplies the singular values of the distortion matrix by $\lambda_i^{-1/2}$, or equivalently by $\lambda^{+1/2}/\lambda_i$. In scaler Wiener filtering we assume a white noise level σ^2 (due to computation error) and modify our multiplication to be given by $\lambda_i^{1/2}/(\lambda_i + \sigma^2)$. This operation is very close to the inverse filter for $\lambda_i \gg \sigma^2$, but for small λ_i (i.e., $\lambda_i \ll \sigma^2$) the filter approaches zero. Thus those singular values that are very small do not reciprocate and cause unstable solution. When this Wiener filtering was used, pictorial images having K values of 61 and 64 gave quite good restorations, whereas without the Wiener modification these images were close to singular. Indeed complete singularity does not occur in the Wiener case until $K = 69$, whereas in the non-Wiener case, this occurs at $K = 63$. There was an obvious improvement obtained by using Wiener filtering, and one could envision an interactive system whereby the viewer could participate in the definition of the computational error σ^2 to optimize the resulting viewed restoration.

Most of the previous discussion was from work done by Andrews and Patterson (1976). However, Huang and Narendra (Applied Optics, 1975) did some work on image restoration using the pseudoinverse matrix in an SVD system with additive noise. They represented the linear degradation of an image by the matrix equation,

$$g = [H] f + n \tag{4.294}$$

where f and g are column matrices containing the samples from the original object and the degraded image, respectively. The numbers of elements in g and f need not be equal. The rectangular matrix $[H]$ was derived from the impulse response (generally spatial varying) of the degrading system. And n is a column matrix containing noise samples. The noise, may, for example be due to the detector.

For an estimate of f they used

$$\hat{f} = [H]^+ g \tag{4.295}$$

where $[H]^+$ is the Moore-Penrose pseudoinverse of $[H]$. In the presence of noise this becomes

$$\hat{f} = [H]^+ [H] f + [H]^+ n \tag{4.296}$$

where the first term on the right-hand side of the equation is the minimum-norm least square estimate in the absence of noise, and the second term represents the contribution due to noise. SVD was used to calculate the pseudoinverse.

The $[H]$ matrix can be decomposed by

$$[H] = \sum_{i=1}^{R} (\lambda_i)^{1/2} U_i V_i^T, \tag{4.297}$$

where R is the rank of $[H]$, U_i and V_i are eigenvectors (considered as column matrices) of $[H][H]^T$ and $[H]^T[H]$, respectively, and λ_i the eigenvalues of either. The pseudoinverse of $[H]$ is given by

$$[H]^+ = \sum_{i=1}^{R} (\lambda_i)^{-1/2} V_i U_i^T. \tag{4.298}$$

The effectiveness of SVD in the presence of noise can be demonstrated by noting that we can trade off between the amount of noise and the signal quality by choosing the number of terms we use in the SVD of the pseudoinverse. Using the SVD of $[H]^+$, the estimate of f becomes

$$\hat{f} = \sum_{i=1}^{R} \lambda_i^{-1/2} V_i U_i^T \{ [H] f \} + \sum_{i=1}^{R} \lambda_i^{-1/2} V_i U_i^T n. \tag{4.299}$$

Each term in the first summation has more or less comparable magnitudes, while the magnitudes of the terms of the second summation increase as $(\lambda_i)^{-1/2}$ (λ_i are in the order of decreasing magnitudes). When we use more and more terms in the summations, the first summation becomes closer and closer to the original object, but the SNR (the ratio of the first summation to the second summation) becomes smaller and smaller. A reasonable balance between the two effects should be achieved. A possible solution would be to look at the result after adding in each new term and stop at the visually best restoration.

Computer simulation was used to determine an optimum number of terms to use for good image restoration in the presence of additive noise. An 8×8 matrix was used that represented the digitized original image. It was found that 38 terms seemed to give the best visual restoration.

SVD Coding Techniques

Simple SVD algorithms are implicitly adaptive. This is true because by decomposing subblocks of an image into their respective singular values and vectors, and by always indexing according to decreasing value, one can then use fixed coding techniques, without worry about using large energy components. In addition, no bookkeeping is necessary for this ordering, as there is no preconceived notion as to the fixed or necessary ordering of the singular vectors.

The simplest algorithm might be to PCM the singular values and singular vectors with a linear quantizer with fixed bit assignment. However, when the distribution of scaler entries that comprise the singular vectors is investigated, a max quantizer becomes more appropriate. The importance (in a least squares sense) of individual singular values warrants variable bit assignment. Because the singular vectors are really one-dimensional sampled waveforms, one might guess that there exists a certain amount of correlation between adjacent scaler entries. Thus a DPCM coding of the scaler components forming the vectors might be in order. For this case an exponential quantizer with variable bit assignment becomes appropriate.

The complex coding techniques are motivated by a general viewpoint, that if a subblock of an image is in a region with similar such subblocks (i.e., common textures, means, variances, etc.) then the SVD of one subblock may be quite similar to that of a neighboring region. Thus one might consider describing classes of SVD's such that a different coding scheme resulted, depending upon whether the subblock represented "busy," "low contrast," "high contrast," etc., regions of an image.

One complex algorithm would envision using the singular vectors obtained from one subblock of a neighborhood as the assumed singular vectors of other neighboring subblocks, and then transmitting only the new singular values of the new blocks. This could have considerable potential in frame-to-frame situations, where adjacent frames in time have a large degree of similarity in SVD.

In order to develop efficient quantizers and coders for the SVD domain, a test image (512×512) was broken into 16×16 subblocks and data gathered over each subblock of the entire frame. For 16×16 subblocks, one obtains 16 monotonically ordered singular values. It was indicated that variable bit coding as a function of the singular value index is needed. Simulation results indicate that coding at 2 bits/pixel can be achieved with about a 1 percent error. This performance is somewhat lower than for conventional transform coding.

Andrews and Patterson also had some comments on computational complexity. For the nonseparable SVPSF case, the point-spread function matrix is $N^2 \times N^2$, and straightforward inversion techniques would require in the order of N^6 computations. However, in the separable situation $N \times N$ matrices need inverting, which requires a computational complexity on the order of $2N^3$. Then pseudoinversion requires on the order of $2KN^2$ computations. If the blur functions were space invariant, Fourier techniques would be applicable and in the order of $2N^2 \log_2 N$ computations would result. However, for the space-variant process the SVD method must be used, because the Fourier domain is not the eigenspace of the distortion. On a CDC 7600 computer, the SVD of a 128×128 matrix requires 3 sec. The subsequent pseudoinversion in iterative fashion for a complex image requires about 30 sec of central processing unit (CPU) time.

4.13.13 Transform Domain Least Squares Image Restoration Using Spline Basis Functions

The problem of achieving good image restoration in the transform domain, has been worked on by many people. Hou and Andrews (1977) approached the problem by using spline basis functions. As time moves along, more and more people are resorting to the use of digital computers to perform image restoration. This arises from the fact that an image in the real world is either corrupted by noise or degraded by various physical phenomena during formation.

An equation for image restoration, which uses a continuous-object-discrete-image model, is the Fredholm integral in two variables

$$G = \int \int_{-\infty}^{\infty} H(\xi, \eta)(\xi, \eta) d\xi \, d\eta + N \qquad (4.300)$$

where G is an $I \times J$ matrix with element g_{ij}, H is an $I \times J$ matrix with element $h_{ij}(\xi, \eta)$, and N is also an $I \times J$ matrix with noise samples n_{ij} as its elements. This model shows that the object scene is a continuous function in two-dimensional space, as it should be in the real world, but the image at the sensor output (or at the input of a processing computer) is discrete and has been sampled into $I \times J$ points. This is certainly a realistic model for digital image-restoration systems. It should be noted that the matrix $H(\xi, \eta)$ is a PSF matrix whose entries are continuous functions of the object plane (ξ, η) as well as the sampling position in the image plane (i, j). Clearly, this point-spread matrix in its most general form can be space variant. The image and object functions that represent light intensities must be nonnegative functions. The PSF is nonnegative because it is the image intensity of a point source. Mathematically we can write

$$\S(\xi, \eta) \geqslant 0 \tag{4.301}$$

$$g_{ij} \geqslant 0 \tag{4.302}$$

$$h_{ij}(\xi, \eta) \geqslant 0. \tag{4.303}$$

This requirement on each individual function is termed "positive restoration."

If the optical system is nonabsorbing (i.e., lossless), the total object energy must be the same as the total image energy. Mathematically this means that

$$\sum_{i=1}^{I} \sum_{j=1}^{J} g_{ij} = \int \int_{-\infty}^{\infty} \S(\xi, \eta)\, d\xi \, d\eta. \tag{4.304}$$

However, from (4.306) it is seen that

$$\sum_{i=1}^{I} \sum_{j=1}^{J} h_{ij}(\xi, \eta) = 1, \qquad \nabla \xi, \eta. \tag{4.305}$$

This is called an equal energy constraint equation. Before proceeding further, a criterion for restoration image quality should be established. The preferred criterion is a modified least squares objective function. By minimizing this objective function we should be able to formulate a linear algebraic equation for solving the unknown parameters. An interpolation scheme must be designed to restore the continuous object from a finite set of discrete data. From a data compression point of view, this interpolation approach is attractive because we need only to solve a linear algebraic equation in finite-dimensional space instead of the integral of Fredholm.

There is a great deal of similarity between data interpolation and curve fitting in the domain of numerical analysis. Reinsch (1967) used the cubic spline function for such interpolation. However, the spline function is not only used for interpolation, but also for smoothing of measured data. The work done here is an extension of that done by Reinsch.

For image-restoration purposes, the estimated object function $\hat{\S}(\xi, \eta)$ will be linearly interpolated by the cubic B-spline basis function. The basis function for one-dimensional B-splines can be written as

$$B_n(\xi, \xi_0, \xi_1, \xi_2 \cdots \xi_{n+1}) = (n+1) \sum_{k=0}^{n+1} \frac{(\xi - \xi_k)_+^n}{\omega(\xi_k)} \tag{4.306}$$

where

$$\omega(\xi_k) = \prod_{\substack{j=0 \\ j \neq k}}^{n+1} (\xi_k - \xi_j) \qquad (4.307)$$

$$(\xi - \xi_k)_+^n = \begin{cases} (\xi - \xi_k)^n, & \text{for } \xi > \xi_k \\ 0, & \text{for } \xi < \xi_k \end{cases} \qquad (4.308)$$

and

$$n = 0, 1, 2 \cdots .$$

We notice the following properties of B_n, for each n:

1 B_n is shift invariant.
2 B_n at one knot is independent from that at the other knot.
3 B_n are strictly positive.
4 $B_1 = B_0^* B_0$, $B_2 = B_0^* B_0^* B_0$, $B_3 = B_0^* B_0^* B_0^* B_0$ where the * denotes convolution.
5 B_n spans finite intervals on the real axis, that is, B_n has the local basis property.

We particularly use B_3, which we have given a special designation $s_k(\xi)$. In two dimensions we use the separable definition as $s_k(\xi)s_1(\eta) = \Phi_{k1}(\xi, \eta)$.

In order to derive a fundamental equation for image restoration in two dimensions, the normal equation of least square analysis, we start from the following objective function, which is minimum at $\hat{\S}(\xi, \eta)$.

$$W(\hat{\S}) = tr\left[(G - \hat{G})^T (g - \hat{G}) \right] + \gamma \int \int_{-\infty}^{\infty} \left[\nabla^4 \hat{\S}(\xi, \eta) \right]^2 d\xi d\eta \quad (4.309)$$

where

$$\nabla^4 = \frac{\partial^2}{\partial \xi^2} \frac{\partial^2}{\partial \xi^2} \qquad (4.310)$$

$$G = \int \int_{-\infty}^{\infty} H\S(\xi, \eta) d\xi d\eta \qquad (4.311)$$

and

$$\hat{G} = \int \int_{-\infty}^{\infty} H\hat{\S}(\xi, \eta) d\xi d\eta \qquad (4.312)$$

is a matrix that has the same size as G, which is $I \times J$. The elements g_{ij}, for $i=1$ to I, and $j=1$ to J, of matrix G are the measured data. H is the point-spread matrix, with elements $h_{ij}(\xi, \eta)$. The second term in (4.315) is the desired smoothing measure for the class of estimated objects $\hat{\S}(\xi, \eta)$. $\hat{\S}(\xi, \eta)$ can be linearly interpolated by a product of 2 one-dimensional cubic B-splines. Hence we write

$$\hat{\S}(\xi, \eta) = \sum_{k=1}^{K} \sum_{l=1}^{L} c_{kl} s_k(\xi) s_l(\eta) = \sum_{k=1}^{K} \sum_{l=1}^{L} c_{kl} \Phi_{kl}(\xi, \eta) \qquad (4.313)$$

where $s_k(\xi)$ and $s_l(\eta)$ are one-dimensional cubic B-splines that were defined earlier.

The column vectors c and $\Phi(\xi, \eta)$ are each $K \times L$ elements long, with the components c_{kl} and $\Phi_{kl}(\xi, \eta) = s_k(\xi) s_l(\eta)$ arranged in lexicographic order, we obtain from (4.319):

$$\hat{\S}(\xi, \eta) = c^T \Phi(\xi, \eta) = \Phi^T(\xi, \eta) c. \qquad (4.314)$$

Now, we can substitute (4.314) into (4.312), and then into (4.309) to get

$$W(c) = tr\left[G^T - c^T \int\int_{-\infty}^{\infty} \Phi(\xi, \eta) H^T d\xi d\eta \right]\left[G - \int\int_{-\infty}^{\infty} \left[\Phi^T(\xi, \eta) d\xi d\eta \, c \right] \right.$$

$$\left. + \gamma \int\int_{-\infty}^{\infty} \left[c^T \nabla^4 \Phi(\xi, \eta) \right]\left[\nabla^4 \Phi(\xi, \eta) c \right] d\xi d\eta. \qquad (4.315)$$

The particular value of c needed to minimize $W(c)$ is found from the derivative of $W(c)$ with respect to c^T, set to zero, or

$$\int\int_{-\infty}^{\infty} \int\int_{-\infty}^{\infty} \Phi(\xi, \eta) tr\left[H^T(\xi, \eta) H(\xi', \eta') \right]$$

$$\cdot \Phi^T(\xi', \eta') d\xi' d\eta' \cdot c - \int\int_{-\infty}^{\infty} \Phi(\xi, \eta) tr\left[H^T G \right] d\xi d\eta$$

$$+ \gamma \int\int_{-\infty}^{\infty} \nabla^4 \Phi(\xi, \eta) \nabla^4 \Phi^T(\xi, \eta) d\xi d\eta \cdot c = 0, \qquad (4.316)$$

which can also be written in short-hand form as

$$(A + \gamma B) c = d, \qquad (4.317)$$

where

$$A = \int \int_{-\infty}^{\infty} \int \int_{-\infty}^{\infty} \Phi(\xi, \eta) \mathrm{tr} \left[H^T(\xi, \eta) H(\xi', \eta') \right]$$

$$\cdot \Phi^T(\xi', \eta') d\xi \, d\eta \, d\xi' d\eta' \tag{4.318}$$

$$B = \int \int_{-\infty}^{\infty} \nabla^4 \Phi(\xi, \eta) \nabla^4 \Phi^T(\xi, \eta) G] d\xi \, d\eta \tag{4.319}$$

$$d = \int \int_{-\infty}^{\infty} \Phi(\xi, \eta) \mathrm{tr} \left[H^T(\xi, \eta) G \right] d\xi \, d\eta. \tag{4.320}$$

We now have some new indexes to define $p = (k, 1)$ and $q = (m, n)$ where matrices A and B are $KL \times KL$ large and the vectors c and d are $KL \times 1$ long. The components of these matrices can be written as follows:

a_{pq} = the (p, q)th element of A

$$= \sum_{i=1}^{I} \sum_{j=1}^{J} \int \int_{-\infty}^{\infty} h_{ij}(\xi, \eta) \Phi_p(\xi, \eta) d\xi \, d\eta$$

$$\cdot \int \int_{-\infty}^{\infty} h_{ij}(\xi', \eta') \Phi_q(\xi', \eta') d\xi' \, d\eta', \tag{4.321}$$

b_{pq} = the (p, q)th element of B

$$= \int \int_{-\infty}^{\infty} \nabla^4 \Phi_p(\xi, \eta) \nabla^4 \Phi_q(\xi, \eta) d\xi \, d\eta$$

$$= \int \int_{-\infty}^{\infty} s_k''(\xi) s_l''(\eta) s_m''(\xi) s_n''(\eta) d\xi \, d\eta, \tag{4.322}$$

d_p = the pth element of d

$$= \sum_{i=1}^{I} \sum_{j=1}^{J} g_{ij} \int \int_{-\infty}^{\infty} h_{ij}(\xi, \eta) \Phi_p(\xi, \eta) d\xi \, d\eta. \tag{4.323}$$

Equation (4.317) deserves special attention because its form is the most general one. We observe from (4.321)–(4.323) that (4.317) can be reduced to the form

$$\left(\tilde{A}^T \tilde{A} + B_x \otimes B_y \right) c = \tilde{A}^T g \tag{4.324}$$

where

$$\tilde{A} = \left(\int\int_{-\infty}^{\infty} h_{ij}(\xi,\eta)s_k(\xi)s_1(\eta)d\xi\,d\eta \right)_{IJ\times KL} \quad (4.325)$$

$$B_x = \left(\int_{-\infty}^{\infty} s_k''(\xi)s_m''(\xi)d\xi \right)_{K\times K} \quad (4.326)$$

$$B_y = \left(\int_{\infty}^{\infty} s_1''(\xi)s_n''(\eta)d\eta \right)_{L\times L} \quad (4.327)$$

and $\otimes$ denotes the Kronecker (direct or tensor) product.

The matrices A and B have the following mathematical properties:

1 A and B are both real, symmetric;
2 $a_{pp} > a_{pq}$ and $b_{pp} > b_{pq}$ for $p \neq q$;
3 A is nonnegative definite with all positive elements; and
4 B is positive definite.

In addition, from (4.318), (4.321), and (4.325) it is clear that A is essentially the amount of correlation or overlap of the PSF represented in spline space. Thus the effect of the blur degradation manifests itself in the rank or degree of A (or $\tilde{A}^T\tilde{A}$). Thus A behaves as a Grammian describing the imaging model, but in spline space. Hence A is nonnegative definite. $\tilde{a}_{pq}$ and a_{pq} are always nonnegative because of the nonnegativeness of $h_{ij}(\xi,\eta)$, $\Phi_p(\xi,\eta)$, and $\Phi_q(\xi,\eta)$. For finite extended $h_{ij}(\xi,\eta)$, the sparseness of matrix $\tilde{A}$ can be expected. Matrix B represents the correlation or overlap of the second derivative of the spline basis functions with themselves. Because of the separability of the basis functions, B can be represented as a Kronecker product. B represents the Grammian of the second derivative basis functions, but is deterministically known because of the expansion and interpolation on cubic B-spline space. The matrices B_x and B_y are nonnegative definite banded matrices as shown by

$$B_x = B_y = \frac{1}{6\Delta^3}\begin{bmatrix} 32 & -9 & 2 & 1 & 0 & & & & \\ -9 & 32 & -9 & 2 & 1 & 0 & & & \\ 2 & -9 & 32 & -9 & 2 & 1 & 0 & & \\ 1 & 2 & -9 & 32 & -9 & 2 & 1 & & 0 \\ 0 & 1 & & & & & & & \\ & & & & & & & & 0 \\ & & & & & & & & 1 \\ & & & & & & & & 2 \\ & & & & & & & & -9 \\ & & & & & 0 & 1 & 2 & -9 & 32 \end{bmatrix}$$

$$(4.328)$$

Obviously, B_x is Toeplitz, almost cyclic, and strictly diagonally dominant. Furthermore, since

$$B = B_x \otimes B_y \tag{4.329}$$

and

$$B^{-1} = B_x^{-1} \otimes B_y^{-1}$$

we conclude that the eigenvalues of B_x, B_y, and B are related by

$$[\lambda(B)] = \lambda(B_x)\lambda^T(B_y),$$

that is, all pairwise combinations of $\lambda(B_x)$ and $\lambda(B_y)$ are strictly positive. Hence B is also positive definite.

We have examined all the mathematical properties of (4.317), let us now turn to the physical meaning of the equation. Physical intuition becomes more clear if we only consider the following three special cases with $\gamma = 0$.

Case 1: The optical system is a perfect imaging system, in which case, the PSF becomes a two-dimensional δ-function, that is,

$$h_{ij}(\xi, \eta) = \delta(\xi - \xi_i, \eta - \eta_i). \tag{4.330}$$

As a result, after subsequent manipulation, the estimated object $\hat{\S}(\xi, \eta)$ at the interpolating grid points turns out to be

$$\hat{\S}(\xi_i, \eta_j) = g_{ij} \qquad \text{for all } ij, \tag{4.331}$$

which is, of course, what we would expect in a perfect imaging system.

Case 2: The optical system is totally blurred without any focus ability, in which case we can expect that the image is completely independent of the object; namely, the PSF is a pure scaler constant for every image point.

$$h_{ij}(\xi, \eta) = h, \qquad \text{for all } \xi, \eta, i, j. \tag{4.332}$$

In other words, the point source position (ξ, η) at the object plane does not change the image. Hence, a_{pq} and the matrix A has rank 1. After further manipulation, we obtain

$$g_{ij} = h\{\hat{\S}(\xi, \eta)\} \tag{4.333}$$

where $\{\ \}$ denotes the spatial average. The one degree of freedom provided by the rank 1 matrix A allows us to measure only the spatial average of the object.

Case 3: The optical system is space invariant, in which case, we make use of (4.324). Here B_x and B_y are space invariant and Toeplitz. If the PSF is space invariant, we have

$$h_{ij}(\xi, \eta) = h(\xi - i\Delta, \eta - j\Delta).\tag{4.334}$$

Then $\tilde{A}$ becomes Toeplitz as well, and the normal equation for $\gamma = 0$ reduces to the least squares formulation of

$$g_{ij} = \sum_{k=1}^{k} \sum_{l=1}^{l} \tilde{a}_{i-k, j-l} c_{k, l},\tag{4.335}$$

which is just a convolution sum resulting in traditional Fourier techniques.

For the more general case, we use the solution provided by the conjugate gradient method. The advantages of adopting an iterative approach are (1) to watch singularity development during iteration and (2) to allow human participation in the convergence process. The conjugate gradient method is useful for the solution of a system of simultaneous linear equations with an arbitrary symmetric positive definite matrix. It is especially advantageous if the matrix is not full, but contains many zero elements.

In our problem as posed by (4.317), the matrix $A + \gamma B$ is symmetric positive definite if γ is positive and nonzero. Matrix A is sparse if the PSF is localized; and matrix B is always sparse. In the case of $\gamma = 0$ and A is singular, the conjugate gradient method will give a least squares solution, in the sense of a generalized inverse, not in the sense of data fitting.

As an introduction to the gradient method in general, let us consider the quadratic form, with the assumption that $\hat{Q}$ is positive definite and symmetric:

$$J(c) = \tfrac{1}{2} c^T \tilde{Q} c - c^T d.\tag{4.336}$$

Take the partial derivative of $J(c)$ with respect to c and let it be r

$$\text{grad } J(c) = r = \hat{Q} c - d.\tag{4.337}$$

The solution of a symmetric positive definite system of equations of the form $Qc = d$ is equivalent to the problem of finding the minimum of a quadratic function. For this reason we call the gradient vector r the residual vector.

In geometrical terms, the contour surfaces of constant $J(c)$ represent concentric ellipsoids in R^n. The gradient r at every c is normal to the surface at that point. To find the minimum of $J(c)$ by varying c is equivalent to moving the vector c in a certain direction for finding the common center of the family of ellipsoids of $J(c) = $ constant.

To find the direction that c should move, let a new trial vector c' be chosen such that it is linearly related to c, so that

$$c' = c + qt\tag{4.338}$$

where t is a nonzero direction vector to be selected and q is a scaler parameter to be specified. From this a new quadratic function is obtained.

$$J(c') = \tfrac{1}{2}(c + qt)^T \hat{Q}(c + qt) - (c + qt)^T d$$

$$= \tfrac{1}{2}q^2(t^T \hat{Q}t) + q(t^T Qc - t^T d) + J(c). \qquad (4.339)$$

By invoking (4.337) we get

$$J(c') = \tfrac{1}{2}q^2(t^T \hat{Q}t) + qt^T r + J(c). \qquad (4.340)$$

For a given fixed c, the parameter q is chosen in such a way that $J(c')$ can attain a minimum. The necessary condition for doing this is to set $[\partial J(c')]/(\partial q) = 0$; that is,

$$q = -\frac{t^T r}{t^T Qt} \qquad (4.341)$$

and to make certain that $J(c')$ is a minimum for this choice of q, we take the second partial derivative of $J(c')$ with respect to q

$$\frac{\partial^2 J(c')}{\partial q^2} = t^T \hat{Q}t, \qquad (4.342)$$

which is always greater than 0 for any nonzero vector t, because $\hat{Q}$ is positive definite by assumption.

This conjugate gradient method, which was originated by Hestenes and Stiefel (1952) is used to determine the new direction vector, based on the gradient at every relaxation step. This iteration process makes good uniform progress toward the solution with a fairly simple algorithm.

The first iteration step is the same as for the steepest descent method; that is, to choose the first direction vector $t^{(1)}$ as the negative gradient $(-r^{(0)})$, so that

$$c^{(1)} = c^{(0)} - q_1 r^{(0)} \qquad (4.343)$$

where $x^{(0)}$ is any trial vector and q_1 is given by

$$q_1 = \frac{r^{(0)T} r^{(0)}}{r^{(0)T} \hat{Q} r^{(0)}}. \qquad (4.344)$$

The next direction of relaxation is in the conjugate direction of $t^{(1)}$, which is stated as

$$t^{(k)T} \hat{Q} t^{(i)} = 0, \qquad \text{for } i \leqslant k - 1.$$

The reason for choosing the new direction in this manner is to expand the directions of search for the center point of the concentric ellipsoids.

Moreover, for further search beyond the initial step, the direction vector $t^{(k)}$ is restricted to the plane spanned by $t^{(k-1)}$ and $r^{(k-1)}$, not merely along the $-r^{(k-1)}$ direction. Therefore,

$$t^{(k)} = -r^{(k-1)} + e_{k-1} t^{(k-1)}, \qquad k = 2, 3, \cdots. \tag{4.345}$$

Finally,

$$e_{k-1} = \frac{r^{(k-1)T} \hat{Q} t^{(k-1)}}{t^{(k-1)T} \hat{Q} t^{(k-1)}}, \qquad k = 2, 3, \cdots. \tag{4.346}$$

Having identified $t^{(k)}$ we can find the minimum point by using the following algorithm

$$c^{(k)} = c^{(k-1)} + q_k t^{(k)} \qquad k = 1, 2, 3, \cdots. \tag{4.347}$$

where

$$q_k = -\frac{t^{(k)T} r^{(k-1)}}{t^{(k)T} \hat{Q} t^{(k)}}. \tag{4.348}$$

Next, we find the final iterative equation r. We multiply both sides of (4.343) by $\hat{Q}$ to get

$$r(k) = r^{(k-1)} + q_k \hat{Q} t^{(k)}, \qquad k = 1, 2, 3, \cdots. \tag{4.349}$$

To summarize, basically, (4.343)–(4.350) constitute the complete algorithm of the conjugate gradient method.

Now, when we put (4.317) in these new settings, we obtain the following analogous relations. By letting

$$Q = A + \gamma B$$

the previous algorithm becomes

$$r^{(0)} = (A + \gamma B) c^{(0)} - d \tag{4.350}$$

$$t^{(1)} = -r^{(0)} \tag{4.351}$$

$$e_{k-1} = \frac{r^{(k-1)T} r^{(k-1)}}{r^{(k-2)T} r^{(k-2)}} \tag{4.352}$$

$$t^{(k)} = -r^{(k-1)} + e_{k-1} t^{(k-1)} \qquad k = 2, 3, \cdots \tag{4.353}$$

$$c^{(k)} = c^{(k-1)} + q_k t^{(k)}, \qquad k = 1, 2, 3, \cdots \tag{4.354}$$

where

$$q_k = \frac{r^{(k-1)T}r^{(k-1)}}{t^{(k)T}(A+\gamma B)t^{(k)}} \tag{4.355}$$

and

$$r^{(k)} = r^{(k-1)} + q_k(A+\gamma B)t^{(k)}. \tag{4.356}$$

From the above algorithm, we can see that no matrix inversion is required. In addition, the conjugate gradient algorithm offers the following advantages:

1 Simplicity of computational procedures.

2 Small storage space needed as compared to the variable metric method.

3 The preservation of the original matrices of coefficients, such as A, B_x, B_y, and G, during the computation.

4 The ability to start anew at any point in the computation.

5 The superiority of each approximation to the preceding ones, in the sense of being closer to the true solution.

The above theory has been borne out in computer simulations, where it was assumed that the PSF was separable, which permitted the use of matrices of $N \times N$, rather than $N^2 \times N^2$ as needed for the nonseparable case. The Gaussian PSF was used during the simulations. For the separable point-spread case we have

$$h_{ij}(\xi, \eta) = h_i(\xi)h_j(\eta) \tag{4.357}$$

where

$$h_i(\xi) = \frac{N}{\sigma_i\sqrt{2\pi}} \exp\left[\frac{-(\xi-x_i)^2}{2\sigma_i^2}\right] \tag{4.358}$$

and

$$h_j(\eta) = \frac{N}{\sigma_i\sqrt{2\pi}} \exp\left[-(\eta-y_i)^2 2\sigma_j^2\right] \tag{4.359}$$

where $\sigma_i = |Kx_i|/\sqrt{2}$ and $\sigma_j = |Ky_j|/\sqrt{2}$ for the space-variant case, and $\sigma_i = \sigma_j = |K|/\sqrt{2}$ are constants independent of image positions for the space-invariant case. In the previous two equations, N is a normalization constant for $h_{ij}(\xi, \eta)$ to satisfy (4.305), and K in σ_i and σ_j is a proportionality constant. The magnitude of K determines the extent of the blur in the images.

In the SVPSF case, the matrix $A + \gamma B$ becomes Toeplitz. Therefore, we should be able to use the scaler FFT algorithm. For computer simulation of large real images, this procedure is very attractive in terms of speed and storage requirements. This simplification is a natural consequence of the use of the spline basis function, along with the fact that the PSF is space invariant. Nevertheless, in our computer simulation of the SIPSF case we have not worked in this direction. Rather, we have concentrated on using one algorithm, that is, the conjugate gradient method, so that both SVPSF and SIPSF images could be processed.

In the computer program we have tracked the errors $\|r\|$, $\|g - \hat{g}\| / \|g\|$, and $\|c - \hat{c}\| / \|g\|$ during each iteration. Naturally, $\|c - \hat{c}\|$ would not be available in normal situations, because f is unknown. These measures are seen to track each other and to decrease monotonically. In addition, $\|g - \hat{g}\| / \|g\|$ and $\|c - \hat{c}\| / \|g\|$ decrease monotonically with decreasing γ. From the pictorial results it is evident that the smaller the γ, the higher the resolution of the restored object.

The computational parameters for these examples included 36-bit floating-point precision on a PDP-K110 computer. Typical computing times on an SVPSF restoration consisted of one iteration every 12 sec of CPU time. Therefore, for 300 iterations, one would expect a one hour computation time, which includes the interpolation and the display phases. These times would be reduced considerably with the use of a larger core memory on the PDP-L110. Much of the total time is due to page and disk swapping, because of the time-share operating system of the computer.

4.13.14 A Finite-Field Number Theoretic Transform

One of a class of number theoretic transforms was reported on by Reed, Truong, Kwoh, and Hall (1977). They have extended the work of Rader (1975) who proposed that the number theoretic transform could be used to accomplish two-dimensional filtering. This has been achieved by using Fourier-like transforms over a finite-field $GF(p)$ of p elements, where p is a prime number. The prime number p of $GF(p)$ was selected for its ability to yield an FFT algorithm, and because it was applicable to the PDP-10 computer. The chosen prime number was $45 \times 2^{29} + 1$, which is large enough to handle a large number of practical applications.

Rader (1974) defined transforms over residue classes of integers modulo, by either a Mersenne prime or Fermat prime p. Agarwal and Burrus (1975) extended Rader's transform for Fermat primes by a factor of 2 by using $\alpha = \sqrt{2}$ as a root of unity. These transforms, now called number theoretic transforms, can be utilized to yield convolutions of the integers without roundoff error. The arithmetic used requires only additions and bit shifts.

Recently, Rader's transform was generalized by Reed and Truong (1975) to transforms over the Galois field $GF(p^2)$, where p is a Mersenne prime.

These transforms also have been extended to operate over a direct sum of Galois fields (Reed and Truong, 1975). If the input data are restricted to $GF(p)$, these number theoretic transforms can be used to compute convolutions of integers. They offer increased transform lengths, but the arithmetic for implementing the transform requires complex integer multiplications. Golomb, Reed, and Truong (1977) proposed that number theoretic transforms could be defined in the Galois field $GF(p)$ to yield circular integer convolutions, where the prime p was of the form $p = k \times 2^n + 1$. A method was developed to perform binary arithmetic, modulo $k \times 2^n + 1$. For this class of primes, the FFT algorithm can be used to realize transforms of integers. These transforms offer a large variety of possible transforms lengths and dynamic ranges. The arithmetic needed requires, in some cases, only slightly modified binary integer additions and multiplications. The computerized implementation of this transform can be programmed to be faster than the conventional radix-2 FFT of real data on the PDP-10 computer. The speed could be considerably improved on computer hardware appropriately specialized to perform modulo p arithmetic.

The largest prime of this form that can be used with a 36-bit word, excluding the sign bit, is the prime $45 \times 2^{29} + 1$. This prime can be used with either the PDP-10 or the Univac 1108 computer, both of which have a word length of 36 bits. A number-theoretic transform using this prime has sufficient dynamic range to compute long-sequence convolutions for many picture processing applications.

Let d, an integer, be divisor of the order $q - 1$ of the cyclic multiplicative subgroup of $GF(p)$, where $q = p^n$, p is a Mersenne prime and n a positive integer. Let G_d be a subgroup of this group with generator r and order d; that is, $G_d = \{r, r^2, \cdots, r^{d-1}, 1\}$. Then, the transform of a sequence $\{a_0, a_1, \cdots, a_{d-1}\}$ of d elements of $GF(p)$ of the form

$$A_k = \sum_{n=0}^{d-1} a_n r^{nk} \tag{4.360}$$

has an inverse of the form

$$a_n = (d)^{-1} \sum_{k=0}^{d-1} A_k r^{-nk} \tag{4.361}$$

where $(d)^{-1}$ denotes the inverse of integer d modulo p.

Let A_k and B_k be the transforms of the sequences $\{a_n\}$ and $\{b_n\}$, respectively, and $C_k = A_k B_k$. The inverse transform of C_k is the circular or cyclic convolution, that is,

$$c_m = (d)^{-1} \sum_{k=0}^{d-1} C_k r^{-km} = \sum_{n=0}^{d-1} a_n (m - n) \tag{4.362}$$

where $(m - n)$ denotes integer, $m - n$, modulo d.

One-dimensional Fourier-like transforms of this type can be extended to two-dimensional Fourier-like transforms over $GF(q)$. If we let the integers d_1 and d_2 divide $q-1$, and the elements r_1 and r_2 in $GF(q)$ generate the cyclic subgroups of d_1, d_2 elements, respectively; $G_{d_1} = \{r_1, r_1^2, \cdots, r_1^{d_1-1}, 1\}$, and $G_{d_2} = \{r_2, r_2^2, \cdots, r_2^{d_2-1}, 1\}$. Then, a two-dimensional transform can be defined by

$$A_{k_1 k_2} = \sum_{n_1=0}^{d_1-1} \sum_{n_2=0}^{d_2-1} a_{n_1, n_2} r_1^{n_1 k_1} r_2^{n_2 k_2}, \quad \text{for } \begin{matrix} 0 \leqslant k_1 \leqslant d_1-1 \\ 0 \leqslant k_2 \leqslant d_2-1 \end{matrix} \quad (4.363)$$

where a_{n_1, n_2} belongs to $GF(q)$ for $n_1 = 0, 1, \cdots, d_1-1$ and $n_2 = 0, 1, \cdots, d_2-1$. However, the inverse of this requires further derivation.

We can let

$$\sum_{x \in G_{d_1}} \sum_{y \in G_{d_2}} x^m y^m = \sum_{i=0}^{d_1-1} \sum_{j=0}^{d_2-1} (r_1^i)^m (r_2^j)^n$$

$$= \sum_{i=0}^{d_1-1} \sum_{j=0}^{d_2-1} (r_1^m)^i (r_2^n)^j. \quad (4.364)$$

Now, if $m \not\equiv 0 \bmod d_1$ or $n \not\equiv 0 \bmod d_2$

$$\sum_{x \in G_{d_1}} \sum_{y \in G_{d_2}} x^m y^m = \frac{1-(r_1^m)^{d_1}}{1-r_1^m} \frac{1-(r_2^m)^{d_2}}{1-r_2^n} = 0 \quad (4.365)$$

and if $m \equiv 0 \bmod d_1$ and $n \equiv 0 \bmod d_2$

$$\sum_{x \in G_{d_1}} \sum_{y \in G_{d_2}} x^m y^m = \sum_{i=0}^{d_1-1} \sum_{j=0}^{d_2-1} 1 = (d_1)(d_2) \quad (4.366)$$

where d_i denotes $d_i \bmod p$ for $(i=1, 2)$. Thus

$$\sum_{x \in G_{d_1}}^{d_1-1} \sum_{y \in G_{d_2}}^{d_2-1} x^m y^m = \sum_{x=0}^{d_1-1} \sum_{y=0}^{d_2-1} (r_1^i)^m (r_2^j)^n$$

$$= (d_1)(d_2) \delta_{d_1}(m) \delta_{d_2}(n) \quad (4.367)$$

where

$$\delta_{d_1}(n) = \begin{cases} 0, & \text{if } n \not\equiv 0 \bmod d_i \\ 1, & \text{if } n \equiv 0 \bmod d_i. \end{cases}$$

Since (d_i) is a nonzero element of the field $GF(q)$ for $(i=1, 2)$, the inverse

$(d_i)^{-1}$ exists in GF(q). Now, multiply $A_{k_1 k_2}$ by $(d_1)^{-1}(d_2)^{-1}r_1^{-k_1 n_1}lr_2^{-k_2 n_2}$ and sum on k_i for $(k_1 = 0, 1, \cdots, d_1 - 1)$. This yields by (4.363) and (4.367)

$$(d_1)^{-1}(d_2)^{-1} \sum_{k_1=0}^{d_1-1} \sum_{k_2=0}^{d_2-1} A_{k_1,k_2} r_1^{-k_1 n_1} r_2^{-k_2 n_2}$$

$$= (d_1)^{-1}(d^2)^{-1} \sum_{k_1=0}^{d_1-1} \sum_{k_2=0}^{d_2-1} \left(\sum_{m_1=0}^{d_1-1} \sum_{m_2=0}^{d_2-1} a_{m_1,m_2} r_1^{m_1 k_1} r_2^{m_2 k_2} \right) r_1^{-k_1 n_1} r_2^{-k_2 n_2}$$

$$= (d_1)^{-1}(d_2)^{-1}(d_1)(d_2) \sum_{m_1=0}^{d_1-1} \sum_{m_2=0}^{d_2-1} a_{m_1,m_2 d_1}(m_1 - n_1)_{d_2}(m_2 - n_2)$$

$$= a_{n_1,n_2}. \tag{4.368}$$

Thus if the two-dimensional transform is

$$A_{k_1,k_2} = \sum_{n_1=0}^{d_1-1} \sum_{n_2=0}^{d_2-1} a_{n_1,n_2} r_1^{n_1 k_1} r_2^{n_2 k_2}$$

the inverse transform is

$$a_{n_1,n_2} = (d_1)^{-1}(d_2)^{-1} \sum_{k_1=0}^{d_1-1} \sum_{k_2=0}^{d_2-1} A_{k_1,k_2} r_1^{-n_1 k_1} r_2^{-n_2 k_2} \tag{4.369}$$

where d_1 and d_2 divide $q-1$, a_{n_1,n_2} and A_{k_1,k_2} are elements of GF(q), and r_1 and r_2 are elements of order d_1 and d_2, respectively.

Now, we need to show the circular convolution property of the two-dimensional transform. We let

$$A_{k_1,k_2} = \sum_{n_1=0}^{d_1-1} \sum_{n_2=0}^{d_2-1} a_{n_1,n_2} r_1^{n_1 k_1} r_2^{n_2 k_2} \tag{4.370}$$

$$b_{k_1,k_2} = \sum_{n_1=0}^{d_1-1} \sum_{n_2=0}^{d_2-1} b_{n_1,n_2} r_1^{n_1 k_1} r_2^{n_2 k_2} \tag{4.371}$$

and

$$C_{k_1 k_2} = A_{k_1,k_2} B_{k_1,k_2}. \tag{4.372}$$

Then, the inverse transform of C_{k_1,k_2} for $(k_1 = 0, 1, \cdots, d_1 - 1$ and $k_2 =$

$0, 1, \cdots, d_2 - 1$) is

$$c_{n_1, n_2} = (d_1)^{-1}(d_2)^{-1} \sum_{k_1=0}^{d_1-1} \sum_{k_2=0}^{d_2-1} C_{k_1, k_2} r_1^{-n_1 k_1} r_2^{-n_2 k_2}$$

$$= (d_1)^{-1}(d_2)^{-1} \sum_{t_1=0}^{d_1-1} \sum_{t_2=0}^{d_2-1} \sum_{l_1=0}^{d_1-1} \sum_{l_2=0}^{d_2-1} a_{t_1, t_2} b_{l_1, l_2}$$

$$\cdot \delta_{d_1}(t_1 + l_1 - n_1) \delta_{d_2}(t_2 + l_2 - n_2)$$

$$= \sum_{t_1=0}^{d_1-1} \sum_{t_2=0}^{d_2-1} a_{t_1, t_2} b_{(n_1-t_1), (n_2-t_2)} \tag{4.373}$$

where $(n_i - t_i)$ denotes the residue of $n_i - t_i$ modulo d_i for $i - 1, 2$. It is straightforward to extend the proof to multidimensional transforms.

If p is a prime of the form $k \times 2^n + 1$, where both k and n are integers, the order of the multiplicative group with generator of $GF(p)$ is given by

$$t = p - 1 = k2^n.$$

Since t has the factor $d = 2^n$, the usual FFT algorithm can be utilized to calculate the transform of as many as $d = 2^n$ points. If $d = 2^m$, $1 \leq m \leq n$ and α is the multiplicative generator of $GF(p)$, then the generator of G_d is evidently $r = 2^{k2^{n-m}}$. Primes of the form $k \times 2_n + 1$ can be found in the table of Robinson (1958). As was mentioned previously, the optimum prime number to use on the PDP-10 computer is $45 \times 2^{29} + 1$. According to Fermat's theorem, $2^{45(2)^{29}} \equiv 1 \bmod p$, where $p = 45 \times 2^{29} + 1$. It can be verified on a computer that $(2^{45})^{2^{29}} \equiv -1 \bmod p$. Thus, $2^{45} \equiv 8589933136 \bmod p$ is an element of order 2^{28}, where $p = 45 \times 2^{29} + 1$. It follows that $r \equiv 2^{45(2)^{28-k}} \bmod p$ is an element of order 2^k, where $0 \leq k \leq 28$. A detailed discussion about how to find the index of order of an element modulo a prime of the form $k \times 2^n + 1$ can be found in Golomb, Reed, and Truong (1977).

Furthermore, if r is the order of d, and d is a power of 2, then r^j is also an element with the order d for $j = 1, 3, 5, \cdots, d - 1$. In order to perform fast multiplications in hardware, it is desirable to choose r with a minimal number of ones from the $d/2$ elements of the order d in $GF(p)$. A program for finding an element of the order 2^k with a minimal number of ones, for $k \leq 28$ was written in assembly language. It was found that 17399515152 and 13958779178 are elements with a minimum number of ones of the order 2^8 and 2^9, respectively, in $GF(p)$.

In order to compute two-dimensional convolutions over $GF(p)$, one needs to constrain the two 2^d-point sequences of integers a_{n_1, n_2} and b_{n_1, n_2} in the dynamic ranges A and B, respectively. In order to avoid overflow, one must also keep the two-dimensional circular convolution c_{n_1, n_2} within the interval

$$\frac{-(p-1)}{2} \leqslant c_{n_1, n_2} \leqslant \frac{p-1}{2}$$

where

$$c_{n_1, n_2} = \sum_{t_1=0}^{d_1-1} \sum_{t_2=0}^{d_2-1} a_{t_1, t_2} b_{(n_1-t_1),(n_2-t_2)}. \qquad (4.374)$$

If a_{t_1, t_2} and $b_{(n_1-t_1),(n_2-t_2)}$ are chosen to be integers such that $0 \leqslant a_{t_1, t_2} \leqslant A$ and $0 \leqslant b_{(n_1-t_1),(n_2-t_2)} \leqslant B$, it is necessary that

$$d^2 A B \leqslant \frac{p-1}{2}$$

where A and B are dynamic ranges of integers a_{t_1, t_2} and $b_{(n_1-t_1),(n_2-t_2)}$, respectively. If $A = B$, the previous scaling constraint becomes

$$A \leqslant \frac{1}{d} \sqrt{\frac{p-1}{2}} . \qquad (4.375)$$

In most picture-processing applications, one of the sequences say b_{n_1, n_2} is known. Therefore, a better bound is

$$A d^2 |b|_{\text{average}} \leqslant \frac{p-1}{2}. \qquad (4.376)$$

Since $d = 2^k$, for $k \leqslant 28$, is a power of 2, one can utilize the FFT, depending on $GF(p)$, to calculate convolutions of 2^9-point sequence of integers.

We can compare the FFT used in the finite-number field with the radix-2 FFT. Let the root of unity ω in the complex number field be replaced by r, the generator of G_d in $GF(p)$. Also, let the arithmetic operations in the complex-number field be replaced by the arithmetic operations in $GF(p)$. Then the FFT algorithm over the finite field is similar in form to the FFT algorithm over the field of complex numbers. The basic operation in the radix-2 FFT algorithm over $GF(p)$ is the butterfly algorithm.

Multiplication modulo the prime number $p = 45 \times 2^{29} + 1$ is straightforward to perform in software on the PDP-10 computer. To perform addition modulo p, let $A + C = A + (C - p)$, where $(C - p) \leqslant 0$. Then if $A + (C - p) \leqslant 0$, the addition is accomplished by the add command; otherwise, it equals $A + (C - p) + p$. To perform subtraction modulo p, if $A - C \leqslant 0$, subtraction is accomplished by the subtract command; otherwise it equals $A - C + p$. If one modifies the FFT algorithm, described in the Fortran code which was given in p. 367 of Rabiner and Gold (1975), one can obtain a radix-2 FFT program for integers over $GF(p)$. One advantage of using the finite-field transform is that it is not complex over $GF(p)$, so that it requires only one-half the memory of the complex number FFT. Also, the finite-field transform does not have roundoff errors.

Now, we should mention the computation speed of the finite-field number-theoretic transform. As implemented on the PDP-10 model KI-10 computer, the measured speed of finite-field transform was approximately 2.4 times faster than the conventional complex FFT, both of which use the radix-2 FFT algorithm. However, since the picture and the filter use only real data, the conventional FFT can be manipulated to save an additional factor of slightly less than 2. In this case, the speed advantage of the finite-field transform over the conventional transform would be 1.2 instead of 2.4. If a mixed radix conventional FFT is employed, these two speeds would be approximately the same, when implemented specifically on the PDP-10 computer, which has faster floating-point multiplication and division operations. However, hardware that has been customized for the finite-field transform would show a definite improvement over the present implementations. Multiplication would be accomplished simply by high-speed shifting and addition operations, thereby taking advantage of the specialized finite arithmetic of the number-theoretic transforms.

4.13.15 Hybrid Coding

Habibi (1974) describes two hybrid techniques that combine the FFT with DPCM. In these systems the strong points of the two methods are combined to obtain improved image coding performance. The transform of the picture elements is encoded into DPCM. The quantized signal is then encoded using either fixed- or variable-length code words and is then transmitted over the digital channels.

Transform coding systems achieve superior coding performance at lower bit rates. They distribute the coding degradation in a manner less objectionable to the human viewer, show less sensitivity to data statistics, are less vulnerable to channel noise. On the other hand, DPCM systems, when designed to take full advantage of spatial correlations of the data, achieve a better coding performance at the higher bit rates. DPCM is relatively simple to implement into hardware. It does not require the large memory, which is a part of transform coding systems. The hybrid system can therefore be designed for real time digital TV usage.

However, DPCM is sensitive to picture statistics or variations from picture to picture, and tend to produce propagation of the channel errors on the transmitted picture.

Habibi describes two hybrid coding systems that use a cascade of unitary transformations and a bank of DPCM systems.

System 1 takes a one-dimensional transform of the image data existing on one TV scan line. Then it operates on each column of the transformed data by using a bank of DPCM systems.

The DPCM systems quantize the signal in the transform domain and takes advantage of the vertical correlation of the transformed data to reduce the coding error.

This hybrid system shows little degradation at small or moderate levels of channel noise, however its performance is degraded significantly at high levels of channel error. This principle can be extended to include interframe coding of TV signals.

The second system uses a two-dimensional unitary transformation on the pictorial data divided into small blocks. The elements of each block in the transformed domain are ordered in a one-dimensional array and are coded by a bank of DPCM systems. The use of a small block size reduces the number of arithmetic operations needed to obtain the transformed data. However, the efficiency of the transformed system is reduced because the elements of various blocks remain correlated in the transformed domain. Ordinary transform coding relies on uncorrelated data in the transformed domain. This would be a defect in the system were it not for the fact that DPCM coding of the elements in the transformed domain takes advantage of this correlation, which actually improves the total system performance.

These two hybrid systems will now be described in more detail.

One-Dimensional Transformation Plus DPCM Coding

Each scan line has M picture elements or pixels. There are N lines in the raster. The data samples are divided into $M \times N$ arrays of pixels $u(x,y)$, where x and y index the rows and columns in each array. The number of samples in a line of image is an integer multiple of M. The mathematical model is set up as follows:

$$u_i(y) = \sum_{x=1}^{M} u(x, y)\Phi_i(x) \qquad \begin{matrix} i=1,2,\cdots, M \\ y=1,2,\cdots, N \end{matrix} \qquad (4.377)$$

$$u(x, y) = \sum_{i=1}^{M} u_i(y)\Phi_i(x) \qquad (4.378)$$

where $\Phi_i(x)$ is a set of M orthonormal basis vectors. The correlation of the transformed samples $u_i(y)$ and $u_i(y+\tau)$ is given by

$$C_i(\tau) = \sum_{x=1}^{M} \sum_{\hat{x}=1}^{N} R(x_1\hat{x}_1 y, \hat{y})\Phi_i(x)\Phi_i(\hat{x}) \qquad (4.379)$$

where $R(x, \hat{x}, y, \hat{y})$ is the spatial covariance of the data.

This equation shows that the correlation of the samples in each column of the transformed array is directly proportional to the correlation of sampled image in the vertical direction. Furthermore, the correlation of samples columns of the transformed array is different. Thus a number of different DPCM systems should be used to encode each column of the transformed data.

A block diagram of the system is shown in Habibi's Fig. 1, and here as Fig. 4.35.

A replica of the original image $u^*(x, y)$ is formed by inverse transforming the coded samples, that is,

$$u^*(x, y) = \sum_{i=1}^{n} v_I(y)\Phi_i(x) \qquad n \leqslant M. \tag{4.380}$$

The MSE is

$$\varepsilon^2 = E\left\{ \frac{1}{MN} \sum_{y=1}^{N} \sum_{i=1}^{M} [u(x, y) - u^*(x, y)]^2 \right\}. \tag{4.381}$$

When it is assumed that $q_i(y)$, the quantization error in the ith DPCM system, is uncorrelated with $u_i(y)$ the coding error is

$$\varepsilon^2 = R(0,0,0,0) - \sum_{i=1}^{n} C_i(0)$$

$$+ E\left\{ \frac{1}{MN} \sum_{y=1}^{N} \sum_{i=1}^{M} [u_i(y) - v_i(y)]^2 \right\} \tag{4.382}$$

where the first two terms are used to compensate for the fact that n rather than M DPCM systems are used.

O'Neal (1966) has shown that for DPCM systems

$$E\left\{ [u_i(y) - v_i(y)]^2 \right\} = Eq_i^2(y) = K(m_i)e_i \tag{4.383}$$

where e_i^2 is the variance of the differential signal in the ith DPCM and $K(m_i)$ is the quantization error of a variate with a unitary variance in a quantizer with $(2)^{m_i}$ levels.

It has been shown that

$$K(m_i) \cong b \exp(-am_i). \tag{4.384}$$

If the variate is Gaussian in nature an optimum value for $a = 0.5 \ln 10$, and $b = 1$ (Habibi, 1971; Habibi and Wintz, 1971; Huang and Schultheiss, 1963). It has also been shown that

$$e_i^2 = C_i(0) - \sum_{j=1}^{m} A_{ij}C_i(j) \tag{4.385}$$

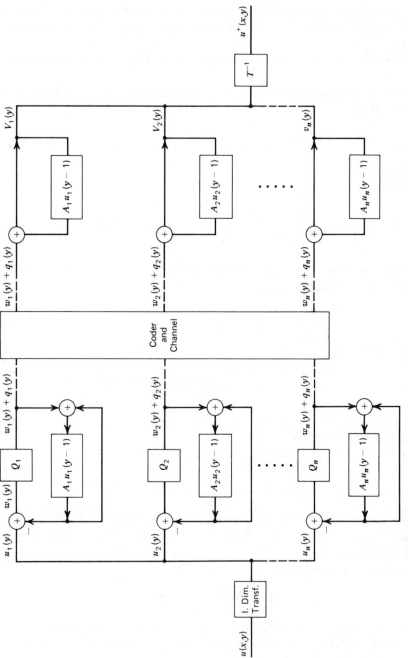

Figure 4.35. Block diagram of the hybrid system using a cascade of one-dimensional transformations and a bank of DPCM systems to encode pictorial data.

where A_{ij} are related to $C_i(j)$ by m algebraic equations. By substitution it can be shown that

$$\varepsilon^2 = R(0,0,0,0) - \sum_{i=1}^{n} C_i(0) + \frac{b}{M} \sum_{i=1}^{n} \exp(-am_i)e_i^2. \qquad (4.386)$$

The optimum number of quantizing bits (m_i) is chosen by

$$m_i = \frac{M_b}{n} + \frac{1}{a}\left[\ln e_i^2 - \frac{1}{n}\sum_{i=1}^{n} \ln e_i^2\right] \qquad (4.387)$$

where n (the number of DPCM systems measured) is chosen to minimize ε^2. Each DPCM quantizer uses 2^{m_i} levels.

The performance of this hybrid system shows little dependence on block size above $M=8$. Habibi's (1974) Fig. 5 shows SNRs versus block size for hybrid systems using Karhunen-Loeve and Hadamard transforms. This is shown here as Fig. 4.36.

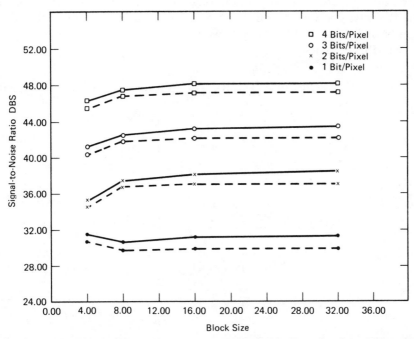

Figure 4.36. Theoretical performance of the proposed one-dimensional encoders versus block size M. The solid and dashed lines refer to the Karhunen-Loeve and the Walsh-Hadamard transforms, respectively.

Two-Dimensional Transformation Plus DPCM Coding

This system requires a large number of computations and a bookkeeping system to keep track of the subblock designations. However, a bank of DPCM systems exploits the correlations between subblocks.

In this system there are $N \times N$ subblocks each divided into $M \times M$ picture elements. A two-dimensional unitary transformation each subblock is obtained and is ordered to form a one-dimensional array. U_i, $i = 1, 2, \cdots, M^2$, where $U(k, 1)$ refers to the ith member of the array, which is obtained by transforming the (k, l)th subblock of the data.

This transformation and its inverse are modeled as

$$U_i(k, l) = \sum_{y=1}^{M} \sum_{x=1}^{M} u(kM + x, lm + y) \Phi_i(x, y) \tag{4.388}$$

and

$$u(kM + x, lm + y) = \sum_{i=1}^{M^2} U_i(k, l) \Phi_i(x, y) \tag{4.389}$$

where $\Phi_i(x, y)$ are a set of orthonormal basis matrices. The elements of $U_i(k, 1)$ arrays for various values of $i(i - 1, 2, \cdots, M^2)$ are correlated, thus could be coded by M^2 DPCM systems.

The MSE becomes

$$\varepsilon^2 = R(0, 0) - \sum_{i=1}^{n} C_i(0) + \frac{b}{M} \sum_{i=1}^{n} \exp(-am_i) e_i^2 \tag{4.390}$$

$$m_i = \frac{M_b}{n} + \frac{1}{a} \left[\ln e_i^2 - \frac{1}{n} \sum_{i=1}^{n} \ln e_i^2 \right] \tag{4.391}$$

where

$$C_i(\tau) = \sum_{\hat{y}=1}^{M} \sum_{y=1}^{M} \sum_{x=1}^{M} \sum_{\hat{x}=1}^{M} R(x - M\tau - \hat{x}, y - \hat{y})$$

$$\cdot \Phi_i(x, y) \Phi_i(\hat{x}, \hat{y}) \tag{4.392}$$

$$e_i^2 = C_i(0) - \sum_{j=1}^{m} A_{ij} C_i(j), \tag{4.393}$$

which has been described previously.

Habibi's Fig. 7 shown here as Fig. 4.37 depicts the theoretical performance of the two-dimensional hybrid system. It depicts SNRs versus sub-

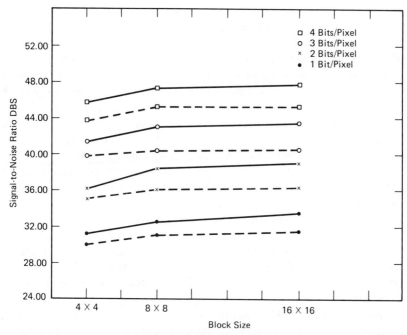

Figure 4.37. Theoretical performance of the proposed two-dimensional encoders versus block size $M \times M$. The solid and dashed lines refer to the Karhunen-Lieve and the Walsh-Hadamard transforms, respectively.

block size; for Karhunen-Loeve and Hadamard transforms plus banks of DPCM systems.

Both of these hybrid systems give good picture quality at a bit rate of 2 bits/pixel, and pictures coded at 1 bit/pixel are acceptable though somewhat degraded in quality. It has been observed that these hybrid systems are affected very little by channel noise corresponding to a bit-error rate of 10^{-4}. However, there is noticeable degradation in picture quality for bit-error rates of 10^{-3} and 10^{-2}. There is some propagation of errors in the vertical direction caused by the DPCM encoders.

When these hybrid systems were compared with the nonhybrid two-dimensional Hadamard transform system, it was found that the subjective quality of the pictures are worse for the hybrid systems, although the hybrid systems produce better picture quality. However, the degradation due to noise does not outweigh the improvement in picture quality of the hybrid system the noise degradation is not that significant.

4.13.16 Transform Image Coding and Noise Immunity

Besides having good bandwidth compression features, transform image coding also provides more tolerance to channel errors. This is due to the error averaging property of the transform coding. To obtain further noise immunity error-correction coding can be resorted to.

The probable reason for improved noise tolerance when transform coding is used, is that the message is transmitted to the frequency domain rather than in the conventional spatial domain. When this is done the noise components are evenly distributed over the entire reconstructed image. Consequently, the noise manifests itself as a low-frequency effect in reconstruction. The human eye is more sensitive to high-frequency noise than to low-frequency noise. Therefore, low-frequency noise is less offensive to the human eye.

In order to take advantage of this phenomenon the quantizing steps should be equally likely to occur. Therefore, each quantized word is equally likely to all the other words, and they are statistically independent. Experimental results bear this out (Hunt, 1969).

Because of this phenomenon the use of error-correcting codes with transform coding, produces better results than does error correction in the normally used spatial domain. However, it appears from experimentation that for error rates of less than 10^{-3} no error correction is needed. But, for low SNRs error correction should be used, even at the expense of not transmitting the entire frequency plane.

According to Hunt (1969) the error-correcting code should have at least six information bits. The two codes mentioned were the Reed-Muller (32, 6) code and the Bose-Chaudhuri-Hocquenghen (BCH) (31, 6) code. Both codes correct 7 or fewer errors. With these codes an error rate of 4×10^{-2} is transferred to a 2.26×10^{-5} error rate.

Hunt (1969) demonstrated that a Fourier transform plus error correction greatly improves the noise immunity of transmitted images.

4.14   CHANNEL-SHARING SYSTEMS

Haskell (1972) investigated, for Picturephone use, the possibility of channel sharing. Simulations were described that tested the feasibility of several conditional replenishment types of coders, sharing the same transmission channel. Channel sharing takes advantage of the fact that many different users are rarely in rapid motion at the same time. Thus when data from several sources are averaged together prior to transmission, it is much more uniform than from a single source. Since the peaks in the averaged data are smaller than with a single source, less buffering is required, and the channel rate required for transmission is much closer to the average data rate generated. The system performs adequately for 12 simultaneous users, and a 2:1 reduction in bit rate is obtainable.

In conditional replenishment systems there are long periods when a single channel is relatively idle. To utilize the channel capacity more efficiently channel sharing can be implemented.

Since separate Picturephone conversations can be assumed to be independent, the combined data are less peaked than from a single source. This not only reduces the required channel capacity, but it also reduces the required buffer size.

In this system, unbuffered data from the various sources are first stored in individual prebuffers, then transmitted to the principal buffer via a multiplexing switch, and finally transmitted over a single light capacity data channel to the receiver where the inverse operation takes place.

For a single source the number of significant changes per field has an exponential probability density function in the region of interest. However, there is a strong field-to-field correlation resulting in large data peaks that can be lost for many fields. It is this characteristic that makes channel sharing so effective in reducing the required bit rate.

4.15 SUMMARY

In this chapter we have discussed many aspects of bandwidth reduction techniques. Some of the basic bandwidth reduction methods were by means of picture degradation, statistical means, direct digital processing, and the use of psychophysical effects. We also discussed bandwidth reduction by the use of various types of dither.

One of the larger sections was devoted to the use of DPCM. DPCM represents a very practical approach to bandwidth reduction. However, it has its problem areas. A DPCM signal is easily degraded by channel noise, and needs to be used with some form of error detection and correction coding. Slope overload is also a problem in that image distortion can result from it. Some method of periodical sync correction must be used to overcome catastrophic loss of sync. However, the use of DPCM in hybrid systems has real potential. DPCM can be combined, for example, with one of the transform systems. Furthermore, delta modulation is very similar to DPCM and has sometimes been called a 1-bit DPCM system. We also discussed the properties of the reduction systems of synthetic highs and lows and bit-plane encoding. Bit-plane encoding is an excellent system to use but has a very high degree of sophisticated hardware complexity.

We discussed various types of systems using frame-to-frame correlation as combined with various other bandwidth reduction schemes. As a rule these systems achieve a large bandwidth reduction, but run into the problem of picture breakup due to the motion of objects within a scene.

Significant bit selection is another practical scheme for a moderate bandwidth reduction.

A large section was devoted to a study of the various transform techniques for bandwidth reduction. A general study of the use of transforms in bandwidth reduction and image restoration was followed by studies of specific systems. The basic transform types studied included: fast Fourier, Haar, Walsh-Hadamard, slant, cosine, and sine. A deterministic type of transform called the SVD transform was also discussed. It was included as a nonreal time technique of digital image processing.

An image restoration method called the least squares spline basis function was also studied. This restoration technique can be combined with other transform methods to achieve faster computational times than are normally available.

The finite-field number theoretic transform was also discussed. This transform can be used to replace the discrete Fourier transform to achieve much faster computation times. However, the available literature did not give a study of its bandwidth compression capabilities.

An application of the DCT was also included in order to show the implementation of this system in realistic hardware.

Table 4.1 Bandwidth Reduction Performance Matrix for Real Time TV

Method Used	Reduction Ratio	Handling of Objects in Motion
Dither	2 : 1	Excellent
DPCM	3 : 1	Excellent
Delta modulation	2 : 1	Excellent
Synthetic highs and lows	4 : 1	Poor
Bit-plane encoding	6 : 1	Excellent
Contour interpolation	4 : 1	Fair
Frame-to-frame correlation (conditional replenishment)	6 : 1	Poor
Frame-to-frame correlation using dot interlace	6 : 1	Poor
Frame-to-frame differential coder	3 : 1	Fair
Frame-to-frame differential coder with dot interlace	3 : 1	Fair
Frame repeating with DPCM	3 : 1	Fair
Channel-shared conditional replenishment	6 : 1	Fair
Combined Intraframe and frame-to-frame encoding	4 : 1	Fair
Combined interframe coding and Fourier transforms	6 : 1	Fair
Significant bit selection	2 : 1	Good
Fast Fourier transform with run-length coding	4 : 1	Good
Haar transform	4 : 1	Good
Walsh-Hadamard transform	6 : 1	Good
Slant transform	4 : 1	Excellent
Cosine transform	4 : 1	Excellent
Sine transform	4 : 1	Excellent
Singular value decomposition (SVD) transform	3 : 1	Not real time
Finite-field number theoretic transform	no experimental data	Excellent
Hybrid coding: DPCM and fast Fourier transform	4 : 1	Excellent
Channel-sharing conditional replenishment systems	2 : 1	Poor

Two forms of hybrid coding using DPCM were included to emphasize that a workable bandwidth reduction scheme may be realized by using combinations with the various other reduction schemes.

The discussion on transform techniques ended with a discussion of the effects of channel noise. The chapter ended with a discussion of a channel-sharing system.

Since bandwidth reduction is the prime consideration of this chapter, a chart has been prepared that compares bandwidth reduction ratios for the various techniques studied. This chart also includes information on the performance of each system with regards to the handling of objects in motion within a TV scene. This information is shown in Table 4.1.

chapter 5

Pseudorandom Coding plus PCM Coding

introduction

Pseudorandom coding can be combined with PCM coding to give improved performance for digital TV communications. This results in reduced transmitter power at the expense of a wider bandwidth. This is called the spread spectrum technique.

However, this system can also make use of bandwidth reduction techniques. The bandwidth reduction method chosen here is a form of dither modulation similar to the one suggested by Roberts (1962).

A good system design starting point is the consideration of the basic TV bandwidth requirements.

5.1 TV BANDWIDTH CONSIDERATIONS

There are several techniques available to minimize the bandwidth requirements without an appreciable loss in picture definition and resolution.

As a design point of reference, consider the bandwidth requirements for analog TV transmission. A television picture consists of continuous two-dimensional brightness distribution. An adequate expression for the cutoff frequency of a low-pass filter designed to handle the bandwidth requirements of analog TV is

$$f_c = \frac{1}{2\sqrt{2}} \left(\frac{W}{h} \right) n^2 N \tag{5.1}$$

where

W the width of a picture frame,

h the height of a picture frame,

n the number of lines of picture elements per frame,

N the frame rate in frames per second.

5.2 DESIGN SYNTHESIS

In practical coding design the first step is the A/D conversion of the video signal. The use of PAM sampling is usually the first step requirement. The total system then becomes a PAM/PCM/PRN/FM system; where the symbol PRN stands for PRN generation; more aptly called pseudorandom code generation.

Such a system has been implemented before, and can be found in the available literature. However, a new method for combining PRN and PCM is presented here. The implementation of this system will be discussed in detail.

5.3 THE TRANSMITTER

The transmitter block diagram is shown in Fig. 5.1. The baseband requirement is 44.1 MHz. This is a large frequency spectrum for frequency modulation to handle. To handle this requires the use of a special technique known as band splitting.

The encoded video signal is fed to the band splitter, where it becomes divided into three separate bands of frequencies. The first segment is from 0 to 500 kHz, and it is used to directly modulate a crystal oscillator. The middle segment is from 500 kHz to 10 MHz, and it is used to indirectly modulate the carrier to FM. Likewise, the third segment is from 10 MHz to 44.1 MHz, and it is used to indirectly modulate the carrier in an FM manner.

The three band segments are smoothly combined in such a manner that the transmitted FM has a baseband from 0 to 44.1 MHz. The transmitted FM two-sided bandwidth is 140 MHz wide, and is deviated by ±46.13 MHz.

The PCM bit rate is 44.1 Mbits/sec, which establishes the maximum baseband frequency.

Of course the encoder is a main part of the transmitter and will now be discussed.

5.4 THE ENCODER UNIT

The encoder unit is shown in the block diagram of Fig. 5.2. It will be noted that the blanked video is fed to the PAM modulator. In the PAM modulator

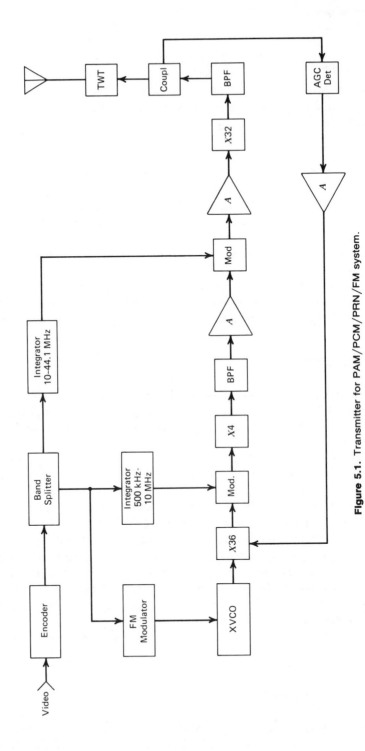

Figure 5.1. Transmitter for PAM/PCM/PRN/FM system.

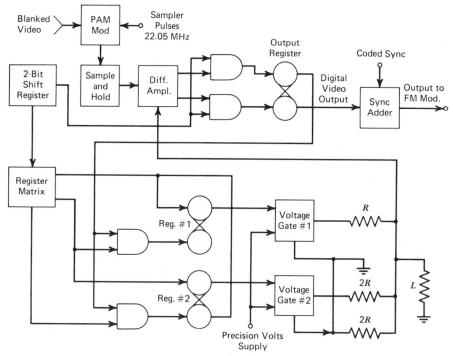

Figure 5.2. Functional diagram of encoder unit.

the video signal is sampled by pulses occurring at a basic 22.05-MHz rate. However, what is not shown here is the fact that these sampler pulses occur at a pseudorandom rate. This is opposed to the usual method of sampling with evenly spaced pulses. More detail will be given on this subject at a later point in the discussion.

Each amplitude-modulated pulse is sent to a sample-and-hold circuit. The function of this circuit is to hold a constant level of voltage for the duration of the sampling period. The voltage level is held equal to the peak value of the input pulse. The sample-and-hold circuit feeds the PCM modulator unit at the input of a differential amplifier.

A 2-bit PCM coding was selected, so that each sampled-and-held level is in turn sampled two times. Thus the basic rate for the system must be two times the PAM sampling rate, or 22.05 MHz times 2 equals 44.1 MHz. This bit rate is a little high, but is feasible to implement, under present state-of-the-art conditions.

However, this 44.1-MHz bit rate cannot be used in a evenly spaced condition, for the system now under consideration. It must be randomized. A little thought will show that each PAM period is different in length of duration than all the others. Therefore, each sampling period must be evenly divided into three equal spaces. To do this requires special synthesis. This calls for a special wave-shaping unit, which is shown in Fig. 5.3.

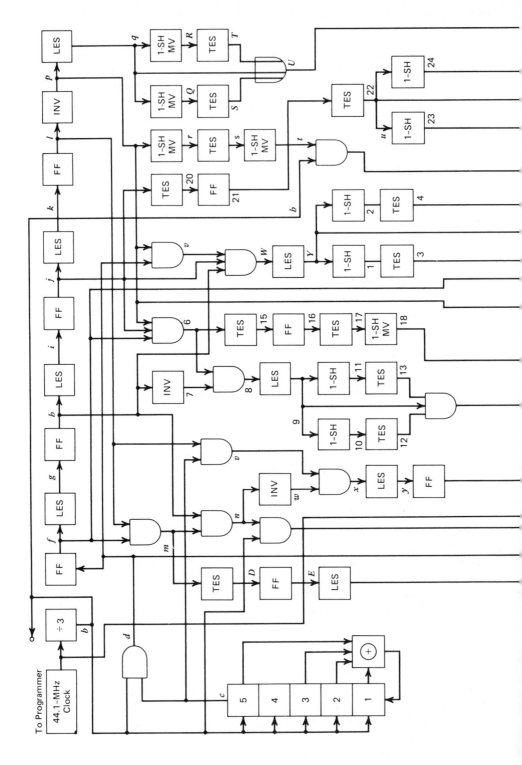

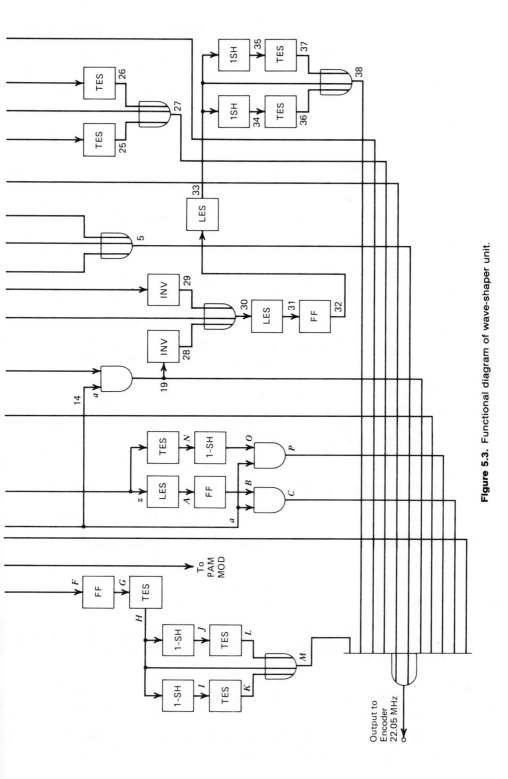

Figure 5.3. Functional diagram of wave-shaper unit.

147

The 2-bit register shown in Fig. 5.2 cares not whether its input is evenly spaced or randomly spaced. The A/D converter shown will work with randomly spaced pulses. Its output feeds an adder, where the TV sync signals are inserted. After the insertion of the coded sync, the signal is sent to the FM modulator unit.

5.5 THE WAVE-SHAPER UNIT

The heart of the wave-shaper unit is the 44.1-MHz clock. Another section of the wave-shaper unit is the PRN. The remainder of the unit deals with synthesis of a waveform at the clock frequency, but which is pseudorandom in output. The labeled points of Fig. 5.3 indicate the corresponding waveforms shown in the waveform chart of Fig. 5.4 (*a*)–(*i*).

Point *a* of Fig. 5.3 has a waveform which is shown in Fig. 5.4(a) on line *a*. It is the clock output. The clock output frequency is divided by three to get waveform *b*, and waveform *b* drives the PRN generator. The waveform output of the PRN generator is shown at point *c* of Fig. 5.4(a).

Waveforms *c* and *b* are AND-gated to obtain the pulse shown at point *d* of Fig. 5.3 and on line *d* of Fig. 5.4(a). These are the PAM sampling pulses and are sent to the PAM modulator unit.

The remainder of the wave-shaper unit functions to synthesize the pulses that will divide the spaces between each pulse pair on line *d* [Fig. 5.4(a)] into three equally spaced pulses. To illustrate the synthesis process, a description of the waveforms from *f* to *o* will be given.

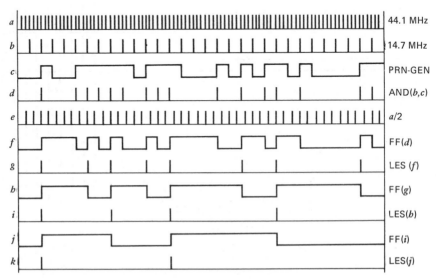

Figure 5.4. (*a*–*i*) Waveform synthesis.

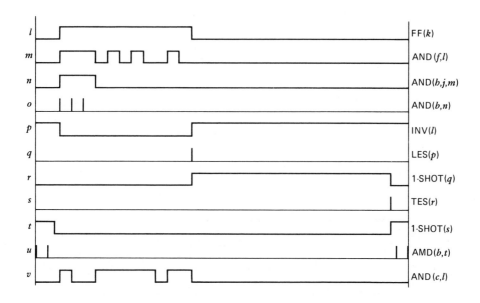

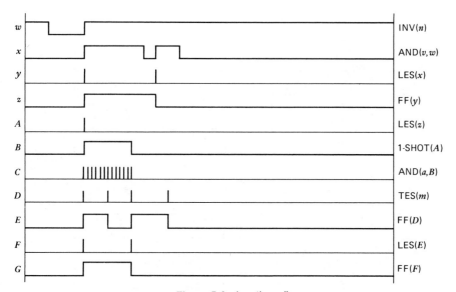

Figure 5.4. *(continued)*

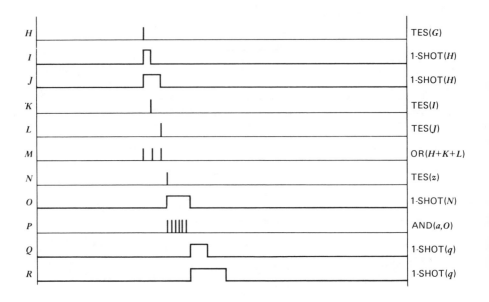

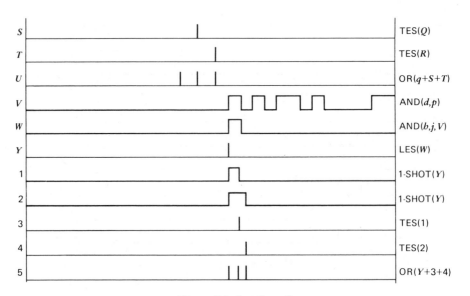

Figure 5.4. *(continued)*

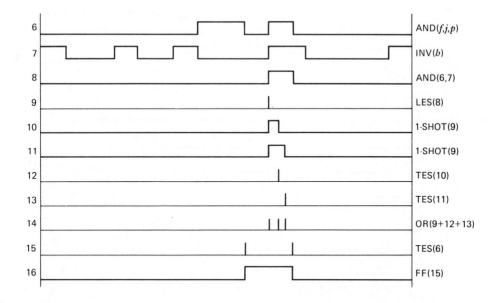

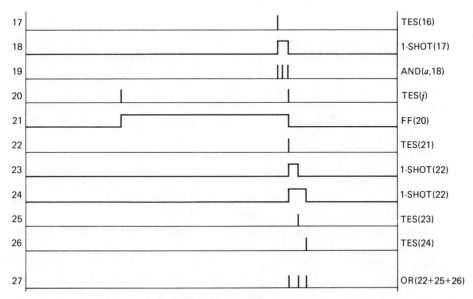

Figure 5.4. (continued)

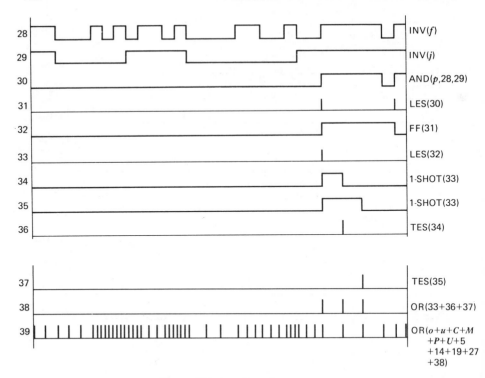

Figure 5.4. *(continued)*

The pulses from d are used to trigger a frequency-divider type of flip-flop (FF), to get the waveform on line f. Waveform f is operated on by a leading edge wave shaper to form the trigger pulses on line g. These pulses, in turn, trigger the FF to get waveform h. Another leading edge shaper (LES) and FF create the waveforms shown on line j. The same procedure is used to get the waveform on line l. Both waveform f and l are AND-gated to get the waveform on line m.

When waveforms h, j, and m are AND-gated, the result is as shown on line n. The pulses on line o are formed by AND-gating lines b and n. Three pulses are shown on line o. The number 2 and 3 pulses on line o divide the number 1 and 2 pulse space of line d into three equal spaces. These three pulses are used later in the final synthesis step.

Similar synthesis procedures are used to form the pulse groups shown on the following lines: Line u of Fig. 5.4(b), line C of Fig. 5.4(c), line M of Fig. 5.4(d), line P of Fig. 5.4(d), line U of Fig. 5.4(e), line 5 of Fig. 5.4(e), line 14 of Fig. 4.5(f), line 19 of Fig. 5.4(g), line 27 of Fig. 5.4(g), and line 38 of Fig. 5.4(i). These are all OR-gated to get the final output waveform shown on line 39 of Fig. 5.4(i). This is the output that will be used to drive the A/D converter of Fig. 5.3. The average bit rate out of the wave-shaper unit is 22.05 Mbits/sec.

5.6   THE PROGRAMMER UNIT

The programmer is shown in Fig. 5.5. It is driven by pulses from line *b* of Fig. 5.4. These are evenly spaced pulses at a rate of 14.7 Mbits/sec. The programmer functions to create the coded sync pulses for use in the transmitted signal, and to create the noncoded pulses for internal use in the camera unit. It also creates the horizontal and vertical blanking gates. The blanking gates are used to blank out the video during the flyback intervals in the TV camera and kinescope tubes. The video after being blanked is sent from the programmer to the encoder, where it enters the PAM modulator unit.

5.7   THE RECEIVER

The receiver is shown in Fig. 5.6. It is a conventional FM receiver using a limiter and discriminator. However, the 140-MHz bandwidth demands an IF center frequency of at least 1000 MHz. When the IF center frequency is in the gigahertz region, a discriminator can be built to handle the 46.13-MHz frequency deviation.

5.8   THE DECODER UNIT

The decoder unit is shown in Fig. 5.7. It functions as a pulse regenerator, a D/A converter, and a bit synchronizer. The demodulated output from the FM receiver is attenuated, sliced, pulse-shaped, and fed to a 1-shot multivibrator (MV).

The waveform out of the MV is cleaned up, but not synchronized. The output from the 1-shot MV is fed to the phase detector of the bit-synchronization unit. The bit sync synchronizer is in the form of a phase-locked loop, having a voltage-controlled oscillator (VCO) at its heart.

The bit-synced VCO output is used to drive the sync separator, in order to derive the horizontal and vertical sync signals. It also drives an integrate and dump circuit. The demodulated and uncleaned signal passes through to the *I & D* circuit to become regenerated and locked in phase to the VCO. The output from the *I & D* is very clean. It undergoes a change of state from digital to analog in the D/A converter, and is PAM demodulated to get back the original video signal.

5.9   THE D/A CONVERTER

The D/A converter is shown in Fig. 5.8. It also contains a VCO at 44.1 MHz nominal frequency. However, in order to convert from digital to analog, the driving waveform cannot be evenly spaced, but must be an exact duplicate of

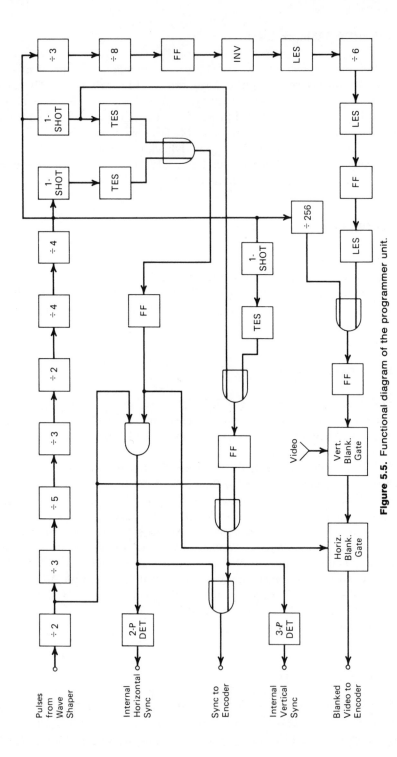

Figure 5.5. Functional diagram of the programmer unit.

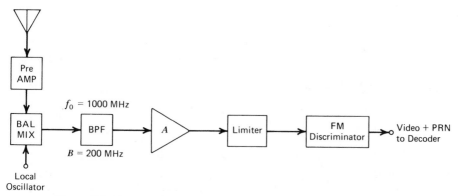

Figure 5.6. Functional receiver diagram for PAM/PCM/PRN/FM system.

the pseudorandom waveform shown on line 30 of Fig. 5.4(i). Therefore, the waveshaper shown in Fig. 5.3 must be used in the receiver system also. This pseudorandom sequence of pulses drives the phase detector shown in Fig. 5.8. The other input to the phase detector is the signal from the integrate and dump circuits of Fig. 5.7. The output from the phase detector is filtered and used to lock the 44.1-MHz VCO unit into phase with the received signal.

The phase-locked output drives the 3-bit shift register, which is the first unit of the D/A converter. The D/A converter is very standard in type, and can handle the PRN type signals.

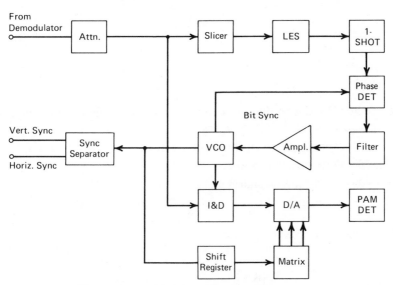

Figure 5.7. Functional diagram of decoder unit.

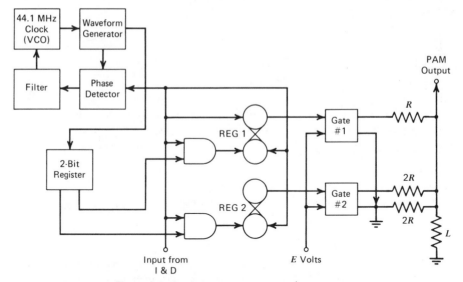

Figure 5.8. Functional diagram of D/A converter.

5.10   THE MATHEMATICAL ANALYSIS

According to Balakrishnan and Abrams (1966), for a **PAM/PCM/FM** system, the error probability can be written as

$$P_b \cong \frac{1}{\sqrt{2\pi(\text{SNR})}} \exp \frac{-\text{SNR}}{2} \qquad (5.2)$$

where

P_b	the bit error probability,
SNR	the SNR due to the internally generated white Gaussian noise and is also written as S/N or E/N_0,
S	the average power in the wanted signal,
N	the noise power in watts,
E	the energy in one signal pulse,
N_0	the noise spectral density, internally generated, watts/hertz.

For a wideband case they state that

$$\text{SNR} = \frac{2}{1 + \left[4\pi^2 f_N^2 / \dot{M}(t_0)^2 \right]} \qquad (5.3)$$

where

ρ^2 the SNR in the IF,

$\dot{M}(t_0)^2$ $4\pi^2 D^2$,

D the maximum one-sided frequency deviation,

         the one-sided IF bandwidth B.

$\mathcal{F}_N$ $B/\sqrt{3}$.

So that the expression for SNR can be written as

$$\text{SNR} = \frac{\rho^2}{1+(B^2/3D^2)} = \frac{\rho^2}{\beta^2}. \tag{5.4}$$

Therefore, expression (1) can be written as

$$P_b = \frac{1}{\sqrt{2\pi\rho^2/\beta^2}} = \exp\frac{-\rho^2}{2\beta^2}. \tag{5.5}$$

The expression for SNR used by Balakrishnan and Abrams (1960) does not take into account the FM improvement factor. We can account for it by multiplying the SNR by the factor $3M^2$, where M is the modulation index. So that, the error probability expression is again modified to

$$P_b = \frac{1}{\sqrt{6\pi M^2 \rho^2/\beta^2}} \exp\frac{-3M^2\rho^2}{2\beta^2}. \tag{5.6}$$

Another modification is needed to account for the use of spread spectrum modulation. So that,

$$P_b = \frac{1}{\sqrt{6\pi M^2 \rho^2 BT/\beta^2}} \exp\frac{-3M^2 BT\rho^2}{2\beta^2}. \tag{5.7}$$

Also Viterbi (1962) in order to account for quantization effects uses a factor of $\log_2 L$, which further modifies the equation:

$$P_b = \frac{1}{\sqrt{6\pi M^2 \rho^2 BT \log_2 L/\beta^2}} \exp\frac{-3M^2 BT\rho^2}{2\beta^2 \log_2 L} \tag{5.8}$$

where

BT   the time bandwidth product,

T the time duration of one signal pulse.

5.11 CHANNEL-SHARING ANALYSIS

As mentioned previously under channel-sharing systems, there are applications where several subscribers may wish digital TV service simultaneously
over the same channel. This has also been called a type of friendly jamming.
Each coding system can be evaluated on the basis of resistance to friendly
jamming. Performance comparison can then be made for each coding type, as
to its merits for use in channel sharing.

An expression can be derived for use in such merit comparisons. There
will be an expression for the combined noise variance:

$$\sigma_T^2 = \sigma_N^2 + \sigma_C^2 \tag{5.9}$$

where

σ_T^2 the total noise variance,
σ_N^2 the variance due to Gaussian noise,
σ_C^2 the variance due to the friendly jammer or clutter.

We can also say that

$$\sigma_N^2 = N_0 B$$

and can let

$$\sigma_C^2 = QC$$

where

C the clutter power.

The factor of Q was given by Blasbalg, Freeman, and Keeler (1964) and is

$$Q = U\lambda_c d$$

where

λ_c the cross-correlation coefficient of the coding system,
d the duty factor of the pulse.

It is easy to see that

$$\frac{\sigma_C^2}{\sigma_N^2} = \frac{QC}{N_0 B} = \rho_C^2 = \frac{QC}{S}\frac{E}{N_0} = \frac{QC}{S}\text{SNR}. \tag{5.10}$$

The expression for error probability can be modified to account for the clutter factor as follows:

$$P_b = \cfrac{1}{\sqrt{6\pi M^2 \rho^2 BT / \beta^2 (1 + \rho_C^2) \log_2 L}}$$

$$\times \exp \frac{-1.5 BT M^2 \rho^2}{\beta^2 (1 + \rho_C^2) \log_2 L} \tag{5.11}$$

where

ρ_C^2	QC/S (SNR)
C	the clutter power (average for 12 total users),
Q	$U\lambda_c d = 0.24$,
U	the number of users $= 12$,
λ_c	the cross-correlation coefficient for the signals of the users $= 0.1$,
d	the duty factor of the pulses $= 0.2$.

5.12 CHANNEL-SHARING PERFORMANCE

The previous parameter values were used to obtain word error probability curves. These results are shown in Fig. 5.9. At a clutter-to-signal power ratio of unity the system has fair performance, but the performance falls off as the C/S ratio increases.

The system parameters were a baseband of 44.1 MHz, a one-sided IF bandwidth of 60 MHz, and an FM deviation of ± 70.0 MHz for a modulation index of about 1.59.

The use of DPCM in the system format of PAM/DPCM/FM results in a baseband of 22.05 MHz versus 44.1 MHz for a PAM/PCM/PRN/FM system. In this case the one-sided IF bandwidth can be set at 30 MHz, and the deviation at ± 35 MHz for a modulation index of 1.59. The channel-sharing word error rate performance characteristics for the DPCM based system are shown in Fig. 5.10.

On a nonchannel-sharing basis the DPCM system is about 3.5 dB better in word error rate performance characteristics than the PAM/PCM/ PRN/FM system described here. For a unity C/S ratio the DPCM based system is much superior to the PCM system. This is assuming identical Q-factors for both systems.

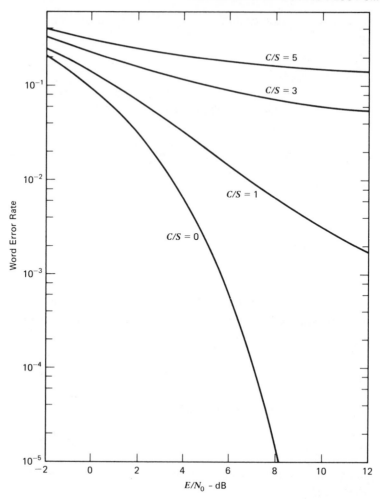

Figure 5.9. Channel-sharing performance of the PAM/PCM/PRN/FM system.

5.13 CONCLUSIONS

The new PAM/PCM/PRN/FM system does not contribute any bandwidth reduction, so that a DPCM based system with half the bandwidth requirements outperforms it in the area of word error rate. However, a new method of using spread spectrum sampling of the video has been demonstrated. A PCM system is a poor vehicle for this new sampling means. In the following chapter a vehicle will be chosen that can take full advantage of spread spectrum sampling.

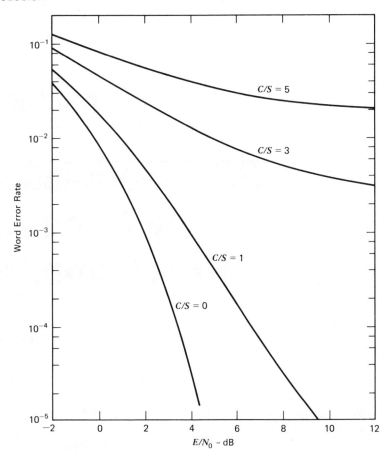

Figure 5.10. Channel-sharing performance of the DPCM/FM system.

chapter 6

PPM Modulation by Means of PRN Sampling

introduction

The principles of spread spectrum sampling can be extended to PPM systems. The resulting system can be in the form of PRN/PPM/FM. In this system pseudo random pulses are generated and then used to directly drive a PPM modulator unit.

Ordinarily, a PPM modulator is driven by regularly spaced clock pulses. In the spread spectrum sampling method the video signal is sampled by pseudo random pulses. Thus, the pulses which are used to sample the video will be irregularly spaced, as opposed to the normal method of equally spaced pulses.

The hardware implementation of this new modulation system will now be described. The implementation of a complete PRN/PPM/FM system will now be discussed. Frequency modulation is used because of the signal-to-noise improvement factor.

6.1 THE TRANSMITTER

The transmitter is FM modulated by the pseudorandom noise-like pulses (PRN/PPM). This creates a combined modulation system which can be called PRN/PPM/FM. There is a problem created by using FM for such a broad baseband spectrum.

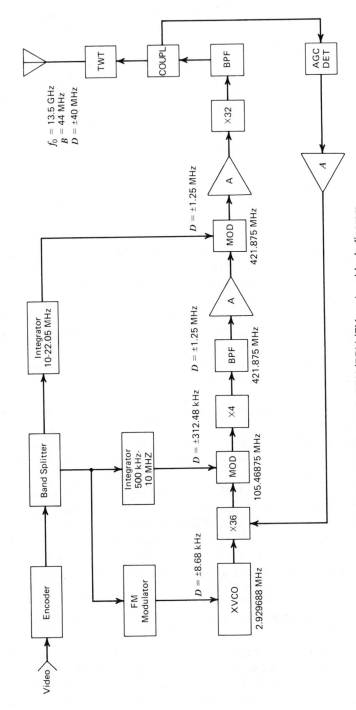

Figure 6.1. Transmitter for PRN/PPM/FM system block diagram.

163

The basic pulse repetition rate of the PRN/PPM pulses must be 22.05 Mbits. Indirect FM must be used in order to FM modulate such a broad spectrum. It cannot be done by direct means. In implementing indirect FM, phase modulation is used, with the baseband having been previously passed through an integrator. This combination creates an FM spectrum through operation on an already existent carrier signal; not by varying the frequency of an FM oscillator. This is the principle used in Fig. 6.1.

However, 22.05 MHz is too broad a spectrum to handle in one step. At least three steps are advisable. A band splitter is used to divide the spectrum into three sections, as follows:

Section no. 1	0	to 500 kHz
Section no. 2	500 MHz	to 10 Mhz
Section no. 3	10 MHz	to 22.05 MHz

The first section makes use of direct FM. The crystal oscillator, at a carrier frequency of 2.929688 MHz is modulated directly. Such an oscillator can be FM modulated by a few kilohertz. In Fig. 6.1, it can be seen that the crystal oscillator will be deviated by ± 8.68 kHz.

The oscillator output is multiplied by 36 times in frequency. Its new center frequency is 105.46875 MHz, and it is deviated by ± 312.48 kHz. At this point the second modulation stage is used. This is indirectly FM modulated by the 500 kHz to 10 MHz band of frequencies. The deviation created by the modulator must equal that of the modulated signal signal from the XVCO, or ± 312.48 kHz.

Thus the first two basebands will be united in the second modulator stage, at equal deviations.

The third modulation takes place after a 4:1 frequency multiplication. The new carrier will be at 421.875 MHz, deviated by ± 1.25 MHz. The indirect FM modulation of the third stage modulates the 10–22.05 MHz baseband section to a ± 1.25-MHz deviation. So that, at the output of the third modulator the center frequency is 421.875 MHz, deviated by ± 1.25 MHz, and having a baseband from 0 to 22.05 MHz.

Next, a multiplier of 32 times is used to get a carrier frequency of 13.5 GHz, deviated by ± 40 MHz, for a baseband from 0 to 22.05 MHz.

Now, a more thorough discussion of the PRN/PPM encoder unit is needed.

6.2 THE PRN/PPM ENCODER UNIT

The heart of any complex encoder is the programmer. The programmer of Fig. 6.2 contains the basic clock, and from it generates the horizontal and vertical synchronization signals, the blanking gates, the coded synchronization signals, and timing pulses for the driving of two random pulse generators.

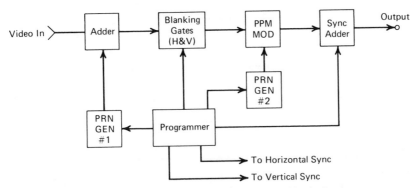

Figure 6.2. encoder for PRN/PPM/FM system block diagram.

The incoming video signal is masked or disguised as random noise by adding PRN to it. The noise generator is driven by the programmer (Fig. 6.3), with evenly spaced pulses counted down from the clock frequency, and having a pulse repetition rate of 122,500 pulses per second. The period of the PRN generator is 16 bits long, which is equal to the length of a horizontal line on the TV raster. Therefore, the PRN signal is synchronized to the horizontal lines of the TV raster. It is easily removed in the receiver.

6.3 PRN GENERATOR NO. 1

The function of the PRN generator has been explained previously. However, a pseudorandom pulse generator sequence period of 16 is an unusual requirement. A maximum length sequence, as generated by a shift register, using linear feedback, is given by the expression:

$$P = 2^n - 1 \tag{6.1}$$

where

P the period length

n the number of bits per word.

When $n = 4$, $P = 15$; not 16. A pseudorandom pulse sequence of 16 is not usually called for.

However, PRN sequences of 16 bits/word length can be generated by a shift register having five stages or more. In this case six stages have been used, as shown in Fig. 6.4.
Mathematically, the polynomial required is

$$a_x = 1 \oplus x_1 \oplus \bar{x}_2 \oplus x_3 \bar{x}_4 \oplus x_5 x_6 \tag{6.2}$$

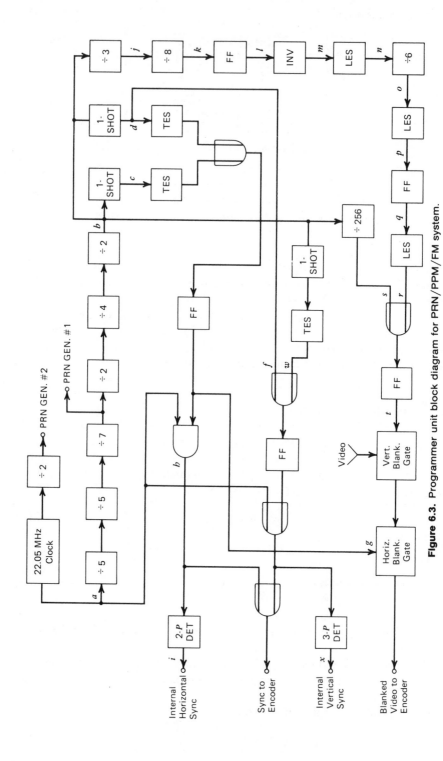

Figure 6.3. Programmer unit block diagram for PRN/PPM/FM system.

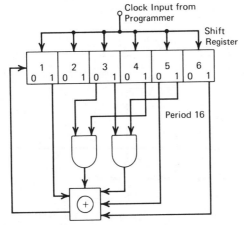

Figure 6.4. PRN generator 1.

and the recursion formula is

$$0 = I \oplus D_1 D_2 \oplus \overline{D}_3 D_4 \oplus \overline{D}_5 \oplus D_6. \tag{6.3}$$

The bars above the symbols (D_3) and (D_5) mean "NOT D_3" and "NOT D_5," or instead of a "one," a "zero." In a shift register the NOT polarity is found on the collector of the shift register transistor, opposite the collector normally used.

6.4   PRN GENERATOR NO. 2

As seen if Fig. 6.2, the PRN generator no. 2 output drives the PPM modulator directly, that is, the blanked video is sampled at a pseudorandom rate, rather than by equidistant pulses. The video is sampled at a rate of 22.05 Mbits/second. The modulator output is PPM at a pseudorandom rate.

A feedback shift register such as is shown in Fig. 6.5 has a period of 22 bits/word. The polynomial required to achieve this period length is

$$0 = I \oplus D_1 D_2 \oplus \overline{D}_3 \oplus D_4 \oplus D_5. \tag{6.4}$$

It produces a series of "ones" and "zeros" as shown in Fig. 6-5(b). The corresponding waveform is shown in Fig. 6-5(c), for the case of 100 percent duty factor pulses. However, in this waveform there are only 5 "ones" and 5 "zeros." In order to obtain narrow sampling pulses, the waveform on line c is AND-gated with narrow clock pulses. This results in line d, in which there are 9 pseudorandomly spaced pulses. Now, there are nine "ones" and thirteen "zeros." Thus the added logic generates low duty factor sampling pulses. The average duty factor is 0.0456.

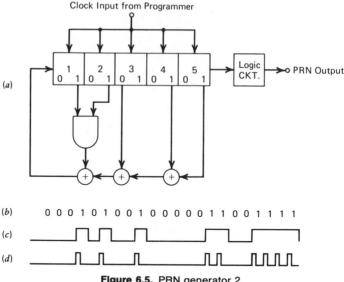

Figure 6.5. PRN generator 2.

6.5 THE PPM MODULATOR

The block diagram for the PPM modulator is shown in Fig. 6.6. It is driven by the blanked video signal, which has been pseudorandomly sampled by the pulses from PRN generator no. 2. The sampling pulses are shown in Figure 6-5(d), and later on line b of Fig. 6.7.

These sampling pulses drive a one-shot MV, which produces 95 percent duty factor pulses, as shown on line c of Fig. 6.7. These widened pulses are used to drive a sawtooth wave generator having a slope duration equal to that of the driving pulses. These sawtooth waves are shown on line d of Fig. 6.7.

These sawtooth waves are summed with the video signal (line e) to give the waveform shown on line f. This waveform is sent to a threshold detector, which removes that portion of the waveform that is below the threshold. The leading edges of the output sawteeth trigger a one-shot MV to give short duty

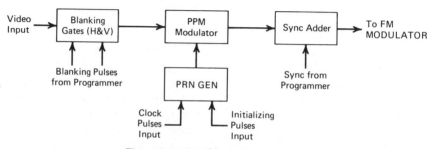

Figure 6.6. PRN/PPm encoder unit.

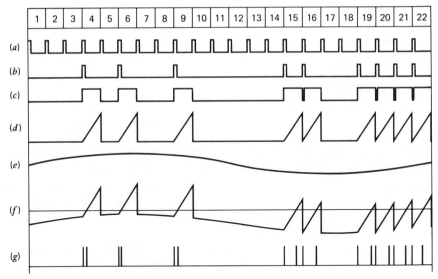

Figure 6.7. The generation of PRN/PPM/FM waveforms.

factor pulses. These pulses are in the form of pseudorandom PPM, or PRN/PPM. The right hand lines on line *g* of Fig. 6.7 represent the PRN/PPM pulses; the left hand lines represent the sawtooth reference points.

6.6 THE RECEIVER

The receiver is shown in Fig. 6.8. It is very straightforward and uses an IF of 500 MHz. An IF this high is needed because the 3-dB bandwidth is set at 72 MHz. An FM discriminator can be made for use on a 72-MHz bandwidth, if the IF center frequency is high enough. In this case 500 MHz should be adequate for the IF center frequency.

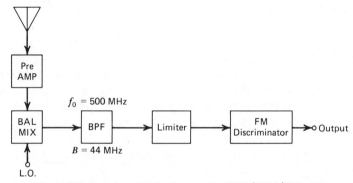

Figure 6.8. Receiver block diagram for PRN/PPM/FM system.

The discriminator output consists of the PRN/PPM pulses, composed of the original video plus PRN signals. The decoder unit recreates the original video waveform.

6.7   THE DECODER UNIT

The decoder unit is shown in Fig. 6.9. The first function of the decoder is to generate the transmitted original waveform. A slicer and a LES start this process. The trigger pulses thus formed are used to trigger a one-shot MV to drive a bit synchronizer unit. The bit synchronizer uses a VCO basically at the frequency of the clock located in the transmitter programmer section. The bit synchronizer locks on to the VCO in the master clock. When locked in the VCO becomes a secondary clock. An integrate and dump circuit is used to produce a very clean wave shape.

A horizontal sync is derived by means of coincident gating, called a two-pulse detector. It can then be used at the kinescope to time the raster lines. The horizontal sync pulses are also used to drive a decoder unit PRN generator no. 1. The purpose of PRN generator no. 1 is to feed a PRN removal device for separating the video from the PRN pulses.

The VCO output, being in a frequency stable condition, is used to drive PRN generator no. 2 of the decoder unit. The AND gate forms trigger pulses for driving a one-shot MV, which drives a sawtooth generator. The sawtooth wave that is generated has irregularly spaced teeth, the rate being controlled by PRN generator no. 2. These teeth drive the PPM demodulator, the output of which consists of inverted video, plus superimposed PRN signal, plus vestiges of the sawteeth. This composite signal is filtered and inverted to obtain the original video, plus the PRN noise. The PRN noise-like signal that was added to the video in the transmitter, prior to sampling, is now removed by additional circuitry, as shown in Fig. 6.9. The waveforms associated with PPM demodulation are shown in Fig. 6.10.

6.8   THE MATHEMATICAL ANALYSIS

In a PPM system the bit-error rate is meaningful. The concept of word error rate does not apply to a pseudorandom sampled PPM system. (PRN/PPM), even though the sampling pulses consist of PRN words. This can be seen by looking at the detection process, where a bit error per word has no more effect upon video distortion than in a pure PPM system. Therefore, only bit-error rates need be considered in a PRN/PPM system of the type used here.

If the signal were merely PRN, the equation for bit-error probability would be

$$P_B = \left(\frac{1}{2\pi E_s / N_0} \right)^{1/2} \exp \frac{-E_s}{2N_0} \tag{6.5}$$

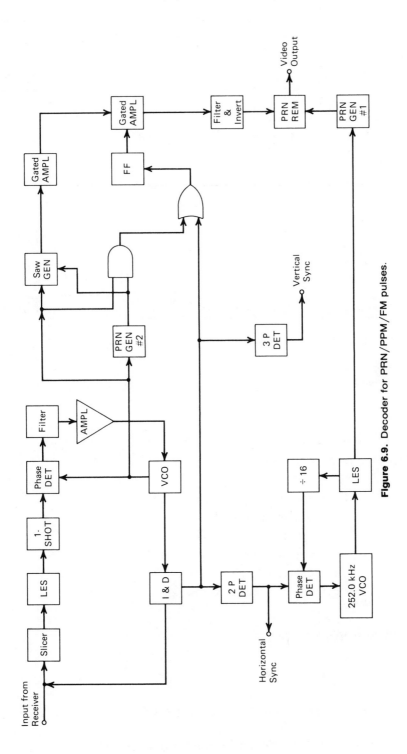

Figure 6.9. Decoder for PRN/PPM/FM pulses.

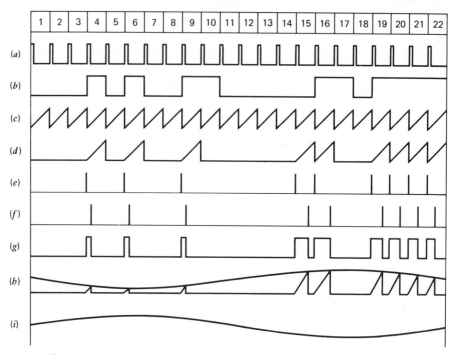

Figure 6.10. Waveforms for the demodulation of PRN/PPM/FM signals.

where

E_s/N_0	$BTE/N_0 = (\text{SNR})BT.$
E_S	$BTE.$
B	IF bandwidth, ± 22 MHz.
T	Time duration of a bit.
E	Energy of the pulse.
N_0	Noise spectral density (watts per hertz).
S	Average received signal power.
N	Noise power in the IF bandwidth.

In order to simplify we can let

$$\rho^2 = \frac{E}{N_0} = \frac{S}{N} = \text{SNR}. \tag{6.7}$$

The FM modulation must be accounted for by using the FM improvement factor of $3M^2$, where M is the modulation index. In the previous example illustrated, $M = 1.82$. The expression for error probability now becomes for a PRN/FM signal

$$P_B = \left[\frac{1}{6\pi M^2 BT\rho^2/2} \right]^{1/2} \exp\frac{-3M^2BT\rho^2}{2}. \tag{6.8}$$

According to Landon (1948), the SNR of PPM/FM has an improvement factor over an AM signal of $(4/9)$ $BT=0.729$. This factor should be used, so that for PRN/PPM/FM

$$P_B = \left[\frac{1}{8\pi M^2 B^2 T^2 \rho^2/3} \right]^{1/4} \exp\frac{-2M^2B^2T^2\rho^2}{3}. \tag{6.9}$$

For the system described above $BT=1.64$ and $(4/9)BT=0.729$. The modulation index is 1.82. Therefore, the numerical evaluation of bit-error probability is

$$P_B = \left(\frac{1}{74.64\rho^2} \right)^{1/2} \exp(-5.939\rho^2). \tag{6.10}$$

6.9 CHANNEL-SHARING PERFORMANCE

This system has excellent performance in channel-sharing application. It is assumed that 12 users of the same channel are being serviced simultaneously. In this case, the bit-error rate equation becomes modified, as was done in Chapter 5 for PCM, to

$$P_B = \left[\frac{1}{8\pi M^2 B^2 T^2 \rho^2/3(1+\rho_c^2)} \right]^{1/2} \exp\frac{-2M^2B^2T^2\rho^2}{3(1+\rho_c^2)} \tag{6.11}$$

where $\rho^2 = \dfrac{QC}{S}$ (SNR).

The Q factor is the product of several other factors:

Q $UD\lambda_C = 0.025$.
U the number of users, 12.
d the pulse duty factor, 0.0456.
λ_c the cross-correlation coefficient for signals of 12 users, 0.0455.

The channel-sharing performance characteristics of the PRN/PPM/FM system using spread spectrum sampling is excellent as shown in Fig. 6.11.

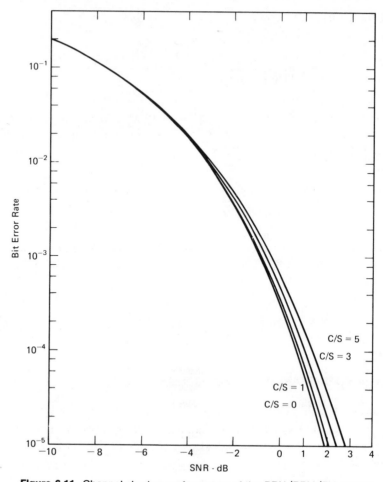

Figure 6.11. Channel-sharing performance of the PRN/PPM/FM system.

chapter 7

Multi-Channel PRN/PPM/FM Television Systems

introduction

In Chapter 6 a new type of PPM modulation was discussed. It was implemented into a PRN/PPM/FM system, in which the PPM pulses were formed from video waveforms, which had been sampled at a pseudorandom rate. The resulting PRN/PPM/FM signal was transmitted in a single channel.

It is possible to simultaneously PPM modulate several different video channels; one channel per TV raster line. This is accomplished by the use of parallel-serial scan techniques. The conventional TV camera can scan only one-line-at-a-time. But, solid-state TV cameras using charge-coupled devices (CCDs) in a focal plane array are now commonplace. Parallel-serial readout of several TV lines simultaneously has been achieved in the present state of the art. This technique is proposed as a means of forming multichannel PRN/PPM/FM systems.

Since CCD technology is a fairly new discipline, we will spend some time discussing their basic properties, and the methods available for applying these properties to a TV camera using parallel-serial readout. The term "readout" is appropriate here, rather than "scan," because no electronic beam is used in solid-state TV cameras using CCDs in a focal plane array.

7.1 AN INTRODUCTION TO BASIC CCD TECHNOLOGY

CCD's are a class of semiconductor structures that come close to being the analog shift registers. In the variable spectrum they are used as photoconductor detectors fabricated in a silicon array chip. Technology uses them in an array at the focal plane of a TV camera lens system.

These arrays take the form of rectangular rows and columns of chips. These rows (or columns) of chips are optimally biased and driven by a clock as shift registers. Thus they continuously sample the various gray shades of the incident light falling upon the focal plane, without the need for the beam scanning used in conventional camera tubes.

The outputs from these analog shift registers can be used to charge detectors or pre-amplifiers with extremely small input capacitances. This leads to amplification at very low noise levels, and to detection of very small numbers of photo-generated carriers.

Very high transfer efficiency and wide dynamic range can be achieved in CCD arrays.

7.1.1 Charge Storage and Transfer

The operation of CCD arrays can be explained in a simplified manner as follows. The CCD array can be thought of as an array of closely spaced metal-insulator-semiconductor (MIS) capacitors in which information in the form of a packet of electric charges, can be stored and transferred, by applying suitable voltage pulses to the metal electrodes.

Charge storage takes place in an MIS capacitor, which could take the form of a p-type semiconductor on which there is an insulating layer and a metal electrode. If then, at zero time, a positive voltage V_g is applied to the electrode, a depletion region is formed in the semiconductor as shown by the energy band diagram of Fig. 7.1(a), as taken from Amelio, Bertram, and Tompsett (1971) in their Fig. 1 (also D. J. Burt, 1975).

When the depletion region is large, the device is not in thermal equilibrium and minority carriers (electrons) begin to accumulate at the interface between the semiconductor and the insulator, where the potential energy is the lowest. The depletion region can be considered as a "potential well." The potential at the interface represents the depth of the well. When carriers accumulate in the well, the surface potential decreases and a large voltage appears across the insulator until equilibrium is reached. This is illustrated in Fig. 7.1(b).

The source of minority carriers is thermal generation within the semiconductor. In high-quality silicon semiconductors the thermal generation current may be so low that equilibrium is reached only after many seconds. This current is known as "dark current" in imaging devices. Minority carriers may be introduced by absorption of incident photons or from forward-biased p-n junctions as an input circuit. In either case, information is represented that can be transferred to a detection circuit and detected.

These information-carrying minority carriers may be considered as a packet of charge, which is stored in a potential well. A linear array of CCDs consists of closely spaced photoconductive detectors, each having gate electrodes on a continuous dielectric film that covers the single-crystal semicon-

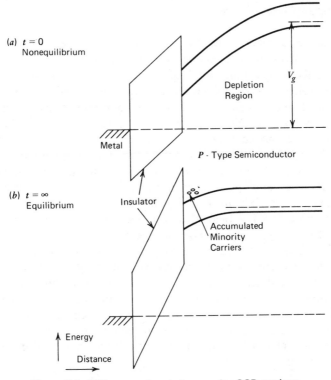

(a) $t = 0$
Nonequilibrium

Depletion
Region

V_g

Metal

P - Type Semiconductor

(b) $t = \infty$
Equilibrium

Insulator

Accumulated
Minority
Carriers

↑ Energy

Distance

Figure 7.1. MIS energy band diagram for CCD analogy.

ductor substrate. The charge-coupled packets may be transferred in serial shift register style from electrode-to-electrode, in a row or in a column of detectors in an the array.

This is done by manipulating the relative depths of the wells under the adjacent gate electrodes, which is accomplished by changing the electrode bias V_g. A higher V_g produces a deeper well; a lower V_g, a lower well. A clock pulse drive to each well electrode moves these packets from well to well, in what is called analog shift register action. The channel region of the device, in and through which the charge storage and transfer takes place, is usually bounded by high concentration "channel-stop" diffusions.

This transfer of charge packets takes place with a very small amount of inefficiency. The inefficiency is caused by the presence of charge trapping states. However, these effects can be minimized by passing a constant background charge called "fat zero" through the detectors. This causes the interface states to remain permanently filled, and minimizes the amount of interface state trapping. Unfortunately, the technique of "fat zero" is not fully effective, due to the presence of edge effects in the channel, and can be difficult to implement in imaging devices where there is no electrical input to the array.

7.1.2 Buried Channel Arrays

An alternative to the surface channel approach is the buried channel approach. The MIS capacitor analogy is presented here in Figure 7.2 and was taken from Kent (1973) and his Fig. 8, which is used here by permission of American Telephone & Telegraph Co. In the buried channel CCD an appropriate doping distribution of polarity opposite to the substrate is introduced over a small region adjacent to the surface. This causes the potential minimum to be moved away from the surface to a location within the impurity layer. Thus the signal charge contact with the interface region is prevented, which eliminates surface trapping effects. This results in high transfer efficiency and speed of charge transfer. This higher speed is due to the fact that the carriers are kept further away from the electrodes and are subject to more fringing field effects. Good charge coupling in the buried channel CCD requires very closely spaced electrodes.

A buried p-channel CCD has a shallow p-type layer between the insulating dielectric material and the n-type silicon substrate of a conventional surface p-channel device structure. The application of suitable potentials to this structure results in the depletion of free carriers from the layer. Signal charge (electrons) can then be stored and transferred in the buried channel without contact with the semiconductor-insulator surface.

The buried channel CCD offers several advantages over the surface channel devices:

1 improved transfer efficiency by elimination of surface trapping mechanisms.

2 lower noise level due to the elimination of surface state "fat zero" noise.

3 higher speed operation resulting from improved carrier mobility and (in deep channels) high fringing fields.

Although the highest operating speeds are obtained in the deeper channels, this is bought at the expense of charge carrying capacity. However, adequate speeds and good capacity can be obtained in relatively shallow channels.

The buried channel CCD has improved dynamic range due to the lower noise levels. Noise in the amplifiers following the CCD matrix is the limiting factor in determining SNR.

However, the buried channel CCD arrays are not free from trapping effects. A phenomenon known as bulk trapping minimizes slightly the charge packet transfer efficiency. This problem has been analyzed by several authors (Barbe, 1975; Mohsen and Tompsett, 1974).

7.1.3 Bulk Trapping in Buried Channel CCD Arrays

The analysis results presented here are from a 1975 Naval Electronic Systems Command report (AD AO22 881). The author of this report is Kenneth

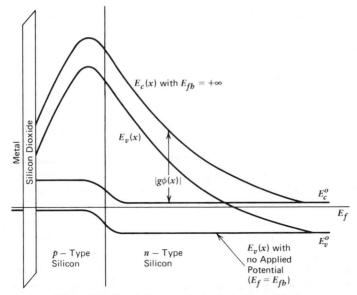

Figure 7.2. Band diagram for buried channel device.

Hoagland (Program Manager, Fairchild Camera & Instrument Corp., Imaging Systems Division) who bases the probability of charge trapping on the number of charges in the well and the distance of the trap from the well minimum. The interaction between charge and trap is not only dependent on trap energy and phase time, but also on the position-dependent energy of the charge with respect to the well minimum. The most effective traps are not necessarily those at the well minimum. Rather, they are those which undergo the largest average increase in the probability of being filled from the dark current condition to the signal condition. A random distribution of traps is assumed.

Calculations were made relative to a set of standard or referenced conditions which include temperature, well volume, trap concentration, trap energy level, and trap cross section.

In its ideal form, the buried channel CCD has very low transfer losses and very low transfer noise. Cumulative transport efficiency can exceed 90 percent after more than 1700 transfers. A high density of bulk traps, however, produces a significant deviation from the ideal structure and is the most serious cause of transfer inefficiency and transfer noise in buried channel CCD low light level image sensors, which operate at reduced temperatures.

This study shows that the determination of rms transient response at bright spot levels of a few electrons requires a microstatistical treatment of trapping probabilities and of geometric trap positions.

The device assumed for analysis was a 500×500 photoelement buried channel CCD array of the interline transfer type, with an area of about 0.360 by 0.480 in. Two-phase operation was assumed with equal times in the two phases. Operation at the conventional TV time rate of 15,750 lines/sec was assumed. This results in a phase time for the vertical register of 31.476 μsec. The phase time for the horizontal register was about 500 times smaller.

Details of the cell geometry and of the buried channel dopant concentration are important only as they affect the shape and size of the potential well, and were discussed briefly.

For the charge levels assumed in this analysis, the mobile charge packets move in a space where fixed charge concentrations exceed mobile charge concentrations by 2 to 4 orders of magnitude, at the point of maximum concentration of mobile charges. The fixed charge concentration and the exposure of this essentially dielectric (rather than semiconducting) space, to the potentials of the surrounding electrodes, result in a strong variation of well potential, and in the effective confinement of the mobile charge to a small region in the vicinity of the potential minimum of the well. Because the transit time from well to well is many orders of magnitudes shorter than the time spent in each vertical transfer well, bulk trapping during transit is inconsequential.

For the trap densities considered, most potential wells have no traps sufficiently close to the potential minimums to trap charges significantly. In those wells that have near the minimum, the traps can be assumed to be randomly distributed in space about the minimum.

The statistically probable trapping behavior of a well with an effective trap is dependent on the probable initial state of the trap, on trap parameters, on the position of the trap with respect to the potential minimum, on the number of mobile charge carriers introduced into the well, and on the time that the charge is held in the well. The wide variations in location of the traps with respect to the well minima must be satisfactorily represented and these locations must be satisfactorily represented and randomized with respect to the transport sequence as they are in the actual device.

The critical portion of the area image sector with respect to bulk trapping is represented by a linear sequence of 500 electrodes, each with a 5 μm-long, 5-V height, ion-implanted or stepped-oxide barrier for the two-phase operation. The potential well under each electrode is 13 μm long by 10 μm wide. The buried channel has a rectangular profile 0.5 μm deep and a net donor concentration of 2×10^{16} atoms/cm^3. Traps are assumed to be distributed randomly throughout the bulk, at a concentration of 10^{11}/cm^3.

Mobile charges are introduced only under the first electrode. They are then transferred through the structure by alternation of phase electrode potentials between 0 and 12 V, which is sufficient for the rapid and complete transfer of all charges not in the bulk traps. Both dark and signal charges are modeled by Poisson series.

For the calculation of the behavior of an average set of electrodes, each potential well is assumed to have one trap within a volume of 10^{-11}cm^3 about the center. The volume is then assumed to be divided into concentric shells, each of volume 2×10^{-14}cm^3, which is 1/500th of the average volume per trap. Each potential well of the set of 500 is assumed to have its trap at the mean radius of a unique concentric shell. This assures that the assigned trap positions are reasonably averaged in space about the potential minima. The order of the wells with respect to radial trap positions is randomized with respect to the order in which the charge moves through the wells, in order to model practical structures more faithfully.

The spatial positions of the trap with respect to the potential minimum must be translated to an energy level with respect to the minimum to allow the calculation of the electron density at the position, which enters into the capture probability. To achieve this the well potentials are assumed to vary quadratically about the minimum. It is further assumed that the ellipsoidal equipotentials can be transported to those of a sphere of equal volume. Computer simulations for a buried channel of a maximum depth of about 0.5 μm, show that well potential changes by 0.020 V in approximately 0.5 μm in both the length direction and the transverse direction of the electrode array. The donor concentration of the buried channel results in a change of 0.020 V in 0.0357 μm in the depth direction. The geometric mean of these three values is approximately 0.208 μm and is the radius of the equivalent sphere at the 0.020 V contour. Its value is about 4×10^{-14} cm^3. The potential can be represented as

$$V_r = V_{r_0} + 0.020 \left(\frac{r}{r_0} \right)^2 \tag{7.1}$$

where

r distance from well minimum,

r_0 0.208 μm.

For this spherical well containing N_w electrons, the electron density is

$$n_r = \frac{N_w}{(\pi/a)^{1.5} r_0^3} \exp\left[-a\left(\frac{r}{r_0}\right)^2 \right] \tag{7.2}$$

where

a $0.020/(kT/q)$
q 1.6×10^{-19}
kT/q 0.01925 V at $-50°$C
kT/q 0.0256 V at $+22°$C

The effectiveness of a trap in modifying charge flow is determined by its characteristic emissions time τ_e and by its capacitance τ_c.

Emission time is given by

$$\tau_\varepsilon = \frac{e^{E_T/kT}}{V_{\text{th}} \sigma N_c} \tag{7.3}$$

where

E_T trap energy below the conduction band.

σ trap cross section.

N_c density of states in conduction band.

V_{th} thermal velocity of the charge carrier.

Capture time is given by

$$\tau_c = 1/V_{th}\sigma N_t$$

where

n_t electron density at the trap location assuming an empty trap and that $n_t = n_r$,

n_r as given previously.

7.1.4 Trap Distribution

The SNR of the output is determined by first assigning trap positions, which are then used to calculate capture time as a function of the number of electrons in a well. After the calculation of emission time, these parameters are employed to determine trap-filling probabilities. From the trap-filling probabilities, cumulative transfer efficiency and SNR are found, by considering in sequence the effects of movement of typical charge packets through 500 wells. Trap positions are assigned as follows for a set of M electrodes (or wells) with a trap concentration of N_T traps/cm^3. One trap is assigned to the volume $1/N_T$ that surrounds the minimum of each potential well. This volume is subdivided into M concentric shells of equal volume. In one of the wells, the trap is assigned to the arithmetic mean radius of a first concentric shell of volume $1/MN_T$. In a second well, the trap is assigned to the arithmetic mean radius of a second concentric shell of the same volume, that is, to a radius equal to the outer radius of the first shell plus half the increment radius of the second shell. The sequence is followed through the Mth well and shell, which is almost $N_T^{-1/3}$ from the well minimum. The electron density in the well n_r is given by

$$n_r = n_{r_0}\exp\left[-a\left(\frac{r}{r_0}\right)^2\right] \tag{7.5}$$

where n_{r_0} is the electron density at the well minimum as was given previously.

7.1.5 Trap Occupancy

The parameters τ_e and τ_c determine the general solution to the probability of trap occupancy f for time t:

$$f = ce^{-t/\tau} + \frac{\tau}{\tau_c} \tag{7.6}$$

where

$$\frac{1}{\tau} = \frac{1}{\tau_e} + \frac{1}{\tau_c}. \tag{7.7}$$

If a trap is initially filled, then $c = 1 - f_{eq}$, where $f_{eq} = e^{-t/\tau}$. If we let f_f represent the probability of trap occupancy for an initially filled well, then $f_f = e^{-t/\tau} + (\tau/\tau_c)(1 - e^{-t/\tau})$. The symbol f_{eq} represents the time asymptote to the trap-filling probability, calculated on the basis of all the electrons in the well.

If a trap is initially empty, then $c = -f_{eq}$, and $f_e = (\tau/\tau_c)(1 - e^{-t/\tau})$, where f_e is the probability of the trap being initially empty.

From these equations, which govern the trapping behavior of a well, for a particular trap location and for a specified number of electrons in the well, the behavior of a chain of wells may be found. The response of a chain of wells to dark current is first considered.

Dark change is assumed to appear at the first well as a Poisson distribution with response time

$$\bar{D}_u = \frac{e^{-u} u^N}{N!} \tag{7.8}$$

where

u the time average value of the number of carriers per pixel.

N any specified number of electrons.

This is a normalized distribution function, that is,

$$\sum_{N=0}^{\infty} \bar{D}_u(N) = 1. \tag{7.9}$$

After many packets of dark charge introduced at the first well have passed through the chain of wells, there is a probable occupancy of the trap at each well P_i. This depends on the trap parameters for the well, on the average u of dark charge, on the distribution function that is not necessarily Poisson after the first well, on the dark charge entering a particular well, and on the phase times. The probable occupancy P_i of the ith well at the time the charge enters, must be the same as that at the end of the transfer interval, because the well is in equilibrium with dark-charge transport.

After many cycles of transport of charge of average value, for example, any individual well will have a probability P_{in} of occupancy of its trap at the time charge is introduced. For each possible quantity N of charge introduced, there will be a probable trap occupancy f at the end of the cycle. This probable occupancy will depend on the initial trap conditions.

For a trap initially full

$$f_f = 1 - \frac{\tau}{\tau_c} e^{-t_h/\tau} + \frac{\tau}{\tau_c} \tag{7.10}$$

where t_h is the time of phase high interval.

For a trap initially empty

$$f_e = \frac{\tau}{\tau_c}(1 - e^{-t_h/\tau}).$$ (7.11)

At the end of a full cycle, these will be

$$f_f = e^{-t_1/\tau_c}\left[\frac{\tau}{\tau_c} + \left(1 - \frac{\tau}{\tau_c}\right)e^{-t_h/\tau}\right]$$ (7.12)

$$f_e = e^{-t_1/\tau_c}\left[\frac{\tau}{\tau_c}(1 - e^{-t_h/\tau})\right]$$ (7.13)

where the additional term takes into account release of charge from the trap during the phase low interval t_1. The expressions given here use a value for τ_c, which assumes the availability of all electrons for trapping at the beginning of the phase high interval. However, if the previous electrode position contains an effective trap, one electron may arrive at some time during the phase high interval, and the capture time may in such a case be significantly larger than the value assumed here. The calculated values for f_e are therefore, in some instances a little larger than the true values, and the effect of the bulk trap on the trapping of low-level signals is, therefore, slightly underestimated. Only 50 out of 500 wells appear to participate significantly in trapping; approximately 10 percent of these or 5 wells are affected in this manner described, because only traps in adjacent wells are involved. It is judged, therefore, that the use of the assumed value for τ_c does not materially affect the conclusion reached for the analysis. In particular, effects of dark-current values of 10 electrons/pixel should be almost negligible; effects at dark current values of 1 electron/pixel are estimated to be less than 20 percent.

For any particular number of electrons N_j, entering a well, the probability of the trap being full at the end of the cycle is

$$F(N_j) = P_{\text{in}}f_f(N_j) + (1 - P_{\text{in}})f_e(N_j)$$ (7.14)

where τ_c in f_e is computed for $N_j + 1$, because of the charge in the trap. Where $\overline{D}_u(N_j)$ is the weighting factor for the entering charge distribution, the average probable occupancy at the end of the cycle is

$$P_{\text{out}} = \Sigma \overline{D}_u(N_j)F(N_j).$$ (7.15)

Since equilibrium is assumed $P_{\text{out}} = P_{\text{in}} = P$ and

$$P = \Sigma \overline{D}_u(N_j)\left[Pf_f(N_j) + (1 - P)f_e(N_j)\right],$$ (7.16)

which can be solved for P to yield

$$P = \frac{\Sigma \bar{D}_u(N_j)f_e(N_j)}{1 + \Sigma \bar{D}_u(N_j)f_e(N_j) - \Sigma \bar{D}_u(N_j)f_f(N_j)}. \tag{7.17}$$

The weighting factor $\bar{E}_v(N_j)$ of the output charge distribution is given by

$$\bar{E}_v(N_j) = P\left[1 - f_f(N_j - 1)\right]\bar{D}u(N_j - 1) + Pf_f(N_j)\bar{D}_u(N_j)$$

$$+ (1 - P)\left[1 - f_e(N_i)\right]\bar{D}_u(N_j) + (1 - P)f_e(N_j + 1)\bar{D}_u(N_j + 1). \tag{7.18}$$

The output distribution is used as the input distribution for the next well. Iteration of this calculation through a chain of wells gives an output distribution from the last well that depends on the input average and distribution and on the trapping behavior and sequence of the wells in the chain.

Calculations for the case of a leading signal charge packet are carried out in the same manner except that P_{in} for the combined signal and dark charge is taken from the output part of the $\bar{D}_u$ results.

Cumulative transfer efficiency n_{cum} is calculated as

$$n_{cum} = \frac{\displaystyle\sum_{j=0}^{\infty} \overline{S + D}_{cum}(N_j)F(N_j)}{\displaystyle\sum_{j=0}^{\infty} \overline{S + D}_{in}(N_j)F(N_j)} \tag{7.19}$$

where

$\overline{S + D}(N_j)$ weighting factor for N_j charges for the combined signal and dark charge.

The SNR voltage ratio, at the output is the square root of the power ratio and is given by

$$\mathrm{SNR} = \left[\frac{\displaystyle\sum_{j=0}^{\infty} (\overline{S+D})(N_j)(N_j - \bar{D})^2 - \sum_{j=0}^{\infty} (\overline{S+D})(N_j)(N_j - \overline{S+D})^2}{\displaystyle\sum_{j=0}^{\infty} \overline{S+D}(N_j)(N_j - \overline{S+D})^2}\right]^{1/2}$$

where the weighting factors and average charges are taken at the output of the chain of wells.

A principle conclusion of this study report was that for the worst case conditions analyzed, the cumulative transfer efficiency for an average dark charge per pixel of 10 electrons and an average signal charge 10 electrons/pixel, is approximately 80 percent; for SNRs of 1.4. Measurement data confirm the theoretical analysis that there is a useful cumulative transfer efficiency after 500 transfers to the extent that 10-electron level signals can be detected.

The major conclusion is that, for typical trap cross sections a usable SNR will be achieved at the 10-electron/pixel signal level, with 10^{10} traps/cm^3, without dark charge being present. About the same result will be achieved with 10^{11} traps/cm^3, if the dark charge averages 10 electrons/pixel.

7.1.6 Dynamic Range in Buried Channel CCD Arrays

The high-light level performance of the buried-channel CCD array is also important. Typical saturation output voltage (V_o) on fabricated units are 10 MV across a 1 Ω source resistor (R_s). If we assume $g_m = 0.6$ MMHOS, the voltage K at the amplifier gate (V_g) is

$$V_g = \frac{V_0(1 + g_m R_s)}{g_m R_s} = 27 \text{ MV}. \tag{7.21}$$

The total capacitance at this node is typically 1.0 pF, so that

$$q_{sat} = 1 \times 10^{-12} \times 27 \times 10^{-3} = 0.027 \text{ pC}$$

$$= 1.7 \times 10^5 \text{ electrons}. \tag{7.22}$$

If we have a dark-current collection area of 234 μm^2 and a dark-current density of 10 nA/cm^2, each photosite generates 2.34×10^{-14} A of dark current at room temperature. We can assume an integration time of 33 μsec (30 frames/sec). The average photosite dark charge is then

$$q_{dark} = 2.34 \times 10^{-14} \times 33 \times 10^{-3} = 7.7 \times 10^{-4} C$$

$$= 4.8 \times 10^{-3} \text{ electrons}. \tag{7.23}$$

The dynamic range is calculated by

$$\frac{q_{sat}}{\sqrt{q_{dark}}} = \frac{1.7 \times 10^5}{69} = 2460 : 1. \tag{7.24}$$

7.1.7 Dark-Current Dependence on Cooling

It has been observed in a variety of buried-channel CCD units that both the average dark signal and the dark signal nonuniformities have a temperature

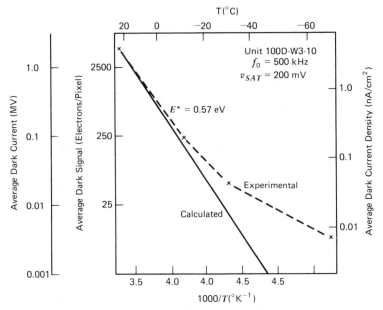

Figure 7.3. Experimental and calculated curves for CCD dark current versus temperature.

dependence that is approximately the same as that of the intrinsic carrier concentration parameter n_i for silicon, over a limited temperature range extending downward from 25°C to −60°C.

Fig. 7.3, which was taken from a NAVAL report (AD-A022 881, Fig. 4.29), shows this temperature dependence. Both theoretical and experimental curves are shown. The deviation between the two curves at lower temperatures, is likely due to the fact that the factors of silicon impurity type and gettering processes were not taken into account in the theoretical calculations.

7.1.8 A Practical In-Line-Transfer Array

Fig. 7.4, which was taken from the NAVAL report (AD-A022 881, Fig. 4.58), shows a practical in-line-transfer (ILT) CCD array layout, using the two-phase shift registers. Vertically arranged shift registers are mounted between detector columns. The vertical columns of CCD detectors are parallel-fed to the shift registers, as shown by the small arrows moving from left to right. Thus one row of detector outputs at-a-time is fed in parallel to the output registers from the vertically moving serial outputs of the array shift registers.

Distributed floating-gate amplifiers (DFGAs) are fed in parallel from the output registers. Inputs to the DFGA are first sensed by FGA1. Then, as they are clocked on down the input channel, they are sensed by the 12 floating-gate structures associated with the DFGA. Finally, they are detected by another single-stage floating gate amplifier (FGA2), and then terminated in a sink diode at the end of the channel.

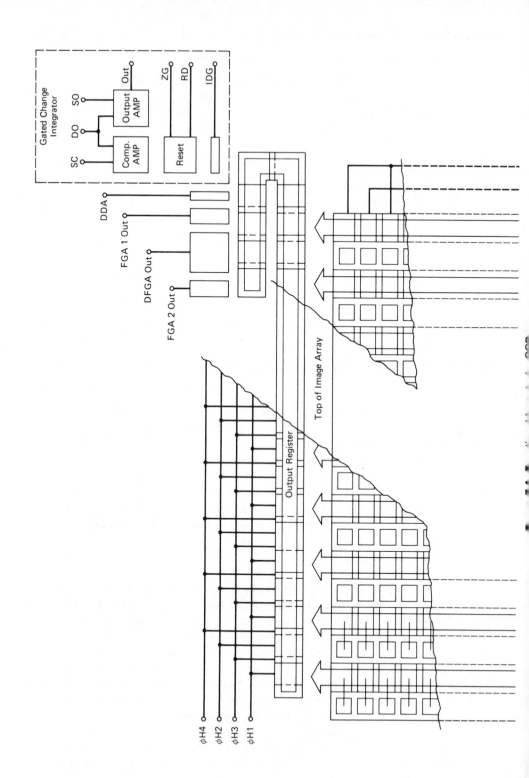

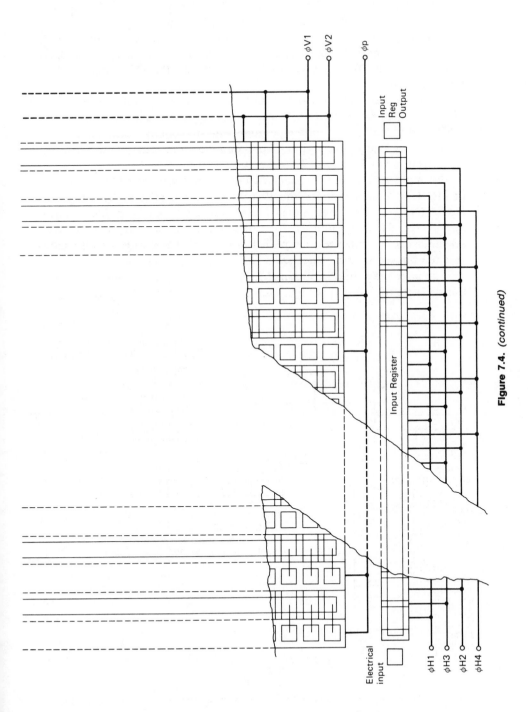

Figure 7.4. (continued)

189

The DFGA system is a very low-noise device and is useful for low-light level CCD TV cameras. However, the tradeoff for low noise is less dynamic range, due to lower overload levels. However, this loss in dynamic range can be corrected by using a gated-charge integrator (GCI) in conjunction with the DFGA. The overload charge from the DFGA can be drained into the DDA, which acts as a DFGA saturation control, where the DDA is a dual-gate differential amplifier.

Experimental results for the DFGA show a noise-equivalent signal (NES) of 10–20 signal electrons for a 50-nsec charge amplifier activation time, and a 3-MHz bandwidth, at room temperature with no cooling.

There are many other types of pre-amplifiers that can be used if there is no interest in low-light level TV operation.

A simplified Interline Transfer CCD array not using the DFGA is shown in Fig. 1 of the appendix of the 1976 Fairchild Imaging Systems reports to NASA (N-77-25429). Their diagram is reproduced here as Fig. 7.5. It shows a serially fed detector-amplifier from the output register. It is a two-phase system using 2:1 interlace. Simplified drive input waveforms are also shown. It is reproduced here by permission from Kenneth Hoagland.

The interlace action is obtained from pulsing (ϕ_p) low, while the adjacent transfer gate (ϕ_{v1} or ϕ_{v2}) is high. During each vertical blanking gate time, ϕ_p,

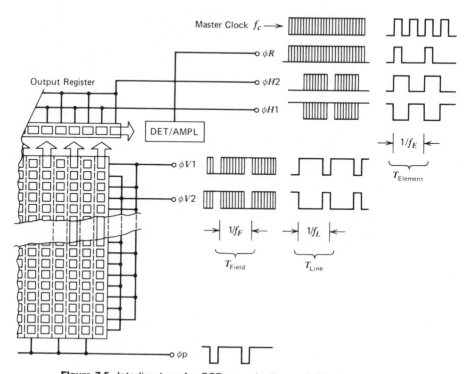

Figure 7.5. Interline transfer CCD organization and drive input waveforms.

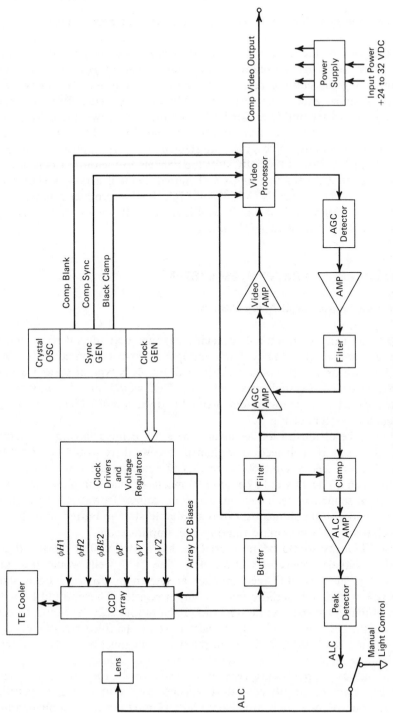

Figure 7.6. CCD TV camera block diagram.

is low and the complementary ϕ_{v1}, ϕ_{v2} waveforms determine which field is used.

At the start of the CCD field readout, elements corresponding to odd-number rows are first shifted in unison into adjacent ϕ_{v1} sites, for row transport along the column registers to the output register. The EVEN field sequence is similar except that the initial shift is into ϕ_{v2} sites. Row transfers at the output register interface (for both ODD and EVEN rows) are effected by holding ϕ_{v1} LOW and ϕ_{v2} HIGH during the horizontal blanking interval. Complementary square wave pulses at element rate are applied to the ϕ_{H1} and ϕ_{H2} transport gates to serially shift the packets to the output detector.

A solid-state TV camera using CCD buried channel arrays is shown as Fig. 3-3 of NASA report N77-25429 (1976). It is reproduced here, with permission of Kenneth Hoagland, as Figure 7.6. It shows the accessory circuits required to complete a TV camera.

7.2 A 25-CHANNEL PRN/PPM/FM SYSTEM

7.2.1 A 25-Channel CCD Array Camera

A CCD array can be structured in such a manner that several rows (lines) may be read out simultaneously. Each multiple channel section could be read out in sequence until a complete field is read out. This type of system would use what is called parallel-serial readout. This would be the basis of a multiple channel TV camera system, a simplified version of which is shown in the 25-channel system of Fig. 7.7.

For the sake of illustration let us assume a TV camera having 450 active lines (CCD rows). If each section contained 25 rows there would be 18 active sections. We will also assume 600 active CCD detectors per row. Synchronization for such a system would be done by means of enabling pulses used for gating ON, in sequence, each section. The 25 channels (rows) of information would be read out simultaneously into a suitable storage buffer, and thence into 25 channels of appropriate filtering and amplification.

The CCD array would be structured so that a row of shift register (light shielded) would lie between CCD registers. The CCD wells would feed the shift registers in a parallel feed manner as shown in Fig. 7.7. Each shift register row would shift serially into an output register. There would be only 25 such output registers; one per channel. Each output register would be 600 bits in length. Each of the 25 output registers would be filled simultaneously. After section 1 has filled the 25 output registers, they must be ready to accept inputs from section 2; and thereafter each section in sequence.

Each output channel shift register is made ready for use of the next section by means of an adjacent shift register. Each output register parallel feeds into one of two such auxilliary registers alternately. Complementary enabling pulses are used to gate these auxiliary registers. The ON register will

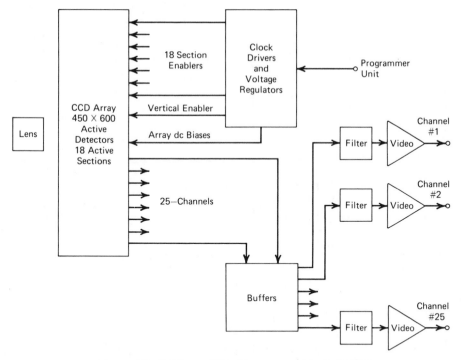

Figure 7.7. 25-Channel TV video camera block diagram.

have clock pulse drive; the OFF register no pulses. The OR gates are fed from these alternating auxiliary registers. There are 25 OR gates; one for each channel.

Fig. 7.8 depicts a single-channel output register structure. The auxiliary registers are enabled on a section-to-section basis. The center register handles each section sequentially, but the auxiliary registers handle the sections on an alternate basis.

Fig. 7.9 shows the CCD array and register structure in greater detail. The shift registers internal to the CCD array are shifted by enabling pulses on a sequential basis. Vertical blanking is done between sections 18 and 1. The vertical blanking time interval is equal to that of two sections. Thus there are 18 active (picturewise) sections and 2 inactive sections for a total of 20 sections per field. For the purpose of simplification, 2 : 1 interlace will not be considered as part of this illustration, although it is certainly possible to attain. Thus synchronization timing will be based on 20 sections. Horizontal blanking must also be provided for between sections. Blanking will be discussed in greater detail later.

We now have a 25-channel CCD focal plane array TV camera and will use it to design a TV communication system. This 25-channel CCD instrumentation is shown in Fig. 7.10. We will now describe a suitable transmitter.

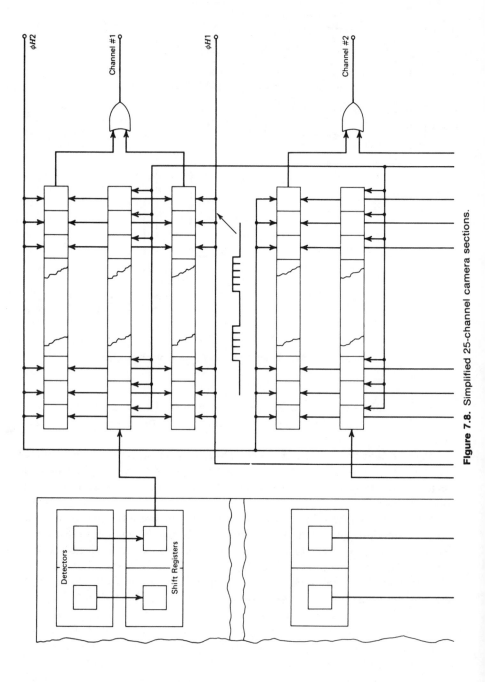

Figure 7.8. Simplified 25-channel camera sections.

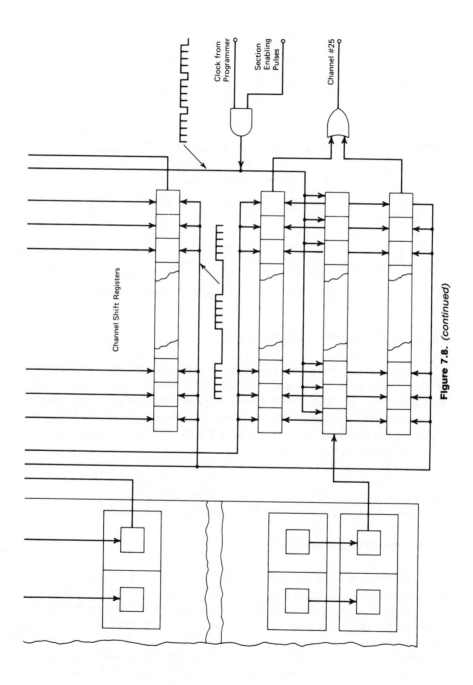

Clock from Programmer

Section Enabling Pulses

Channel #25

Channel Shift Registers

Figure 7.8. *(continued)*

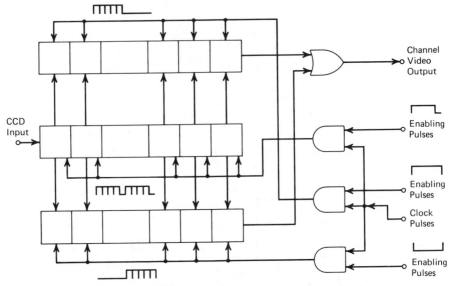

Figure 7.9. CCD output shift registers for each channel.

7.2.2 The Transmitter plus Modulator

The transmitter is frequency-modulated by pseudorandom noise-like pulses (PRN/PPM), as in Chapter 6. This creates a combined modulation system that can be called PRN/PPM/FM. However, the system described in this chapter differs from the one in Chapter 6, because of the use of 25 orthogonal channels (to be described later) that are simultaneously transmitted over a one-carrier transmitter.

The band-splitting technique described in Chapter 6 is used here as shown in Fig. 7.11. We use a transmitted carrier of 4.2 GHz, with a modulation baseband of ±1.6 MHz and a ±3.2-MHz FM deviation. Now, a more thorough discussion of the PRN/PPM encoder unit is needed.

7.2.3 The Channel Encoders

The encoder unit is shown in Fig. 7.12. The 25-channel TV camera feeds the channel encoders. The video signal to each channel is blanked during section and vertical blanking periods. These blanking pulses are generated in the programmer unit, which are described later. After blanking the channel video is fed to a PPM modulator, which is sampled by the pseudorandom technique described in the preceding chapters. The PRN generators furnish PRN sampling pulses to the PPM modulators. After PRN/PPM modulation is accomplished, sync pulses are added to the signal at appropriate points. Then each of the 25 channels is fed into an OR gate, which in turn feeds the frequency modulator.

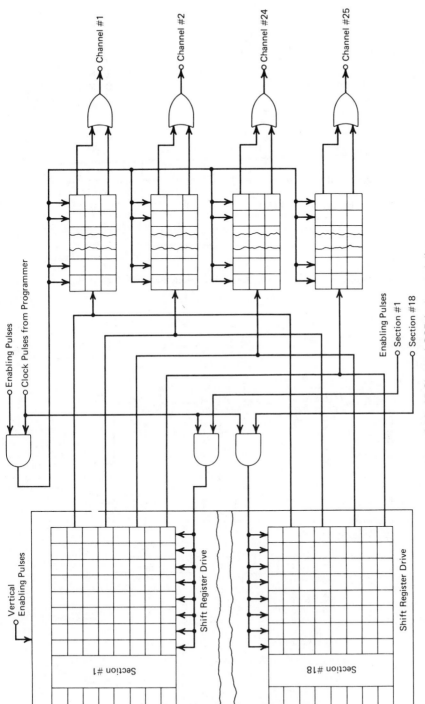

Figure 7.10. 25-Channel CCD instrumentation.

197

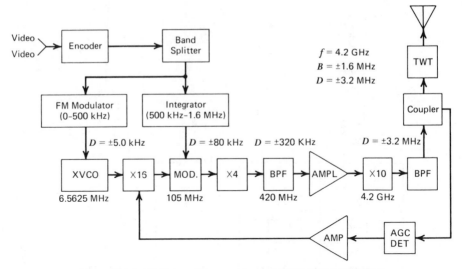

Figure 7.11. 25-Channel frequency modulator plus transmitter.

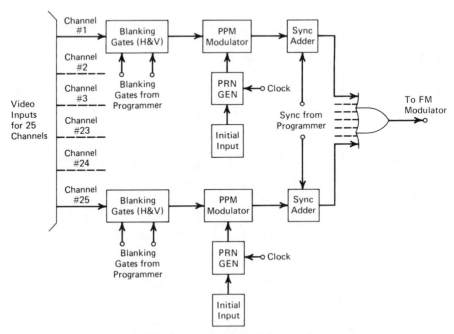

Figure 7.12. The 25-channel encoders.

It should be mentioned here that each of the 25 channels is encoded so that it is orthogonal to every other channel. Thus only one FM carrier is needed in the transmission. The coding means is described later.

7.2.4 The Programmer Unit

The programmer unit is shown in Fig. 7.13. The basic clock rate is 3.2 MHz. It was derived as follows. The system readout is parallel-serial as was mentioned previously. The bandwidth for such a system is

$$BW = (\text{aspect ratio}/2)(\text{number of active lines})(\text{number of sections})(\text{frame rate}).$$

$$(7.25)$$

For clean synchronizing reasons we use an aspect ratio of $5:3$. This is easier to synchronize than a $4:3$ ratio. Since we are not concerned with interlace, we need a reasonably high readout rate; 40 Hz per field, or in this case per raster. Therefore, the video bandwidth from the TV camera is

$$BW = 5 \times 450 \times 18 \times 40/6 = 2.7 \times 10^5 \text{ Hz} = 270 \text{ kHz} \qquad (7.26)$$

As in Chapter 6 we use a pseudorandom readout rate of $6:1$ to get the digital pulse rate of the baseband to be transmitted:

$$B = 6 \times 2.7 \times 10^5 = 1.62 \text{ MHz}. \qquad (7.27)$$

We choose a basic clock rate of 3.2 MHz, a nice round number. This number divides by 4000 to get a section sync rate of 800 pps. We assume 800 horizontal elements, of which 750 are active horizontal elements.

To get the clock rate drive for the CCD registers we use $800 \times 800 = 6.4 \times 10^6$ pps. which point is available on the divider chain. The time interval for each section is $1/800 = 1.25 \times 10^{-3}$ sec. The optimum blanking time is $0.08 \times 1.25 \times 10^{-3} = 1 \times 10^{-4}$ sec. The nearest point to this on the divider chain is at 1.28×10^4 pps, and the time interval between pulses is 78.125 μsec. Thus the section blanking gate will be 78.125 μsec wide. This represents a loss of 50 clock pulses at the CCD register drive during the section blanking interval, and leaves 750 active clock pulses to drive each section.

The vertical blanking gate will be 2.5 msec wide, or two sections wide, as stated previously.

In addition to sync signals the programmer unit furnishes the 1.6-MHz clock pulses to the PRN generator unit. The function of the PRN generator unit is to furnish 25 separately coded and orthogonal PRN pulses to the PPM modulators. The PRN coding is described next.

7.2.5 Orthogonal Code Synthesis for the 25-Channel Code System

We need 25 channels of orthogonal code structure. The nonlinear shift register can be the basic unit for each code. First, we establish the length of

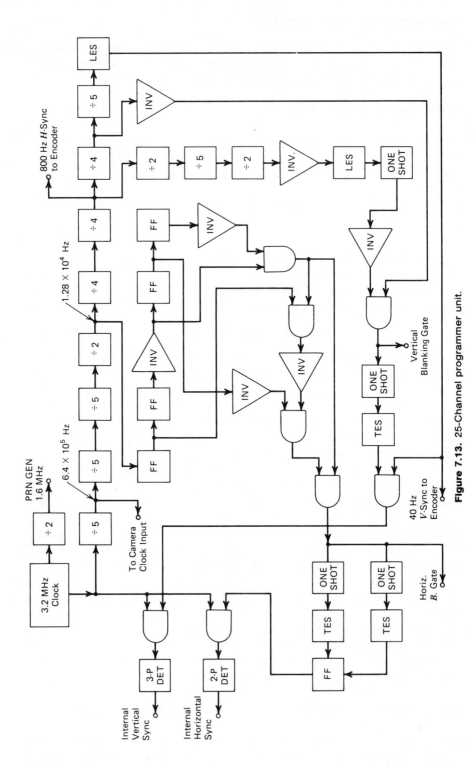

Figure 7.13. 25-Channel programmer unit.

the code in number of bits. If we use an even number of bits, it is possible to achieve perfect orthogonality. Let us choose a bit length of 14 per pseudorandom period. Nonlinear codes of 4 bits/word will not give a total bit length of 14. It takes 5 bits/word to do this.

There are several feedback shift register configurations that will give 14 bits/period. The mathematical expression for a 14-bit code period is

$$p = \text{period} = 2^{k-1} - 2. \tag{7.28}$$

When we let $k = 5$, $p = 14$.

One of the recursion equations that will give 14 bits/period is

$$O = I + D_2 + D_4 + D_5 + D_1D_2 + \overline{D_2D_4}. \tag{7.29}$$

The basic shift register derived from this recursion is shown in Fig. 7.14. When the initializing word is 01111, the following code results:

$$01111101010001 \ (\text{Code} \ \# \ 1). \tag{7.30}$$

This code has 8 ones and 6 zeros. It is usable as one of the 25 codes that are needed. However, better orthogonality results when there are 7 ones and 7 zeros. Logic circuitry at the output of the shift register can be used to remove a "one" from this code. If we remove the second "one" we get

$$01011101010001 \ (\text{7 ones and 7 zeros}), \tag{7.31}$$

which is a usable code.

Another useful recursion is

$$O = I + D_2 + D_4 + D_5 + \overline{D_1D_3} + D_1D_4 + D_3D_4, \tag{7.32}$$

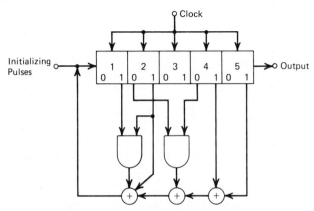

Figure 7.14. a 14-bits/period PRN code generator.

which gives 01111001101100. This code can also be manipulated to give several other code words. A short list of useful recursions and their basic code words is as follows.

$$O = I + D_2 + D_3 + D_5 + D_1 D_2 + \overline{D_1 D_4} + D_2 D_4 \tag{7.33}$$

$$01111000110110$$

$$O = I + D_3 + D_5 + D_1 D_3 + D_2 D_3 + D_2 D_4 \tag{7.34}$$

$$01111101100101$$

$$O = I + \overline{D_2} + D_3 \, \overline{D_4 D_5} + D_5 \tag{7.35}$$

$$01111101010111$$

$$O = I + D_2 + D_3 \, \overline{D_4} \, D_5 + D_6 \tag{7.36}$$

$$01111110011000.$$

These basic codes can be manipulated and shifted to give 25 orthogonal 14-bit codes. Such a code set is shown is Fig. 7.15. The criterion for code

	1	2	3	4	5	6	7	8	9	10	11	12	13	14
1	0	1	1	1	1	1	0	1	0	1	0	0	0	1
2	0	0	0	1	1	1	1	0	0	1	1	0	1	1
3	0	1	1	0	0	1	1	1	1	0	0	0	1	1
4	0	0	1	0	1	0	1	1	1	1	1	0	1	1
5	1	0	1	0	1	1	1	0	1	1	1	1	1	0
6	1	0	1	0	1	1	1	0	1	0	1	0	0	0
7	0	0	0	1	0	1	1	1	0	1	0	1	0	1
8	0	1	0	0	0	1	0	1	1	1	0	1	0	1
9	0	0	1	0	1	1	1	1	0	0	1	0	1	0
10	0	1	0	1	1	0	0	1	1	0	1	1	0	0
11	0	0	1	1	0	1	0	0	1	1	0	1	1	0
12	0	0	0	1	1	1	0	0	0	1	1	0	1	1
13	0	1	1	0	0	1	0	1	1	0	0	1	0	1
14	1	0	1	0	0	1	1	0	1	1	0	0	1	0
15	0	1	0	0	1	1	1	0	0	1	1	0	0	1
16	1	0	1	0	1	1	0	1	0	0	1	1	0	0
17	0	1	0	1	0	1	0	1	0	1	0	1	1	0
18	0	0	1	0	1	0	0	1	0	1	1	0	1	1
19	0	0	1	0	1	0	0	1	0	1	0	1	1	1
20	1	0	1	0	0	0	1	0	1	0	1	0	1	1
21	1	1	0	0	1	1	0	0	0	1	0	1	0	1
22	0	1	1	0	1	1	1	0	0	1	1	0	0	0
23	0	0	1	1	1	0	1	1	0	0	1	1	0	0
24	0	0	0	1	1	1	1	0	1	0	0	1	1	0
25	0	0	0	0	1	0	1	1	1	1	0	0	1	1

Figure 7.15. The 25-channel code set.

Figure 7.16 — Code set cross-correlation analysis work table (triangular cross-correlation matrix). Each cell shows a code pair (e.g. 1/2) with its correlation value below; the parenthesized value at the end of each row is the row total.

```
1/2  +2   1/3  -2   1/4  -4   1/5  +6   1/6  +4   1/7  -4   1/8  -4   1/9  0    1/10 0    1/11 -2   1/12 -4   1/13 -4   1/14 +4   1/15 -4   1/16 -2   1/17 -4   1/18 -4   1/19 -4   1/20 +8   1/21 -4   1/22 -4   1/23 +2   1/24 +4   1/25 0

2/3  +2   2/4  +4   2/5  -2   2/6  0    2/7  -4   2/8  -4   2/9  -4   2/10 +4   2/11 0    2/12 -12  2/13 +6   2/14 +6   2/15 -8   2/16 +4   2/17 +6   2/18 -4   2/19 0    2/20 0    2/21 0    2/22 -6   2/23 0    2/24 -4   2/25 -4   (-34)

3/4  -4   3/5  +2   3/6  0    3/7  0    3/8  -4   3/9  -4   3/10 +4   3/11 0    3/12 +2   3/13 -4   3/14 -8   3/15 +4   3/16 +4   3/17 0    3/18 0    3/19 0    3/20 -4   3/21 +10  3/22 +4   3/23 +4   3/24 +4   3/25 -4   (-14)

4/5  -4   4/6  -2   4/7  +4   4/8  +2   4/9  -6   4/10 +2   4/11 +2   4/12 +2   4/13 +2   4/14 -2   4/15 -2   4/16 +2   4/17 +6   4/18 -8   4/19 -6   4/20 -6   4/21 +6   4/22 -2   4/23 -2   4/24 +2   4/25 -10  (-24)

5/6  -8   5/7  +4   5/8  +4   5/9  -4   5/10 +4   5/11 -4   5/12 +2   5/13 +4   5/14 -8   5/15 0    5/16 -4   5/17 0    5/18 +4   5/19 0    5/20 -2   5/21 0    5/22 -4   5/23 -4   5/24 -4   5/25 -4   (-22)

6/7  +6   6/8  +6   6/9  -6   6/10 +6   6/11 +2   6/12 +2   6/13 +2   6/14 -6   6/15 -2   6/16 -6   6/17 +2   6/18 +2   6/19 +10  6/20 -6   6/21 -6   6/22 -10  6/23 -2   6/24 -2   6/25 +2   ( 0 )

7/8  -6   7/9  +2   7/10 +2   7/11 -2   7/12 -4   7/13 -2   7/14 -6   7/15 -2   7/16 +2   7/17 -2   7/18 +2   7/19 -2   7/20 +6   7/21 -2   7/22 +2   7/23 -2   7/24 -2   7/25 -2   (-18)

8/9  +6   8/10 -2   8/11 -2   8/12 +2   8/13 -10  8/14 +2   8/15 -2   8/16 -6   8/17 +6   8/18 +6   8/19 -2   8/20 +6   8/21 -6   8/22 +2   8/23 +6   8/24 +2   8/25 -2   (+2)

9/10 +2   9/11 +2   9/12 -2   9/13 -2   9/14 -2   9/15 -2   9/16 -6   9/17 -6   9/18 -2   9/19 -2   9/20 -4   9/21 +6   9/22 -6   9/23 -6   9/24 -2   9/25 -2   (-28)

10/11 +2  10/12 -2  10/13 -2  10/14 +10 10/15 +2  10/16 +2  10/17 -2  10/18 +2  10/19 +2  10/20 +6  10/21 +2  10/22 +2  10/23 -6  10/24 -2  10/25 +2  (+18)

11/12 -2  11/13 -6  11/14 -6  11/15 +6  11/16 +2  11/17 -6  11/18 -2  11/19 -2  11/20 +2  11/21 +2  11/22 +2  11/23 -6  11/24 +2  11/25 -4  (-4)

12/13 +6  12/14 -6  12/15 -6  12/16 +2  12/17 +2  12/18 -4  12/19 -2  12/20 -2  12/21 +2  12/22 +2  12/23 +2  12/24 -2  12/25 -2  (-4)

13/14 +2  13/15 +2  13/16 +2  13/17 +2  13/18 +2  13/19 +2  13/20 +2  13/21 +2  13/22 +6  13/23 +2  13/24 +2  13/25      (+22)

14/15 +2  14/16 +2  14/17 +2  14/18 +2  14/19 +2  14/20 -6  14/21 +2  14/22 -2  14/23 +6  14/24 -2  14/25 -2  (+6)

15/16 +2  15/17 +2  15/18 -2  15/19 +2  15/20 +2  15/21 +2  15/22 -8  15/23 -6  15/24 +2  15/25 -2  (-6)

16/17 +2  16/18 -2  16/19 +2  16/20 +2  16/21 -2  16/22 -2  16/23 -2  16/24 +6  16/25 +6  (-2)
```

Figure 7.16. (a) and (b) Code set cross-correlation analysis work table.

```
17/18 17/19 17/20 17/21 17/22 17/23 17/24 17/25
+2    -2    +10   -4    +2    +2    -2    +2    (+10)

18/19 18/20 18/21 18/22 18/23 18/24 18/25
-10   -2    +2    -2    -2    +6    -6    (-14)

19/20 19/21 19/22 19/23 19/24 19/25
+2    -2    +2    -2    +2    -6    (-4)

20/21 20/22 20/23 20/24 20/25
+6    +2    +2    +2    -2    (+10)

21/22 21/23 21/24 21/25
-2    +6    +2    +2    (+8)

22/23 22/24 22/25
-2    +2    +2    (+2)

23/24 23/25
-2    +2    ( 0 )

24/25
-2    (-2)
```

$$\lambda_c = -120/4200 = 0.0286$$

Figure 7.16. (continued)

word selection is a low cross-correlation coefficient. The cross-correlation λ_c between any two code words of equal length is

$$\lambda_c = \frac{\text{number of agreements} - \text{number of disagreements}}{\text{number of agreements} + \text{number of disagreements}}. \tag{7.37}$$

To correctly assay λ_c for the entire code set requires the correlation between a given line and the other 24 lines taken one by one. This was done in tabularized form as shown in Fig. 7.16. It was found that

$$\text{number of agreements} - \text{number of disagreements} = -120. \tag{7.38}$$

To save the tedium of counting the total number of line-by-line comparisons, one can use the summation

$$N = L \sum_{n=1}^{M} (x - n) \tag{7.39}$$

where

N total number of comparisons,
L number of bits per period,
M number of codes per code set minus one $= 24$,
x total number of codes per code set $= 25$.

For the 25 line code set shown:

$$N = \sum_{n=1}^{24} (25 - n) = 4175. \tag{7.40}$$

Therefore,

$$\lambda_c = \frac{-120}{4175} = -0.0286. \qquad (7.41)$$

It may prove useful to show here the code syntheses used to obtain lines 1, 6, 7, and 9. Codes 6, 7, and 9 are variations of code # 1. The timing chart is shown in Fig. 7.17, and the block diagram is shown in Fig. 7.18. This method of combining common code elements saves on the number of components needed per hardware card.

The same synthesis method was used to obtain the remainder of the code set. Fig. 7.19 is the timing chart for codes 5, 8, 19, 20, and 21, while Fig. 7.20 is the block diagram for codes 5, 8, 19, 20, and 21. Fig. 7.21 is the timing

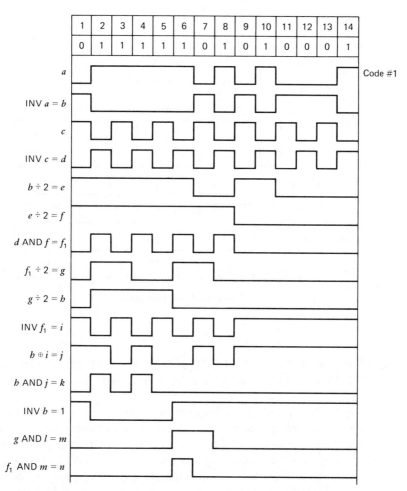

Figure 7.17. (a) and (b) Timing chart for codes 1, 6, 7, and 9.

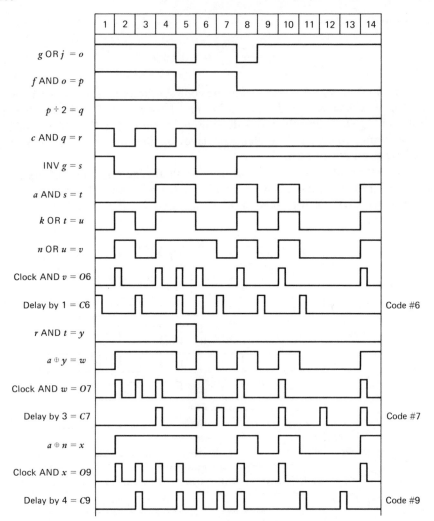

Figure 7.17. *(continued)*

chart for codes 4, 13, 14, 15, 16, 17, and 18. Fig. 7.22 is the timing chart for codes 22, 23, 24, and 25. Fig. 7.23 is the timing chart for codes 2, 10, 11, and 12.

By using the preceding method for synthesizing an orthogonal code set, we can use one only initializing waveform to handle every code word in the code set. The timing chart needed for obtaining the initialization word is shown in Fig. 7.24, and the accompanying block diagram in Fig. 7.25. Thus initialization is used continually, and for this particular code set the initializing word is 01111000000000.

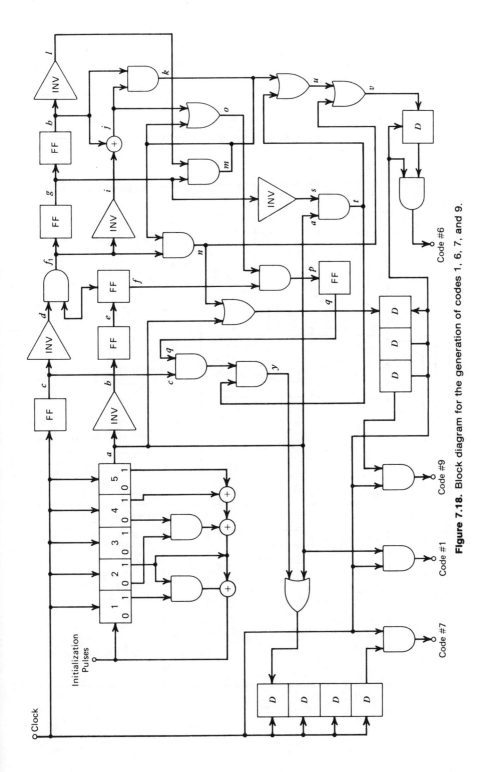

Figure 7.18. Block diagram for the generation of codes 1, 6, 7, and 9.

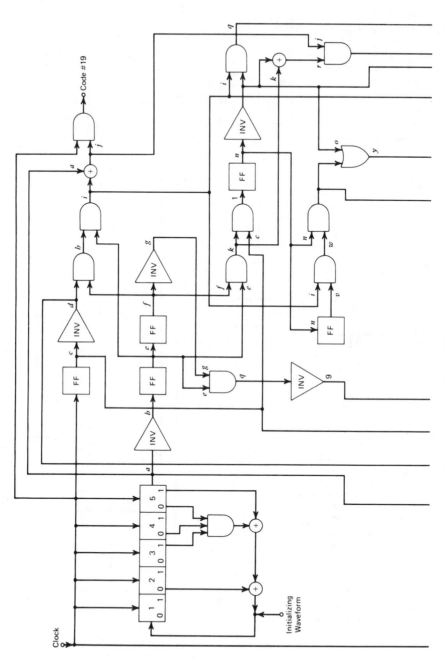

Figure 7.19. Block diagram for the generation of codes 5, 8, 19, 20, and 21.

208

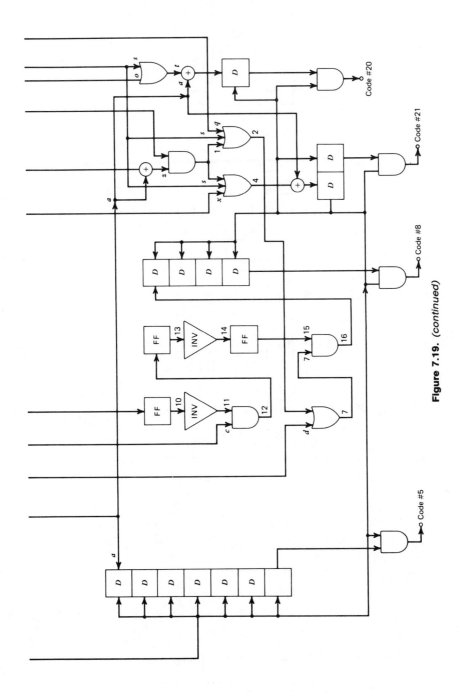

Figure 7.19. *(continued)*

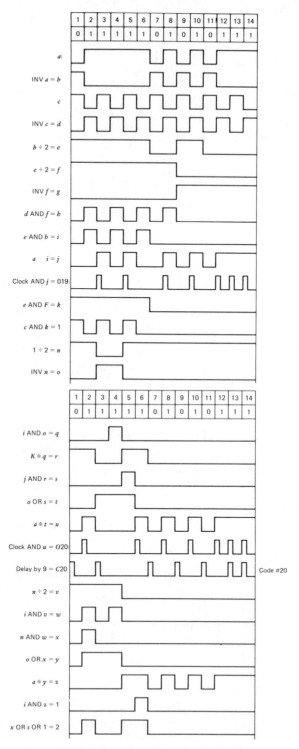

Figure 7.20. Timing chart for codes 5, 8, 19, 20, and 21.

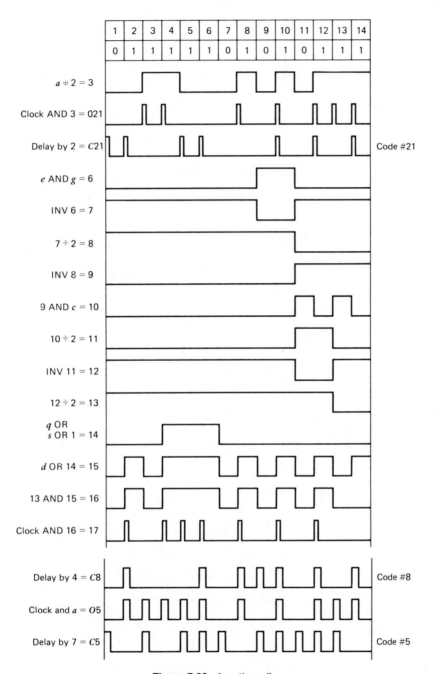

Figure 7.20. *(continued)*

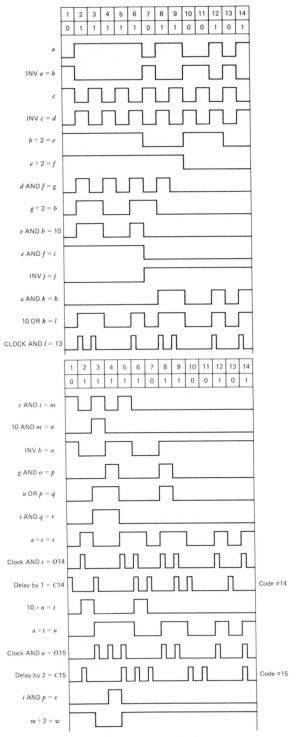

Figure 7.21. Timing chart for codes 4, 13, 14, 15, 16, 17, and 18.

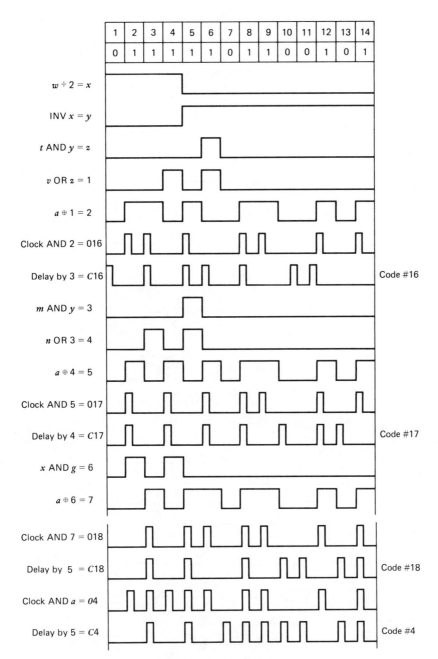

Figure 7.21. *(continued)*

213

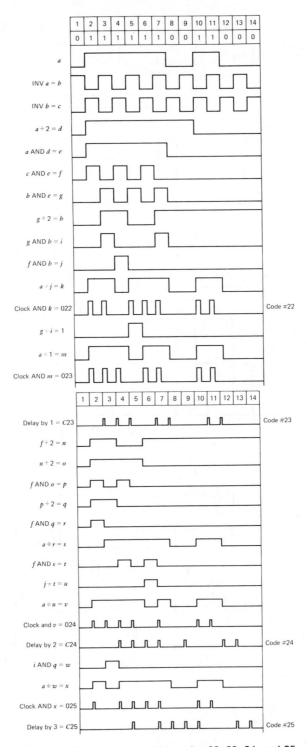

Figure 7.22. Timing chart for codes 22, 23, 24, and 25.

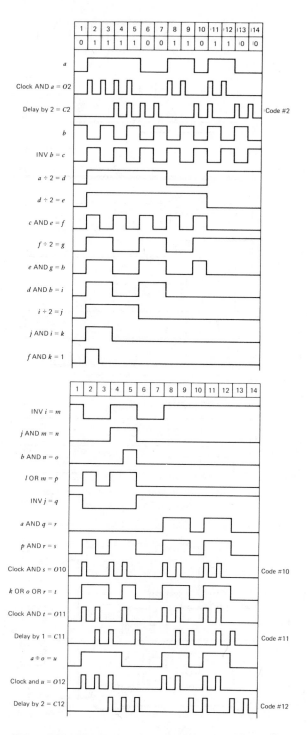

Figure 7.23. Timing chart for codes 2, 10, 11, and 12.

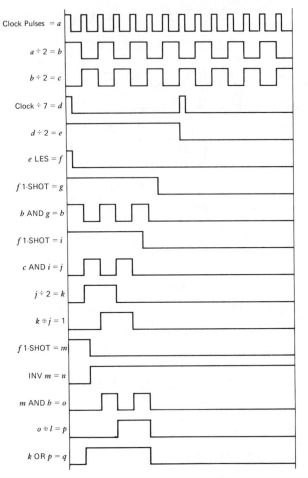

Figure 7.24. Timing chart for initializing pulses.

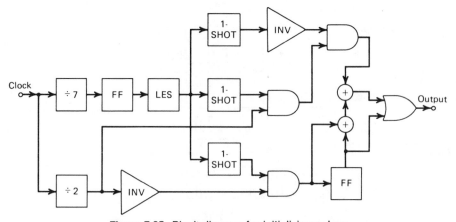

Figure 7.25. Block diagram for initializing pulses.

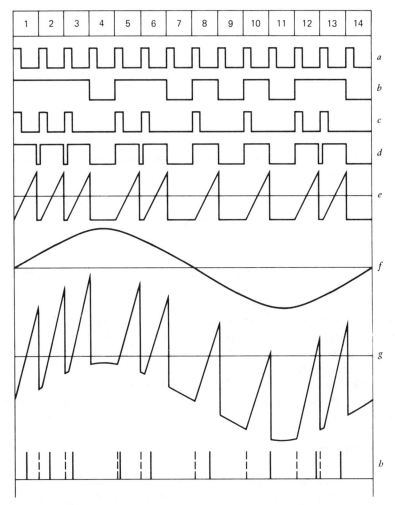

Figure 7.26. PPM by means of pseudorandom sampling.

7.2.6 The PRN/PPM Modulators

Fig. 7.26 shows the timing waveforms needed to create PPM modulation, when sampled in a pseudorandom manner. Since the usual PRN output from a shift register is in the form shown on line *b*, and it is desirable to use narrow sampling pulses, we AND gate line *b* with the clock pulses from line *a* to get the sampling pulses on line *c*. Now, we generate 95 percent duty cycle pulses by triggering a 1-shot MV with pulses from line *c*. These wide pulses are shown on line *d* and are in turn used to drive the sawtooth generator having a slope duration equal to that of the driving pulses, as shown on line *e*. A sine wave representing a typical video frequency waveform is shown on line *f*. The

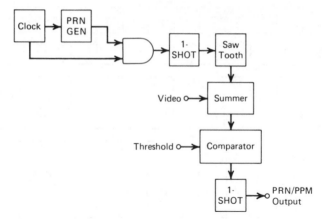

Figure 7.27. PRN/PPM modulator unit.

video of line f is summed with the sawtooth of line e to get the waveform of line g. A comparator is used to separate and use that portion of the sawtooth waves that are above the threshold, as shown on line g. The leading edges of the saved sawtooth waves are used to create the outgoing PRN/PPM narrow pulses as shown in solid notation on line h. The dashed notation shows the references pulses. Even though the reference pulses are shown in the timing chart they are not transmitted.

A block diagram of a PRN/PPM modulator unit is shown in Fig. 7.27. Of course, there is one modulator unit used in each of the 25 channels, as is shown in Fig. 7.12.

7.2.7 The Receiver

The receiver is shown in Fig. 7.28. It is very conventional, having an IF with a 60-MHz center frequency, and a 3-dB bandwidth of ± 2.6 MHz. The discriminator output consists of the PRN/PPM pulses, composed of the

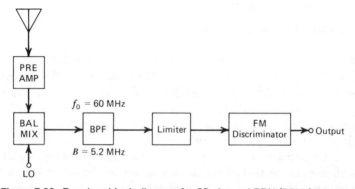

Figure 7.28. Receiver block diagram for 25-channel PRN/PPM/FM system.

original video plus PRN signals. The decoder unit recreates the original video waveform. However, prior to decoding, some form of pulse regeneration and sync separation is needed.

7.2.8 The Pulse Generators and Sync Separators

The discriminator output drives the channel separators, which are in the form of integrate-and-dump units combined with phase-locked oscillators. The I & D circuits are used to regenerate the noisy pulses, and the phase-locked oscillators give a form of coherent decoding, as shown in Fig. 7.29. The

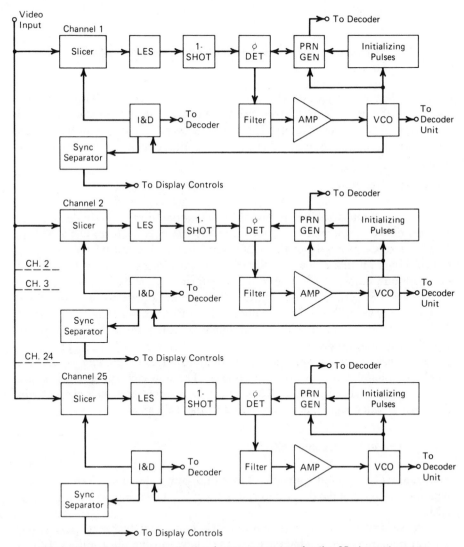

Figure 7.29. Pulse regenerators and sync separators for the 25-channel system.

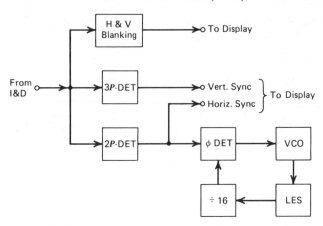

Figure 7.30. The sync separators for the 25-channel system.

coherent decoding separates the video into 25 separate channels. There are outputs to the decoder units from the VCO circuits, the PRN generator, and the integrate-and-dump circuits. The separated sync signals go to the display unit controls. Fig. 7.30 shows the sync separator circuits in block diagram detail.

7.2.9 The Decoder

The decoder unit is shown in Fig. 7.31. The operation of a single decoder channel can be explained by referring to the timing waveform chart of Fig. 7.32. The clock pulses from the VCO units are used to create the sawtooth waves shown on line c. The PRN generator waveforms are used to gate the sawtooth output in order to obtain the waveforms of line d.

The PRN generator output pulses are also used to gate some of the clock pulses as shown on line e. The incoming PRN/PPM pulses (line f) from the I & D are OR gated with the pulses from line e, to obtain those shown on line g. The OR gated output drives a FF to obtain the waveforms of line h. The line h pulses are used to gate a portion of the sawtooth waves of line d, to obtain the amplitude-modulated waveform of line i. These waveform envelopes are obtained in the wrong polarity. These must be inverted and filtered to obtain the original video signal, as it originated in the TV camera.

7.2.10 The Video Display Unit

The video display unit consists of a multi-gun kinescope, driven by the separated video signals; one video channel per TV raster line as shown in Fig. 7.33.

The separated sync signals contain the horizontal blanking pulses and vertical blanking pulses, which are used to blank out the retrace lines. It also

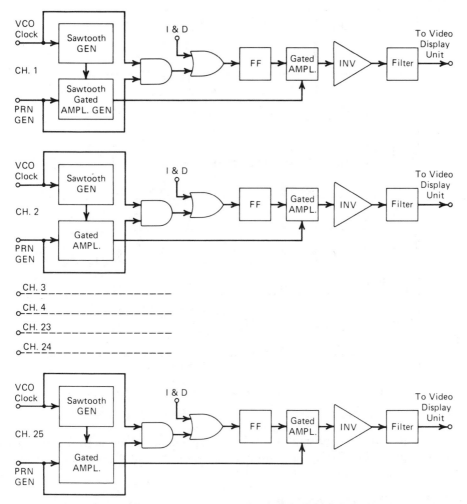

Figure 7.31. Block diagram for the 25-channel decoders.

contains the horizontal and vertical sync pulses, which are used to create the sweep circuits needed for the kinescope display.

The sweeping format is the same as that of the transmitter camera section. Twenty-five lines of video will be displayed simultaneously for one section; then the next 25 lines will be displayed in the appropriate area of the kinescope screen, which is just below that of section one. This procedure is continued until the 18 active sections of video have been displayed. This is followed by the vertical blanking period; and then back to number 1 again.

At the present state of the art, 25-gun kinescopes are not available. Perhaps, with a little effort, a 25-gun kinescope will be possible to obtain in the near future. The reward for developing such a video tube is the use of narrow channel digital TV communications in real time.

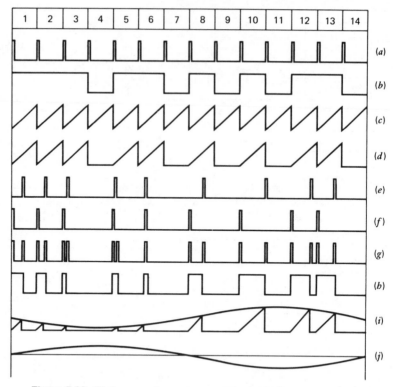

Figure 7.32. Timing waveforms for the PRN/PPM/FM demodulators.

An alternate to the multi-gun kinescope would be a matrix of LED diodes. Such LED display units are undergoing R & D at the present time. A breakthrough could occur at any time.

7.2.11 Mathematical Analysis

The basic equation for bit-error rate in a PRN/PPM/FM system are given in (6.7) of Chapter 6. We repeat it here:

$$P_B = \frac{\exp(-2M^2B^2T^2\rho^2/3)}{(8\pi M^2B^2T^2\rho^2/3)^{1/2}}. \tag{7.42}$$

This equation handles the single channel case. But, the 25-channel case calls for modifications to it. We now have

$$P_B = \frac{\exp(-2M^2B^2T^2\rho^2/3(1+\rho_c^2))}{\left[8\pi M^2B^2T^2\rho^2/3(1+\rho_c^2)\right]^{1/2}} \tag{7.43}$$

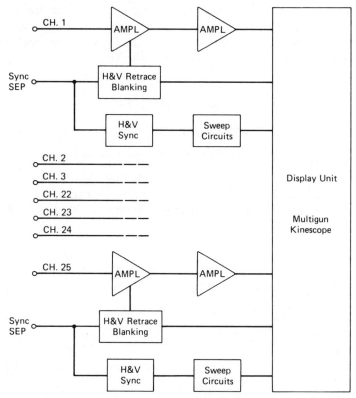

Figure 7.33. The 25-channel kinescope display block diagram.

which accounts for the crosstalk effects between channels. The symbol

$$\rho_c^2 = QC \, (\text{SNR})/S \tag{7.44}$$

where

C crosstalk interference level,
S the signal level,
Q $Kd\lambda_c = 0.01098$,
K number channels in the code set
d pulse duty factor
λ_c channel-to-channel cross correlation.

We can set $C/S \cdot$ so that

$$\left(1 + \rho_c^2\right) = \left[1 + Q \, (\text{SNR})\right] = (1 + 0.01098 \, \text{SNR}). \tag{7.45}$$

However, the 25-channel bandwidth is much less than that for a single-channel system. The bit-error rate should reflect the resulting improvement in SNR. If we let B_0 be the transmission bandwidth of a single-channel system, then we can use the ratio of B_0/B times the SNR. This modifies the equation to

$$P_B = \frac{\exp\left[-2M^2BT^2B_0\rho^2/3(1+\rho_c^2)\right]}{\left[8\pi M^2BT^2B_0\rho^2/3(1+\rho_c^2)\right]^{1/2}}. \qquad (7.46)$$

The following design constants were used to determine the bit-error rate plot of Fig. 7.34: $M=1.82$, $BT=1.64$, $1+\rho_c^2=1+(0.01098\ \text{SNR})$, $B_0=22$ MHz, and $B=2.624$ MHz. There are four curves shown in Fig. 7.34. The curve labeled $C/S=0$ is that for the 25-channel system with no transmission interference.

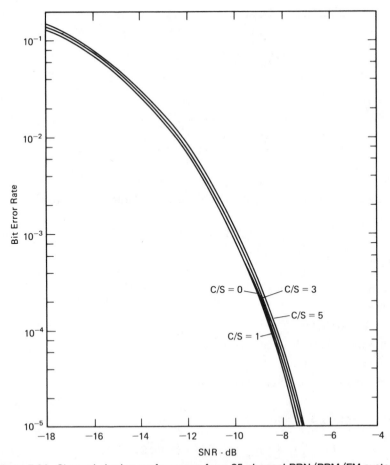

Figure 7.34. Channel-sharing performance for a 25-channel PRN/PPM/FM system.

However, the system will probably be used on a channel-sharing basis. We have assumed 12 simultaneous, 25-channel systems being transmitted over the same carrier frequency. Effectively, this changes the value of Q in the bit-error rate equation to

$$Q = (K + U) d\lambda_c \qquad (7.47)$$

where U is the number of simultaneously users of the 25-channel system.

We have assumed C/S ratios of 1, 3, and 5 to obtain three of the curves shown in Fig. 7.34. It can be seen that channel sharing at the chosen C/S levels produces very little performance degradation.

7.2.12 Channel Capacity, Information Rate, and Efficiency for the 25-Channel System

Channel capacity was analyzed by Shannon (1948) for band-limited signals. He obtained

$$C = \tfrac{1}{2} \log_2(1 + \text{SNR}) \text{ bits/sample.} \qquad (7.48)$$

He assumed the Nyquist sampling rate of twice the baseband $(2W)$ to get

$$C = 2W/2 \log_2(1 + \text{SNR})$$

$$= W \log_2(1 + \text{SNR}) \text{ bits/sec.} \qquad (7.49)$$

However, pseudorandom sampling demands a sampling rate of 6 times the baseband, which makes

$$C = 3W \log_2(1 + \text{SNR})$$

$$= \frac{B}{2} \log_2(1 + \text{SNR}) \text{ bits/sec.} \qquad (7.50)$$

We wish to compare the 25-channel capacity to that of a single channel. The $6W$ bandwidth is already accounted for, but the SNR should be modified to account for the ratio B_0/B. Therefore,

$$C = \frac{B}{2} \log_2\left[1 + (B_0/B) \text{SNR}\right] \text{ bits/sec.} \qquad (7.51)$$

Moreover, we are dealing with a PRN/PPM/FM system, which modifies the SNR just as it did in the bit-error rate equation. Now, we obtain

$$C = \frac{B}{2} \log_2\{1 + \left[4M^2B^2T^2/3(1 + \rho_c^2)\right](B_0/B)\text{SNR}\} \text{ bits/sec.} \quad (7.52)$$

In addition to these modifications we must modify due to bandwidth occupancy. We are using a correlated PRN/PPM decoder. The bandwidth occupancy due to PRN is $\log_2(M)/2T$, while that for PPM is $1/2T$. The bandwidth occupancy for PRN/PPM is $\log_2(M)/4T$, where $M = 2^{n-2} - 2 = 14$, and $\log_2 M = 3.81$. The channel capacity is, now, modified to

$$C = \frac{B \log_2 M}{8} \left\{ 1 + (4/\log_2 M)[4M^2 B^2 T^2/3(1+\rho_c^2)](B_0/B)\,\text{SNR} \right\} \text{ bits/sec.}$$

(7.53)

We are dealing with bit-error rate and in a PRN/PPM/FM system the ratio of bit-error rate to word-error rate for $M = 14$ is

$$\frac{P_B}{P_W} = \frac{2^{n-2}-1}{2^{n-1}-2} = \frac{7}{14} = 0.5.$$

(7.54)

From this ratio it can be seen that the probability of 1 bit being in error is $P_B/(2^{n-2}-1)$. Apply Bayes' rule to the entropy function to obtain

$$H_x(y) = -\sum_i P(x_i) \sum_i P(y_i|x_i) \log_2 P(y_i|x_i)$$

$$= -\sum_i P(y_i|x_i) \log_2 P(y_i|x_i)$$

$$= -[1 - P_B(n)] \log_2 P(1 - P_B(n) - P_B(n))[\log_2 P_B(n) - \log_2(2^{n-2}-1)].$$

(7.55)

Since the rate of transmission is $1/T$ bits/sec, the uncertainty factor is $H_x(y)/T$. We subtract this from the transmission rate to get

$$H = B \left\{ 1 + [1 - P_B(n)] \log_2[1 - P_B(n)] \right.$$

$$\left. + P_B(n)[\log_2 P_B(n) - \log_2(2^{n-2}-1)] \right\} \text{ bits/sec.}$$

(7.56)

The channel efficiency is simply

$$\eta = \frac{H}{C}.$$

(7.57)

A plot of channel capacity versus SNR in decibels is shown in Fig. 7.35. A plot of information rate versus SNR in decibels is shown in Fig. 7.36, while channel efficiency versus SNR in decibels is plotted in Fig. 7.37.

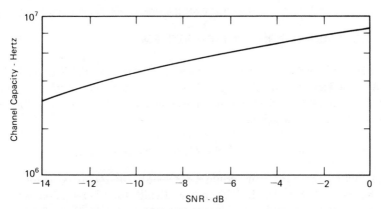

Figure 7.35. Channel capacity for a 25-channel PRN/PPM/FM system.

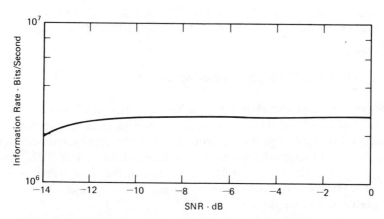

Figure 7.36. Information rate for a 25-channel PRN/PPM/FM system.

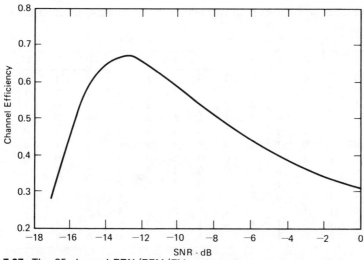

Figure 7.37. The 25-channel PRN/PPM/FM system channel efficiency versus signal-to-noise ratio SNR.

227

7.3   A 13-CHANNEL PRN/PPM/FM SYSTEM

For the sake of comparison we now investigate a 13-channel PRN/PPM/FM system. The camera has 39 active 13-channel sections, and two inactive sections, for a total of 41 sections. There are $13 \times 39 = 507$ active TV lines. The base video bandwidth is

$$W = 5 \times 507 \times 39 \times 40/6 = 6.591 \times 10^6 \text{ Hz}. \qquad (7.58)$$

The calculated sampling rate would be 3.9546×10^6 Hz. It is better to use a sampling rate of 4×10^6 Hz. When we divide 4×10^6 by 2500 we obtain 1600-pps section rate, which divides down, very nicely, to give a frame rate of 40 pps.

When we use a time-bandwidth (BT) product of 1.64, we obtain a transmission bandwidth of $\pm 6.56 \times 10^6$ Hz, or $13.12 = $ MHz two-sided bandwidth.

7.3.1   13-Channel Orthogonal Code Synthesis

Again we use nonlinear shift registers having a period of 14 ($M = 14$). The detailed procedure for generating such codes was given above for the 25-channel system, and need not be repeated here. An orthogonal code set was found for the 13-channel system, and is shown here in Fig. 7.38. Analysis shows that this code set is completely orthogonal, the cross correlation being equal to 0. However, for the purpose of assessing the effect of channel sharing upon performance, we use

$$\lambda_c = -\frac{1}{78} = -0.0128. \qquad (7.59)$$

This is an arbitrary choice and it might be more realistic to use a lower value of λ_c than this.

	1	2	3	4	5	6	7	8	9	10	11	12	13	14
1	0	1	0	0	0	1	0	1	1	1	0	1	0	1
2	0	0	0	1	1	1	0	1	0	1	1	0	1	1
3	1	0	1	0	0	1	1	0	1	1	0	0	1	0
4	0	1	0	0	1	1	1	0	0	1	1	0	0	1
5	1	0	1	0	1	1	0	1	0	0	1	1	0	0
6	0	1	0	1	0	1	0	1	0	1	0	1	1	0
7	0	0	1	0	1	0	0	1	0	1	1	0	1	1
8	0	0	1	0	1	0	0	1	0	1	0	1	1	1
9	0	1	0	1	1	0	0	1	1	0	1	1	0	0
10	1	1	0	0	1	1	0	0	0	1	0	1	0	1
11	0	1	1	0	1	1	1	0	0	1	1	0	0	0
12	0	0	1	1	1	0	1	1	0	0	1	1	0	0
13	0	0	0	1	1	1	1	0	1	0	0	1	1	0

Figure 7.38. A 13-channel code set.

We now assume pulsewidths of 1.54×10^{-8} sec. Since the value of T is 2.5×10^{-7} sec, the duty factor is

$$d = \frac{1.54 \times 10^{-8}}{2.5 \times 10^{-7}} = 0.616. \tag{7.60}$$

The Q factor is found by

$$Q = 13 \times 0.0616 \times 0.0128 = 0.001025. \tag{7.61}$$

The bandwidth ratio is $B_0/B = 22/6.56 = 3.3537$. The bit-error rate is numerically

$$P_B = \frac{\exp\{(-5.94 \times 3.3537 \times \text{SNR})/[1 + 0.001025(C/S)]\}}{(2\pi \times 11.88 \times 3.3537 \times \text{SNR})/[1 + 0.001025(C/S)]^{1/2}}. \tag{7.62}$$

But $C/S = 1$ internal to the system, so that

$$P_B = \frac{\exp[-19.921\,\text{SNR}/(1 + 0.001025\,\text{SNR})]}{\{(250.3\,\text{SNR})/[1 + (.001025\,\text{SNR})]\}^{1/2}}. \tag{7.63}$$

The channel capacity is numerically

$$C = \left(\frac{3.81}{8}\right) \times 6.56 \times 10^6 \log_2\left[\frac{1 + (4 \times 11.88 \times 3.3537 \times \text{SNR})}{3.81 \times (1 + 0.001025\,\text{SNR})}\right]$$

$$= 3.12 \times 10^6 \log_2\left[1 + \frac{159.4\,\text{SNR}}{3.81 + (1 + 0.001025\,\text{SNR})}\right] \tag{7.64}$$

$$H = 6.56 \times 10^6[(P_B - 1)\log_2(P_B - 1) + P_B(\log_2 P_B - \log_2 7) + 1] \text{ bits/sec.} \tag{7.65}$$

THe channel efficiency is, of course, H/C.

7.4 SINGLE-CHANNEL PRN/PPM/FM

In Chapter 6 we discussed a single-channel PRN/PPM/FM system using pseudorandom sampling. However, in Chapter 6 we used a nonlinear word period of 22. If we wish to compare the performance of single channel PRN/PPM/FM to multi-channel PRN/PPM/FM systems, we need to use equal word lengths. We will use the period of 14 here also.

The object is to compare channel capacities, information rates, and channel efficiencies. In this case $QC/S = 0$.

7.5   PERFORMANCE COMPARISON OF PRN/PPM/FM SYSTEMS

Performance comparisons have been made for single channel, 13-channel, and 25-channel PRN/PPM/FM systems, having 14-bit word periods.

First, we compared these systems for bit-error rate. The results are shown in Fig. 7.39. It is easy to see from these curves that the 25-channel has superior performance to the other two systems. This is due, of course, to its smaller bandwidth, which gives a better SNR.

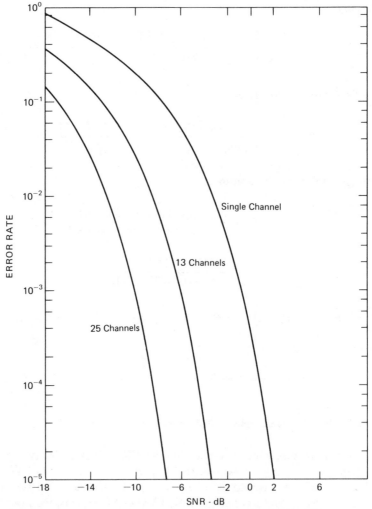

Figure 7.39. Performance comparison of multi-channel PRN/PPM/FM systems versus signal-to-noise ratio SNR.

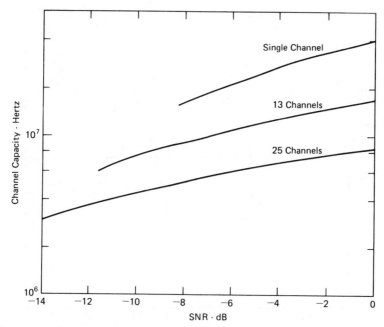

Figure 7.40. Multi-channel PRN/PPM/FM systems channel capacity comparison.

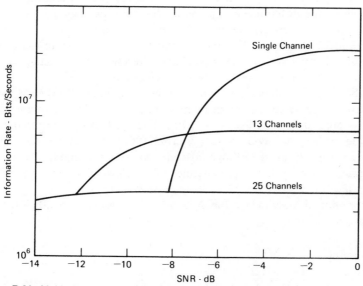

Figure 7.41. Multi-channel PRN/PPM/FM systems information rates comparison.

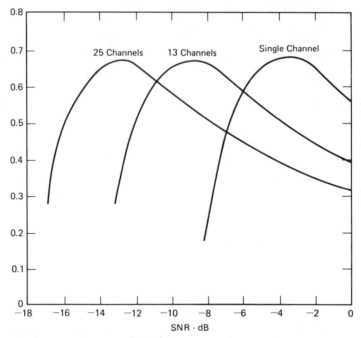

Figure 7.42. Multi-channel PRN/PPM/FM systems channel efficiency comparison.

Channel capacity comparison is found in Fig. 7.40, while information rates are compared in Fig. 7.41. Channel efficiency is compared in Fig. 7.42. There is very little difference in maximum channel efficiency among the three systems, but it is reverse to the trend found in the bit-error rates curves. The single channel has the best efficiency. This can be correlated to the crosstalk in the multi-channel systems. However, the difference can be made negligible by improving the cross correlation between channels.

The bandwidth reduction in the multi-channel PRN/PPM/FM systems can be used to great advantage in many applications. The channel efficiency of the pseudorandom sampled PRN/PPM/FM systems is higher than for the block-type orthogonal code sets (Viterbi, 1961). He shows the channel efficiency for $n = 20$ block orthogonal codes to be a maximum of 0.65, while the channel efficiency for single channel PRN/PPM/FM systems ($M = 14$), is 0.675, and for 25-channel PRN/PPM/FM it is 0.672.

As mentioned previously, the following transmission bandwidths were obtained (two-sided)

Single channel	44 MHz
13-channel	13.3 MHz
25-channel	5.2 MHz

7.6 SUMMARY

In this chapter the use of multi-channel PRN/PPM/FM systems was discussed. These systems make use of parallel-serial readout of CCDs focal plane array cameras. In order to familiarize the reader with the application of CCD techniques, the first portion of this chapter was devoted to a theoretical discussion of basic CCD theory. This was followed by a discussion on the practical implementation of the CCD into a usable focal plane array. Next, a detailed discussion was presented on how to implement a multi-channel PRN/PPM/FM system. A detailed discussion was presented on the fabrication of a 25-channel parallel-serial readout PRN/PPM/FM system. This discussion included a mathematical analysis of the performance characteristics of the 25-channel system. This was followed by a short description of a 13-channel PRN/PPM/FM system, and a single channel PRN/PPM/FM system.

It should be noted here that a 25-channel PRN/PPM baseband system is capable of a little greater than 12:1 bandwidth reduction to about 0.5 bits/pixel. If instead of FM we use Single Sideband (SSB), and assume external channel-sharing numbering 12, the 2.64-MHz band would support 12 simultaneous multi-channel PRN/PPM/SSB systems. Therefore, for a 5.28-MHz communication transmission spectrum, 24 simultaneous users could be accomodated. This means that for commercial TV use, twenty-four 25-channel PRN/PPM/SSB digital TV channels could be placed in the spectrum now allotted to one commercial TV channel, a gain of 24:1 in spectrum utilization. In digital TV terms 5.28 MHz of bandwidth is close to being equivalent to a 6:1 bandwidth reduction. Therefore, on a channel-sharing basis of channel occupancy improvement ratio of 144:1 is feasible. This efficiency in the use of the spectrum, however, would be diminished by the use of needed guard bands. Also, multi-channel PRN/PPM/FM systems would require roughly twice the bandwidth of multi-channel PRN/PPM/SSB systems.

This represents a very large improvement in the utilization of TV spectrum. This is achieved without deterioration of picture quality. This type of system should be immune to picture breakup from objects in fast motion, or from fast panning.

Although, at present, cathode ray tube manufacturers have not produced any 25-gun kinescopes, one such manufacturer has stated in a unpublished conversation that it should be possible to make 20-gun tubes without crowding the state of the art. This makes it feasible to start R & D on multi-channel PRN/PPM modems now. However, the proposed flat plate display for the kinescope is not far from being possible. According to the latest information, the Westinghouse Research Laboratories have achieved practical electroluminescent thin film flat plate displays for TV use, and are not far from achieving commercial TV quality. Such displays could be adapted to multi-channel PRN/PPM systems displays. They also have the necessary phosphors for color TV.

The circuit complexity of the multi-channel PRN/PPM systems is relatively large. However, when one looks at the poor results of other competitive systems in conquering breakup due to fast moving objects, it can be seen that conquering such defects could lead to complex circuit solutions. It could be that multi-channel PRN/PPM systems could be produced in a shorter time limit than it would take for its practical competitors, and at comparable circuitry complexity. The R & D efforts on a multi-channel PRN/PPM system should be relatively straightforward.

chapter 8

Bi-Phase Transorthogonal Block Type Coding

introduction

The system discussed here uses reverse-phase modulated transorthogonal coding.

Transorthogonal coding is used to sample the analog video signal. The modulated carrier is then transmitted and received in the telecommunication sense. This system uses a block code of seven words, each having a length of 7 bits. This constitutes an M-ary system, with the modulated carrier being in the form of bi-phase modulation.

The following discussion will include both mathematical analysis and the schematical implementation of the system. Implementation will be discussed first.

8.1 THE TRANSMITTER

The RF portion of the transmitter, which is shown in Fig. 8.1 is very simple. A phase modulator is used to change the phase of the carrier $\pm 180°$ at each pulse edge. This method is used more than any other for transmitting transorthogonal codes.

8.2 THE ENCODER

The heart of the encoder is the block code modulator, assisted by the programmer unit, as shown in Fig. 8.2. The input video signal is first blanked

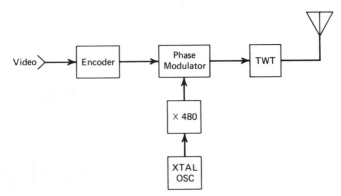

Figure 8.1. Bi-phase transorthogonal code transmitter.

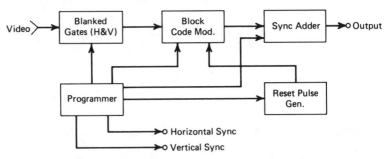

Figure 8.2. Encoder for bi-phase transorthogonal codes

out, at points in time coincidental with the horizontal and vertical synchronizing pulses. After modulation the sync pulses are added for transmission.

8.3 THE BLOCK CODE MODULATOR

The block transorthogonal code is usually formed by the Kroneker product construction of the Hadamard matrix of the order of 2. It has $2^k - 1 = M - 1$ bits/word. It can, however, be generated in various ways. The method chosen here is to use pseudorandom code generation, plus external logic circuitry. The resulting code block is as shown in Table 8.1 (Everitt, 1964).

In this type of code, the number of disagreements between any pair of code word symbols, exceeds the number of agreements by one. Hence, the cross-correlation coefficient is

$$\lambda = \frac{-1}{M-1}. \tag{8.1}$$

The generation of this type of code is shown in Fig. 8.3. The numbered triangles at the left-hand side of the figure are differential amplifiers. They are

Table 8.1

Block Code
1 0 1 0 1 0 1
0 1 1 0 0 1 1
1 1 0 0 1 1 0
0 0 0 1 1 1 1
1 0 1 1 0 1 0
0 1 1 1 1 0 0
1 1 0 1 0 0 1

fed inputs from the blanked video and the reset pulses. The reset pulses occur at the end of every $M - 1$ digit code word.

There are seven amplifiers for the sampling of seven different gray levels in the video signal. For example, assume a gray level between 2 and 3. The bottom two differential amplifiers pass the signal because their reference voltages are set at such levels that pass or reject various input video levels. Each differential amplifier feeds a set/reset type of FF. A signal from the differential amplifier cause its FF to set. In the example given both of the two lower FFs become set; being normally in the reset position. While they are set their polarities are reversed, and read 10 rather than 01.

In the example given AND gate number 2 is the only one fired during a set period. At the end of each code word a reset pulse causes all of the FF's to reset.

A gate is formed for the duration of one word. This gate passes pulses from the clock. It will pass only the correct number of pulses to the PRN generators, which will form the correct code word in the output. Each code word is triggered on sequentially. No two code words exist at the same time. The PRN generator outputs are OR gated and sync is added prior to the output.

The generation of the reset pulses and the PRN generators is explained in detail later. We now go to the programmer.

8.4 THE PROGRAMMER

The programmer is shown in Fig. 8.4. It has at its heart the 22.05-MHz clock. The clock pulses are divided down to form the horizontal and vertical sync pulses, and to operate the blanking gates. The blanking gates pass blanked video output to the block code modulator unit. The clock itself directly feeds the block code modulator and the reset pulse generator.

8.5 THE RESET PULSE GENERATOR

The reset pulse generator is in the form of a feedback shift register. This register has six stages, nonlinear feedback, and modulo-2 addition in the

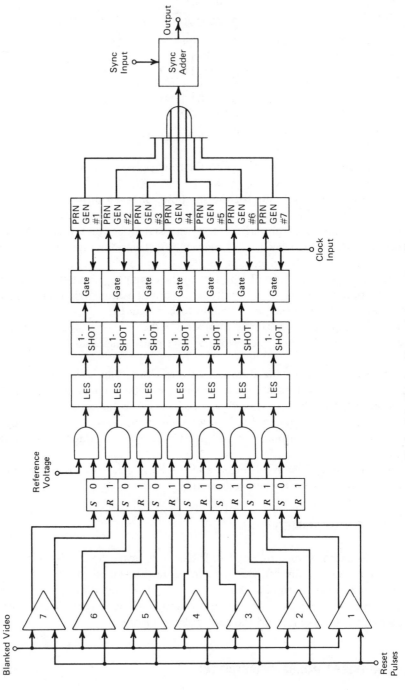

Figure 8.3. Block code modulator unit diagram.

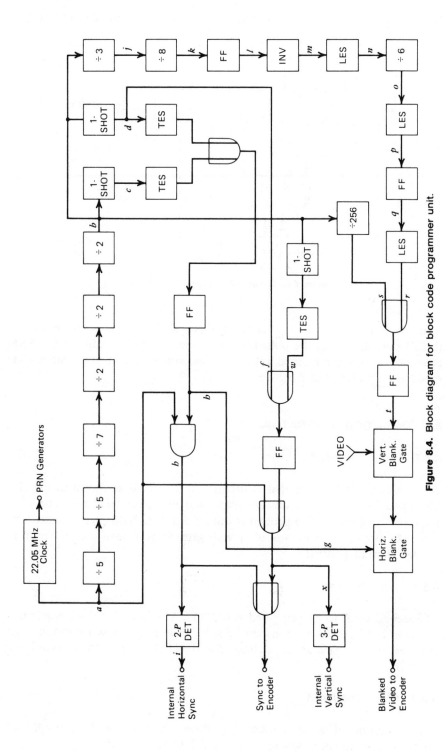

Figure 8.4. Block diagram for block code programmer unit.

239

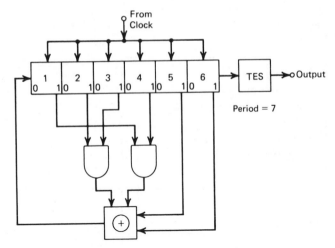

Figure 8.5. Reset pulse generator diagram.

feedback path. The word generated is 0111111. This is a nonreturn to zero (NRZ) output, so that the pulses formed from the trailing edge of the PRN generator output, can be used as reset pulses, which are sent to the block code modulator unit (see Fig. 8.5).

8.6 THE WORD GENERATORS

8.6.1 Word Number 1

The first word to be generated is 1010101. It is formed as shown in Fig. 8.6. The basic PRN generator has six stages and nonlinear feedback. The PRN sequence is 0111111. The generator output is NRZ, but is transformed into return to zero (RZ) by the use of an AND gate. Logic is then used to form the needed word of 1010101.

8.6.2 Word Number 2

The second word to be generated is 0110011. The same PRN generator that was used in Fig. 8.6 is again used in Fig. 8.7. But, in this case the output logic includes an AND gate at the output. The 0111111 word is transformed into 0110011.

8.6.3 Word Number 3

The generation of word number 3 is shown in Fig. 8.8. It is simply the modulo-2 addition of words number 1 and 2, to get 1100110.

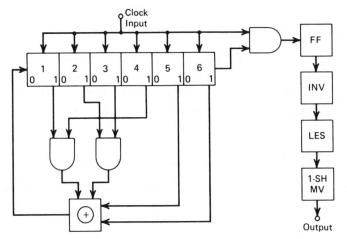

Figure 8.6. Diagram for PRN generator 1.

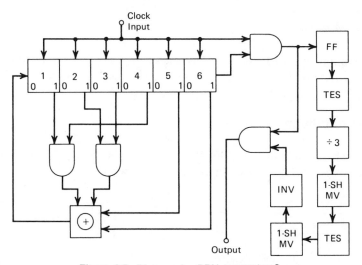

Figure 8.7. Diagram for PRN generator 2.

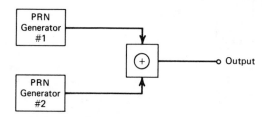

Figure 8.8. Diagram for PRN generator 3.

8.6.4   Word Number 4

The generator used to generate word number 4 is complex. The PRN generator has five stages and puts out the (RZ) word of 0111100. The feedback portion of the generator has six feedback paths; three linear and three nonlinear. Logic circuitry is used to obtain the word 0001111, which is a phase shifted version of the generator output. This is shown in Fig. 8.9.

8.6.5   Word Number 5

The generation of word number 5 is shown in Fig. 8.10. It is simply the modulo-2 addition of words number 1 and 4 to get the word 1011010.

8.6.6   Word Number 6

The generation of word number 6 is shown in Fig. 8.11. It uses the basic PRN generator of Fig. 8.9, and needs no output logic to obtain the word 0111100.

8.6.7   Word Number 7

The generation of word number 7 is shown in Fig. 8.12. It is simply the modulo-2 addition of words number 1 and 5. The resultant word is 1101001.

8.7   THE RECEIVER

The receiver is shown in Fig. 8.13. It is a very simple one, with an IF center frequency of 500 MHz. It has a limiter and a discriminator output.

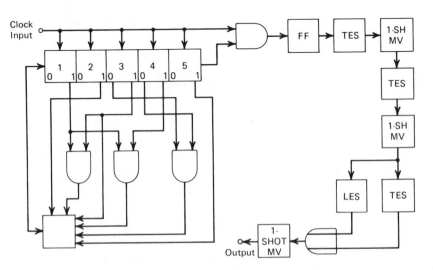

Figure 8.9. Diagram for PRN generator 4.

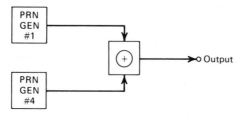

Figure 8.10. Diagram for PRN generator 5.

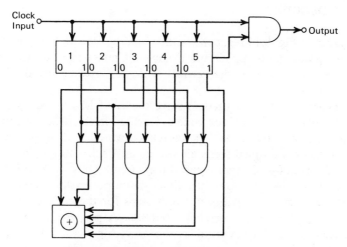

Figure 8.11. Diagram for PRN generator 6.

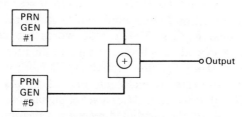

Figure 8.12. Diagram for PRN generator 7.

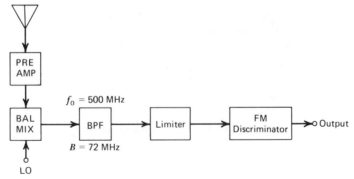

Figure 8.13. Receiver diagram for a bi-phase transorthogonal coding system.

8.8 THE DECODER

The decoder of Fig. 8.14 regenerates the signal by use of bit synchronization and an integrate-and-dump circuit. Coincident gates form the sync outputs. A three-pulse detector forms the vertical sync; and a two-pulse detector forms the horizontal sync.

The phase-lock loop is used to slave the 22.05-MHz clock, which drives the word generators. The block code is repeated in the decoder and synchronized to the transmitter code. The correlated words feed seven separate AND gates. The incoming signal feeds all AND gates. Each AND gate output feeds a specialized form of an integrate-and-dump circuit. Only the correlated output will have four pulses per word. The other six outputs will have two pulses per word. Staircase circuitry will build four steps for the correlated output, but only two steps for the uncorrelated outputs. A properly set threshold will select only the correlated output as shown in Fig. 8.15. Thus the system is M-ary correlated.

8.9 MATHEMATICAL ANALYSIS

As explained previously, transorthogonal codes have a family of M words with M bits/word, where $M = 2^n$. If $n = 3$, then $M = 8$, and an orthogonal code block would consist of 8 code words with 8 digits per word.

In transorthogonal coding $(2^n - 1)$ code words are used, with $(2^n - 1)$ bits/word. This type of code has been analyzed in the literature. They form a class of codes called M-ary systems. Nuttall (1962) has a very good approach to the problem. We will start by including some of his analysis.

Assume that $(s_k(t))$, $k = 1, 2, \dots, M$, is a set of signals used for transmission, and $n(t)$ is additive noise, the combined received waveform being

$$s_k(t) + n(t), \qquad (8.2)$$

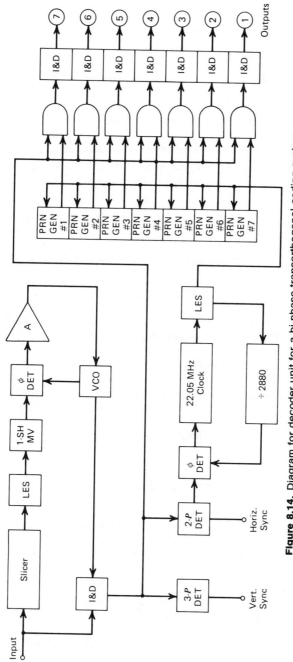

Figure 8.14. Diagram for decoder unit for a bi-phase transorthogonal coding system.

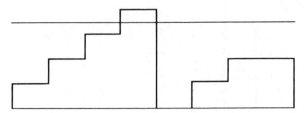

Figure 8.15. Demodulation threshold action.

If the kth signal of the set was sent. When we assume that the signal number 1 is sent we can see that the receiver computes the quantities

$$x_k = \int_{s_k} (t)\big[s_1(t) + n(t) \big] dt \qquad \text{for } k+1,2,3,\dots,M. \tag{8.3}$$

The optimum correlated receiver decides that the signal k was sent, if

$$x_k = \max(x_1, x_2, \dots, x_m). \tag{8.4}$$

We assume that signal number 1 was sent, so that the probability of a correct decision P_c is the probability that $x_1 > x_2 > \cdots > x_m$); or

$$P_c = P_r(x_1 > x_2 > \cdots > x_m). \tag{8.5}$$

Now, we assume equal cross-correlation coefficients in the signal set:

$$\frac{\int s_j(t) s_k(t) dt}{E} = \lambda_{jk} = \lambda_{(j \neq k)} \tag{8.6}$$

where E is the common received signal energy

$$E = \int s_k^2(t) dt \qquad k = 1, 2, 3, \dots, M. \tag{8.7}$$

There is a restriction on the value of

$$\int \left(\sum_{k+1}^{M} s_k(t) \right)^2 dt = \sum_{j=1}^{M} \sum_{k=1}^{M} \int s_j(t) s_k(t) dt$$

$$= ME + (M^2 - m) \qquad \lambda E \geqslant 0 \tag{8.8}$$

so that

$$\lambda = \frac{-1}{M-1}. \tag{8.9}$$

In the case of the transorthogonal code set $\lambda = -1/M - 1$. For general λ we have

$$x_1 = E + y_1 \tag{8.10}$$

$$x_k = E\lambda + y_k \qquad k = 2,3,4,\ldots,M \tag{8.11}$$

$$y_k = \int s_k(t)n(t)\,dt \qquad k = 1,2,3,\ldots,M. \tag{8.12}$$

By substituting (8.10) into (8.5) we have

$$P_c = \int_{-\infty}^{\infty} dy_1 \int_{-\infty}^{E(1-\lambda)+y_1} \cdots \int dy_2 \cdots \int dy_m p(y_1,y_2,\ldots,y_M). \tag{8.13}$$

Since the input noise is zero-mean Gaussian, and (8.12) constitutes a linear operation, the variables (y_k) are also zero-mean Gaussian. Therefore,

$$p(y_1,y_2,\ldots,y_m) = (2\pi)^{-M/2}|Ma|^{-1/2}\exp\left(-\tfrac{1}{2}y^T Ma^{-1}y\right) \tag{8.14}$$

where y is a column matrix and Ma is the matrix of cross-correlation coefficients

$$|Ma| = \left(\overline{y_j y_k}\right), \tag{8.15}$$

where the overbar denotes a statistical average over the noise.

Using (8.11), (8.6), and (8.7), we have

$$\overline{y_j y_k} = \int\int s_j(t_1)s_k(t_2)\,\overline{n(t_1)n(t_2)}\,dt_1\,dt_2$$

$$= \begin{cases} \dfrac{N_0 E}{2}, & j = k \\[2mm] \dfrac{N_0 E\lambda}{2}, & j \neq k \end{cases}. \tag{8.16}$$

So that

$$|Ma| = \left(\frac{N_0 E}{2}\right)^M (1-\lambda)^m[1+(M-1)\lambda] \tag{8.17}$$

and the co-factors of Ma are given by

$$M_{jk} = \begin{cases} \left(\dfrac{N_0 E}{2}\right)^{M-1}(1-\lambda)^{M-2}[1+(M-2)\lambda], & j = k \\[3mm] -\left(\dfrac{N_0 E}{2}\right)^{M-1}(1-\lambda)^{M-2}\lambda & j \neq k \end{cases}. \tag{8.18}$$

When we substitute (8.17) and (8.18) into (8.14), we obtain, after regrouping,

$$p(y_1, y_2, \ldots, y_M) = \left[\pi N_0 E(1-\lambda) \right]^{-M/2} \left[\frac{1-\lambda}{1+(M-1)\lambda} \right]^{-1/2}$$

$$\times \exp\left\{ -\frac{1}{N_0 E(1-\lambda)} \left[\sum_{k=1}^{M} y_k^2 - \frac{\lambda}{1+(M-1)\lambda} \left(\sum_{k=1}^{M} y_k \right)^2 \right] \right\}.$$

$$(8.19)$$

By substituting (8.19) into (8.13) and defining

$$u_k = \frac{y_k}{\left[N_0 E(1-\lambda)/2 \right]^{1/2}}, \tag{8.20}$$

we obtain

$$P_c = \int_{-\infty}^{\infty} du_1 \int_{-\infty}^{(2E(1-\lambda)/N_0)^{1/2}} \cdots \int du_2 \cdots \int du_M$$

$$\cdot (2\pi)^{-M/2} \left[\frac{1-\lambda}{1+(M-1)\lambda} \right]^{1/2}$$

$$\times \exp\left\{ \frac{-1}{2} \left[\sum_{k=1}^{M} u_k^2 - \frac{\lambda}{1(M-1)\lambda} \left(\sum_{k=1}^{M} u_k^2 \right)^2 \right] \right\}. \tag{8.21}$$

Nuttall gives a solution to this in the form of

$$P_c = \int_{-\infty}^{\infty} dy (1-\lambda)^{1/2} \phi\left[y(1-\lambda) \right]^{1/2} \int_{-\infty}^{\infty} du_1 \phi\left(u_1 - \sqrt{\lambda y} \right)$$

$$\times \left[\int_{-\infty}^{[2E(1-\lambda)/N_0]^{1/2} + u_1} \phi\left(u_2 - \sqrt{\lambda y} \right) du_2 \right]^{M-1} \tag{8.22}$$

where

$$\phi = \frac{1}{\sqrt{2\pi}} \exp\frac{-x^2}{2} \tag{8.23}$$

and

$$\Phi(x) = \int_{-\infty}^{x} \phi(y)\, dy. \tag{8.24}$$

Allowing for the fact that $(\lambda)^{1/2}$ may be imaginary, we manipulate the integrals on u_1 and u_2 by defining a new variable of integration

$$x = u_2 - \sqrt{\lambda y} \,, \qquad dx = du_2 \qquad (8.25)$$

and expression (8.22) can be rewritten as

$$P_c = du_1 \phi\left(u_1 - \sqrt{\lambda y}\right) \left[\int_{-\infty + \sqrt{\lambda y} - u_1}^{[2E(1-\lambda)/N_0]^{1/2} + u_1 - \sqrt{\lambda y}} \phi(x)\,dx \right]^{M-1}. \qquad (8.26)$$

This can be written as

$$P_c = \int_{-\infty}^{\infty} du_1 \phi\left(u_1 - \sqrt{\lambda y}\right) \phi^{M-1}[2E(1-\lambda)/N_0]^{1/2} + \left(u_1 - \sqrt{\lambda y}\right). (8.27)$$

Now we let

$$z = u_1 - \sqrt{\lambda y} \,, \qquad dz = du_1, \qquad (8.28)$$

so that

$$P_c = \int_{-\infty - y}^{\infty - y} dz \phi(z) \phi^{M-1}\left\{ z + \left[2E(1-\lambda)/N_0 \right]^{1/2} \right\} \qquad (8.29)$$

$$= \int_{-\infty}^{\infty} dz \phi(z) \phi^{M-1}\left\{ z + \left[2E(1-\lambda)/N_0 \right]^{1/2} \right\}. \qquad (8.30)$$

Nuttall proceeds to simplify his equations, but the solution of (8.30) is given more precisely by Birdsall and Green (1958).

The general form of Nuttall's equation (8.30) is

$$P_c = \int_{-\infty}^{\infty} \phi(z) \phi^{M-1}(z + a)\,dz. \qquad (8.31)$$

An appropriate solution is given by Birdsall and Green (1958). Their approach, which gives an equation similar to (8.30) is

$$P_e = T_{M-1}(x) \phi(x + d')\,dx \qquad (8.32)$$

where

$$d' = \left(\frac{2E}{N_0} \right)^{1/2}$$

$$T_{M-1}(x) = \text{Tippett's distribution,}$$

where

$$T_{M-1}(x) = \frac{\phi(x+m)_{M-1}}{\sigma_{(m-1)}} \tag{8.33}$$

and the probability of error can be written as

$$P_e = \frac{\int_{-\infty}^{\infty} \Phi(x+m)_{M-1}}{\sigma_{(m-1)}\phi(x+d')dx} \tag{8.34}$$

$$= \frac{\Phi(d'+m_{(M-1)})}{\sqrt{1+\sigma_{(M-1)}^2}}. \tag{8.35}$$

Peterson, Birdsall, and Fox (1954) give the following equations:

$$m_{(M-1)} = 1 - \frac{1}{M-1} + \frac{0.67722}{\sigma_{(M-1)}} \tag{8.36}$$

and

$$\sigma_{(M-1)} = M-2, \tag{8.37}$$

so that

$$m_{(M-1)} = \frac{M-2}{M-1} + \frac{0.57722}{M-2}. \tag{8.38}$$

They give

$$\sigma_{(M-1)} = \frac{\pi(M-1)^{1/2}}{M-2} \tag{8.39}$$

or

$$\sigma_{(M-1)}2 = \frac{\pi^2(M-1)}{(M-2)^2} \tag{8.40}$$

and

$$\sqrt{1+\sigma_{(M-1)}^2} = \frac{\sqrt{(M-2)^2 + \rho^2(M-1)}}{M-2}. \tag{8.41}$$

Now, an expression for the probability of error can be written as

$$P_e = \frac{\sqrt{(M-2)^2 + \pi^2(M-1)}}{(M-2)\sqrt{2\pi\left(\dfrac{2E}{N_0} + \dfrac{M-2}{M-1} + \dfrac{0.57722}{M-2}\right)}}$$

$$\times \exp\left[-\frac{1}{2} \frac{\dfrac{2E}{N_0} + \dfrac{M-2}{M-1} + \dfrac{0.57722}{M-2}}{\dfrac{(M-2)^2 + \pi^2(M-1)}{(M-2)^2}}\right]. \tag{8.42}$$

Tippett's solution does not include the cross-correlation coefficient of Nuttall, which should be included.

$$P_e = \frac{\sqrt{(M-2)^2 + (M-1)\pi^2}}{(M-2)\sqrt{2\pi}\sqrt{\dfrac{2E(1-\lambda)}{N_0} + \dfrac{M-2}{M-1} + \dfrac{0.57722}{M-2}}}$$

$$\times \exp\left\{-\frac{1}{2} \frac{\left[\dfrac{2E(1-\lambda)}{N_0} + \dfrac{M-2}{M-1} + \dfrac{0.57722}{M-2}\right]}{\dfrac{(M-2)^2 + \pi^2(M-1)}{(M-2)^2}}\right\}. \tag{8.43}$$

Viterbi (1961) included the term $\log_2 M$, therefore this term should also be included in the equation:

$$P_e = \frac{\sqrt{(M-2)^2 + \pi^2(M-1)}}{\sqrt{2\pi}\,(M-2)\sqrt{\dfrac{M-2}{M-1} + \dfrac{0.57722}{M-2} + \dfrac{2E(1-\lambda)}{N_0\log_2 M}}}$$

$$\times \exp\left[-\frac{1}{2} \frac{\dfrac{2E(1-\lambda)}{N_0\log_2 M} + \dfrac{M-2}{M-1} + \dfrac{0.57722}{M-2}}{\dfrac{(M-2)^2 + \pi^2(M-1)}{(M-2)^2}}\right]. \tag{8.44}$$

Also, the time-bandwidth product BT should be included, so that

$$P_e = \frac{\sqrt{(M-2)^2 + \pi^2(M-1)}}{\sqrt{2\pi}\,(M-2)\sqrt{\dfrac{M-2}{M-1} + \dfrac{0.57722}{M-2} + \dfrac{2EBT(1-\lambda)}{N_0 \log_2 M}}}$$

$$\times \exp\left[-\frac{1}{2}\frac{\dfrac{2EBT(1-\lambda)}{N_0 \log_2 M} + \dfrac{M-2}{M-1} + \dfrac{0.57722}{M-2}}{\dfrac{(M-2)^2 + \pi^2(M-1)}{(M-2)^2}}\right]. \qquad (8.45)$$

Thus far, nothing has been said about the effect of the bi-phase modulation, which is equivalent to double sideband AM with suppressed carrier. Both are superior to ordinary AM by a power factor of 3; or, for equal transmitted powers, both have three times the SNR (power ratio) than AM has. So that

$$P_e = \frac{\sqrt{(M-2)^2 + \pi^2(M-1)}}{\sqrt{2\pi}\,(M-2)\sqrt{\dfrac{M-2}{M-1} + \dfrac{0.57722}{M-2} + \dfrac{6EBT(1-\lambda)}{N_0 \log_2 M}}}$$

$$\times \exp\left[-\frac{1}{2}\frac{\dfrac{6EBT(1-\lambda)}{N_0 \log_2 M} + \dfrac{M-2}{M-1} + \dfrac{0.57722}{M-2}}{\dfrac{(M-2)^2 + \pi^2(M-1)}{(M-2)^2}}\right]. \qquad (8.46)$$

At this point we might simplify by setting

$$A = \frac{\sqrt{(M-2)^2 + \pi^2(M-1)}}{M-2} \qquad (8.47)$$

and

$$B = \left(\frac{M-2}{M-1} + \frac{0.57722}{M-2}\right) \qquad (8.48)$$

and

$$\rho^2 = \frac{E_s}{N_0} = \frac{6EBT(1-\lambda)}{N_0 \log_2 M}. \qquad (8.49)$$

Then

$$P_e = \frac{A}{\sqrt{2\pi(\rho^2 + B)}} \exp\left(-\frac{1}{2}\frac{\rho^2 + B}{A^2}\right). \tag{8.50}$$

8.10 CHANNEL-SHARING CONSIDERATIONS

We, again, consider the case where there are 12 simultaneous users on the same frequency channel. In this case there will be one wanted channel and 11 unwanted signals. In order to evaluate the error probability, (8.50) can be modified to

$$P_e = \frac{A}{\sqrt{2\pi\dfrac{\rho^2}{1+\rho_c^2} + B}} \exp\left[-\frac{1}{2}\frac{\dfrac{\rho^2}{\rho_c^2+1} + B}{A^2}\right]. \tag{8.51}$$

Equation (8.53) should be used for low SNR conditions, and (8.54) for high SNR conditions. As explained in the previous two chapters, $\rho_C^2 = QC/S(E/N_0)$. For the bi-phase transorthogonal coding system described in this chapter,

Q $Ud\lambda_C$,
U 12 users,
d 0.2 duty factor.

The evaluation of the cross-correlation coefficient term λ_C will be considered separately.

8.10.1 Evaluation of λ_C

The RF mixer in the receiver has modulo-2 addition qualities. In modulo-2 addition there are four possible states of interference. In state 1, the wanted signal is a one and the unwanted signal is a zero for a probability of zero. In state 2, the wanted signal is a one and the unwanted signal is a zero for a probability of zero. In state 3, the wanted signal is a zero and the unwanted signal is a one for an interference probability of one. In state 4, the wanted signal is a one and the unwanted signal probability is also one for a probability of zero. Since there are four states, the resultant probability is 0.25. This condition holds for only a small fraction of the total time available.

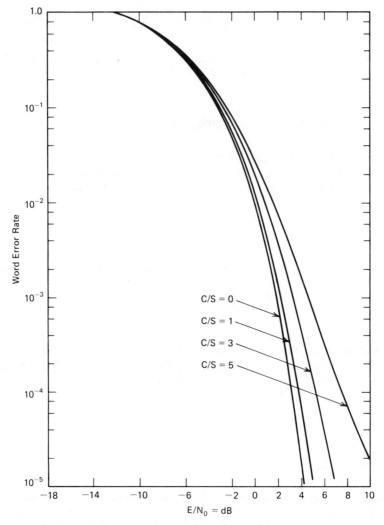

Figure 8.16. Channel-sharing performance for the bi-phase transorthogonal system.

Due to the threshold setting; if the wanted signals and the unwanted signals are in phase, no error can occur. When they are the opposite in phase there is unity probability of error. However, the actual phase is seldom in either condition, but is somewhere in between the two-phase positions; in or out. For about $\frac{1}{3}$ of the time, or for 120° phase angle, interference will occur. So, the error probability is $0.33 \times 0.25 = 0.083$. So the summation of error probability resulting from the modulo-2 addition is equal to $0.25 + 0.083 = 0.333$, or

$$P_{m2} = 0.333. \tag{8.52}$$

In order to find λ_C, we multiply P_{m2} by the cross-correlation coefficient for a transorthogonal code, or

$$\lambda_C = P_{m2} \frac{-1}{M-1} \tag{8.53}$$

$$= \frac{-P_{m2}}{7} = \frac{-0.333}{7} = -0.0476. \tag{8.54}$$

8.10.2 Channel-Sharing Performance

From the above information the value of Q can be obtained:

$$Q = Ud\lambda_C = 12 \times 0.2 \times 0.0476 = 0.114. \tag{8.55}$$

This value of Q was used to obtain the curves shown in Fig. 8.16. The results are reasonably good, but do not compare favorably with those of a PRN/PPM/FM system.

chapter 9

The Convolutional Codes

introduction

A class of error-correcting codes called the convolutional codes give excellent performance when used in real time digital television. While they outperform block codes, they are not to be considered as such. There are many binary versions of these codes, but also q-ary codes exist.

Fig. 9.1 depicts an encoder for a general (Mn, Mk) convolutional code. The encoder accepts k information symbols [elements of GF(q)] at its input and produces $n > k$ code symbols at its outputs.

These output symbols are linear combinations over GF(q) of the input symbols in the preceding M blocks.

Convolutional encoders are based on the use of feedback shift registers. A basic example is shown in Fig. 9.2. Each unit of time, a new binary source digit u_n enters the encoder and each preceding source digit moves one place to the right in the shift register. The new source digit is transmitted directly on the channel as $x_n^{(1)} = u_n$. Following each such information digit is a check digit. The output switch samples source digits and shift register digits in succession. These output digits are clearly labeled in the figure.

For this particular example

$$x_n^{(2)} = u_n \oplus u_{n-3} \oplus u_{n-4} \oplus u_{n-5}$$

$$= x_n^{(1)} \oplus x_{n-3}^{(1)} \oplus x_{n-4}^{(1)} \oplus x_{n-5}^{(1)}. \tag{9.1}$$

9.1 THE BINARY SYMMETRICAL CHANNEL

The transmission characteristics of the binary BSC depict the propagation mode for the convolutional codes. Additive white Gaussian noise, of course,

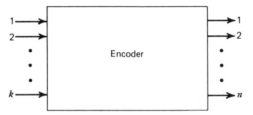

Figure 9.1. Simplified form for $A(n, k)$ convolutional encoder.

exists on this channel giving rise to transition probabilities as shown in Fig. 9.3.

In this figure the signal vectors (S_i) lie on the vertices and quantize the components of the received vector into two levels. Input and output alphabets are denoted by the discrete channel symbols $(0, 1)$. Probability is denoted by the symbol p. It is the probability that the transmitted vector is received incorrectly.

Communication over this BSC is as shown in Fig. 9.4. The $(0, 1)$ components are fed to the BSC from the parity check coder (in this case, convolutional). The BSC output feeds the decoder.

The effects of the transmitter modulator, the Gaussian channel noise, and the receiver quantizer are all combined into the probability parameter p.

When one of M equally likely messages is communicated over the BSC by means of N component binary vectors (y_i), the optimum receiver compares the N component received vector r' with each of the (y_i) and determines for which i the probability $P(r'|y_i)$ is maximum. The probability that a transition occurs with any single use of the BSC is p and successive uses of the channel are statistically independent. Therefore, whenever r' and y_i differ in d_i coordinates,

$$P(r'|y_i) = p^{d_i} q^{N-d_i} = q^N \left(\frac{p}{q} \right)^{d_i}; \qquad q \triangleq 1-p. \qquad (9.2)$$

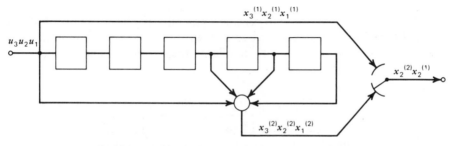

Figure 9.2. Diagram for a sampled convolutional encoder.

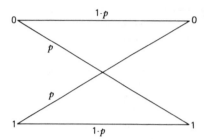

Figure 9.3. Representation of the binary symmetrical channel transitional probabilities.

The quantity d_i is called the Hamming distance between r' and y_i.

The right side of (9.2) is a monotonically decreasing function of d_i, for $p < \frac{1}{2}$. Accordingly, the optimum BSC decoder may determine $\hat{m}$ by computing the set of Hamming distances (d_i), $i = 0, 1, \ldots, M-1$, and setting $\hat{m} = m_k$ whenever d_k is the smallest member of the set.

Since the vector $r' \oplus y_i$ contains a "1" only in coordinates in which y_i differs from r', the Hamming distance d_i is conveniently obtained by forming $r' \oplus y_i$, and counting the resulting number of 1's. By convention, the number of 1's in any binary vector "a" is called its weight, denoted by $W(a)$. With this notation

$$d_i \triangleq W(r' \oplus y_i). \tag{9.3}$$

Convolutional decoders make use of this Hamming distance weighting.

9.2 CONVOLUTIONAL DECODING THEORY

As mentioned previously convolutional codes are not block codes, but rather belong to the family of tree codes. The vectors $(0, 1)$ in the tree codes are as shown in Fig. 9.5.

The input to the convolutional encoder is of the form

$$X = (X_1, X_2, \ldots, X_L). \tag{9.4}$$

We can initially set the shift register to an all zero state. The first V digits of y, obtained by shifting the first component of X into stage 1 and sampling the V adders, depend only on X_1. Similarly, the second V digits depend only

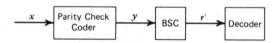

Figure 9.4. Diagram for simplified BSC communications.

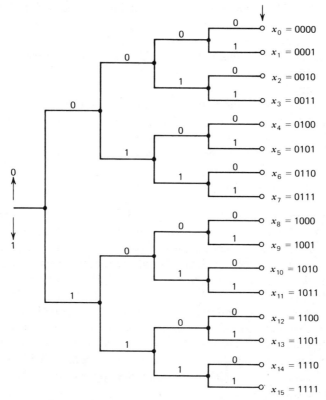

Figure 9.5. Convolutional tree code structure.

on X_1 and X_2. In general, the V digits of y obtained immediately after shifting component X_h into the X register depend only on X_h and the $(K-1)$ components of X preceding X_h. This implies that if two-input vectors agree in their first $(h-1)$ coordinates, the corresponding output vectors agree in their first $(h-1)V$ coordinates.

As depicted in the code tree of Fig. 9.5, the set of 2^L L-component input vectors (X_i) is diagrammed by adopting the convention that the upper branch diverging from any node of the tree corresponds to shifting a 0 into the shift register and the lower branch to shifting in a 1. Thus each input vector X_i designates a distinct path all the way through the tree to one of the 2^L terminal nodes. The association of the (X_i) and the paths is indicated in the figure.

Next consider any intermediate node of the input tree: the path leading up to this node designates the contents of the shift register just prior to the shifting in of a digit, and the contents immediately thereafter determine the next V digits of y. Along each branch of the code tree is written the V digits of y associated therewith. In this example $K=4$, and $V=3$ for the convolutional encoder. The branch vectors are not independent of each other, but depend on all prior branch vectors.

The convolutional encoder is a parity check device due to the use of a modulo-2 adders. When the number of 1's into the adder are even the output is 0; if odd, the output is a 1. This is similar to the parity check system of block encoders.

There are several methods for decoding convolutional codes. Wozencraft and Reiffen (1961) use a method called suboptimum decoding. A discussion of this is found in Wozencraft and Jacobs (1965) and is used in that which follows.

9.2.1 Suboptimum Decoding

According to the authors, the decoder decides on each of the L components $X_1, X_2, \ldots, X_L$ of the encoder input vector in turn. Thus there is a sequence of decisions $\hat{X}_1, \hat{X}_2, \ldots, \hat{X}_L$. Each decision is based on (1) the previous decisions $\hat{X}_1, \hat{X}_2, \ldots, \hat{X}_L$, and (2) the KV digit span of the received vector that is directly affected by X_h. They refer to the KV vectors $(_h r)$, and the tree node under consideration specified by $\hat{X}_1, \hat{X}_2, \ldots, \hat{X}_L$ as the hth starting node. Each X_h is decoded in turn by determining which of the 2^K K-branch code-word segments that diverge from the hth starting node is the most probable cause of $_h r$. The decoder calculates the Hamming distance between each KV digit code-word segment and $_h r$. If the code-word segment with the smallest distance leaves the hth starting node along the upper branch, the decoder sets $\hat{X}_h = 0$; otherwise $\hat{X}_h = 1$.

The error probability is $P(E)$, that at least one error will be made by the suboptimum decoder in the decision sequence $\hat{X}_1, \hat{X}_2, \ldots, \hat{X}_L$. Let $P(E_h)$ be the conditional probability of an error on the hth decision, given that the hth starting node is correct. They have bounded the unconditional probability $P(E)$ by deriving the sequence of equations

$$P(E) \leqslant \sum_{h=1}^{L} P(E_h) \tag{9.5}$$

$$P(E_h) = P(E_1) \tag{9.6}$$

$$P(E_1) = P(E_1 | X_0) \tag{9.7}$$

in which the condition on the right hand of (7) indicates that the all-zero message sequence is transmitted. Therefore,

$$P(E) \leqslant L P(E_1 | X_0). \tag{9.8}$$

The average probability of error was found by averaging (9.8) over an ensemble of communication systems, each with a different convolutional code:

$$\overline{P(E)} < L2^{-N(R'_0 - R_N)}. \tag{9.9}$$

This equation is a bound on the mean probability that one or more errors will be made in decoding an L-bit message input sequence with convolutional coding, in a BSC, and the suboptimum decoder. The coder constraint length is $K = NR_N = N/V$, and R'_0 is the BSC error exponent of (9.9):

$$R'_0 = 1 - \log_2\left[1 + 2\sqrt{p(1-p)}\,\right] \tag{9.10}$$

where

$$p = \Phi\left(\sqrt{\frac{2E_b R_N}{N_0}}\,\right) \tag{9.11}$$

$$R_N = \frac{M}{N}. \tag{9.12}$$

9.2.2 Sequential Decoding

The suboptimum decoder becomes very complex in its implementation when the length of the encoding shift register increases. This complexity increases exponentially with the number of shift register stages used (K). To get around this complexity problem Fano (1963) created a sequential decoding algorithm.

Basically the sequential decoder proceeds in much the same way as the suboptimum decoder of Wozencraft and Reiffen (1961). In both cases the decoding of X_h is equivalent to the decoding of X_1, provided the hth starting node is correct. Fano changed the method of deciding $\hat{X}_h$.

We can make use of the code tree to explain the Fano decoding algorithm. Let $_1y$ represent the first $N = KV$ digits encountered along a particular tree path and let $_1n$ denote the first N noise digits.

If we assume initially that the BSC is noiseless, so that, $_1n = 0$, then $_1r = {_1y} \oplus {_1n} = {_1y}$. In this case a decoder that has been provided with a replica of the encoder as a point of reference, can trace out the first K branches of the path designated by X. The decoder starts at the first node of the code tree, generates both branches diverging therefrom, and follows the branch that agrees with the first V digits of $_1r$. At node 2 the decoder again generates both divergent branches and follows again the correct path. Thus the decoder progresses from node to node. The decoder determines the first K digits of X. Decisions at each node can be made if the two possible paths of choice differ by one or more digits. This condition can be guaranteed by connecting the first stage of the shift register to the first modulo-2 adder.

However, the BSC is more often than not, noisy and $_1n \neq 0$. To solve this problem a slight procedural modification is needed. If neither branch going away from a particular node matches the reference encoder replica, the decoder chooses that branch that agrees more closely with the replica. If more than $V/2$ transitions occur in the transmission of a branch, the decoder proceeds to an incorrect node. Having once made such a mistake, however, the decoder will not find any agreements upon proceeding further.

When this situation occurs the Fano algorithms tell the decoder to go back to the last prior known correct node. Once there, the decoder selects the branch that was before rejected and proceeds in the normal manner. The object of the decoder is trace a path through the code tree to one of its 2^K terminal nodes. As soon as this path has been traversed $\hat{X}_1$ is recorded in an accumulator, and the whole process repeated to determine $\hat{X}_2, \hat{X}_3, \ldots, X_L$.

Let us assume that the decoder has penetrated l branches into the code tree, such that $0 \leqslant l < K$. Let $d(l)$ denote the total number of differences (Hamming distance) observed by the decoder between the path it is following, say $y^*(l)$ and the corresponding l-branch segment of the received sequence $r(l)$;

$$d(l) \triangleq W(y^*(l) \oplus r(l)). \tag{9.13}$$

As the sequential decoder penetrates deeper into the code tree, it keeps a running count of $d(l)$. After each successive penetration the decoder compares $d(l)$ against a discard criterion function $k(l)$. If $d(l)$ exceeds $k(l)$, the tentative path is discarded as improbable. The decoder then backs up to the nearest unexplored branch for which $d(l) - k(l)$ and again proceeds as far along the tree branches as $k(l)$ permits. The decoder keeps track of the branches it has explored and thereby avoids retracing false steps. The previous description of the Fano algorithm is simplified. A detailed analysis may be found in Wozencraft and Jacobs (1965). Precise analysis is complicated by the fact that the size of the decoding register is finite. Meaningful insight into decoder performance, however, is gained by assuming that the shift register is so large that the search position pointer is never forced to the output end of the register. Using this assumption they found the average error probability to be:

$$\overline{P(E)} = \left[LA_0 2^{-(K/2)[(R'_0/R_N) - 1]} \right] \qquad \text{for } R_N < N_0 \tag{9.14}$$

where

$$A_0 = \frac{2}{1 - 2^{1/2[(R'_0/R_N) - 1]}}. \tag{9.15}$$

The main drawback to the use of the Fano decoder is the large storage space required for the memory bank as K grows large.

9.2.3 The Viterbi Decoding Algorithm

To get around the problem of a large storage buffer needed in the Fano decoding system, Viterbi (1967) created a decoding algorithm (see also, Peterson and Weldon, 1972). If the number of code words in a convolutional code is reasonably small, the decoding procedure by Viterbi is useful. In this procedure all code word replicas are generated successively (q^K words). Each replica word is compared with the received word. Consider a convolutional code with rate $R_N = K/N$ and the basic block length N. Let the encoding constraint length of the code be denoted by $N_e = M_e N$ and let $K_e = M_e K$ denote the number of information symbols in a constraint length. In the code tree the K_e-tuple to be encoded determines one of q^{K_e} paths of M_e branches through the tree. Each node of that tree has q^K branches, and with each branch is associated an N-tuple. Let $a_0, a_1, \ldots$, denote the input information sequence; a_i is a K-tuple of elements of $GF(q)$. Let

$$C_0 = C_{00} \quad C_{01} \cdots C_0(M_e - 1)$$

$$C_1 = C_{10} \quad C_{11} \cdots C_1(M_e - 1) \tag{9.16}$$

$$C_q^{K_{E-1}} = C_{(q^{K_e}_{-1})0} \quad C_{(q^{K_e}_{-1})} \cdots C_{(q^{K_e}_{-1})}(M_e - 1)$$

denote the q^K code words of length N_e associated with the paths of the tree. The symbol C_{ij} denotes a g-ary N-tuple.

If the received sequence is $r = r_0, r_1, \ldots$, the decoder calculates the q^{Ke} distances

$$d(C_{ij}r) = \sum_{j=0}^{M_e - 1} d(C_{ij}, r_j) \qquad i = 0, 1, \ldots, q^{K_e} - 1 \tag{9.17}$$

where r_j denotes a q-ary N-tuple. Again the Hamming distance may be used. The distances associated with the q^K paths for which the last $M_e - 1$ information blocks $a_1, a_2, \ldots, a_{M_e - 1}$ are identical are compared. The path having the smallest distance is called the survivor. This constitutes step 1.

At step 2, the distances between the received sequence r and $q^{K_e}N(M + 1)$ -tuples formed by lengthening each survivor by one branch are calculated. Again, the distances associated with the q^K paths for which the last $M_e - 1$ information blocks are identical are compared and in each case the path that is closest to the received sequence is denoted the survivor.

The procedure continues with q^{K_e} calculations being performed at each step. The decoder must store the information sequences associated with $q^{K_e - K}$ survivors, as well as the distances for these paths. Let m denote the number of branches that the decoder can store and let $n = mN$. Typically n, the decoding constraint length, is several times larger than the encoding constraint length.

At the $(m - M_e + 1)$th step, paths of length m are being considered and the storage capacity of the decoder has been reached. Now if all $q^{K_e - K}$ survivors have the same information symbols in the 0th block, then the 0th block is decoded as these symbols. If all survivors do not agree, the decoder can either signal a detected error or simply guess at a_0, presumably by taking a majority vote of the survivors. After decoding, if errors have been corrected, the distance functions for the survivors must be reduced by the number of errors corrected.

The probability of incorrectly decoding a code word is

$$P(E) \sum_{i=[(d+1)/2]}^{d} \binom{d}{i} P^i Q^{d-i} \qquad d \text{ odd} \qquad (9.18)$$

$$= \frac{1}{2} \binom{d}{d/2} P^{d/2} Q^{d/2} + \sum_{i=[(d+1)/2]}^{d}$$

$$\binom{d}{i} P^i Q^{d-i} \qquad d \text{ even} \qquad (9.19)$$

where

$$Q = P - 1$$

and

$$P = \Phi\left(\sqrt{\frac{2 E_b R_N}{N_0}}\right).$$

The great advantage of the Viterbi maximum likelihood decoder is (Heller and Jacobs, 1971), that the number of decoder operations performed in decoding L bits is only $L2^{(K-1)}$, which is linear in L. However, Viterbi decoding is limited to relatively short constraint length codes due to the exponential dependence of decoder operations per bit decoded on K. Fortunately, excellent decoder performance is possible with good short constraint length codes.

9.3 BURST ERROR-CORRECTING CODES

The convolutional codes described previously are able to correct random type errors occurring in singles. There are a class of convolutional codes that are able to correct errors that occur in bursts. These bursts may last two or more bit periods in length. The first burst error-correcting convolutional codes were devised by Hagelbarger (1959; 1960). These codes were also studied by Kilmer (1960). Wyner and Ash (1963) define Type $B2$ codes. Their results were used by Berlekamp (1964) and Preparata (1964) to construct burst error-correcting convolutional codes. Massey (1965) independently discovered this class of codes. The class of codes is described as the BPM codes.

Burst error-correction convolutional codes have a burst error length symbolized by b. The upper bound for b is given (Peterson and Weldon, 1972) as

$$b \leqslant \frac{(m-1)(N-K)}{\left(1 + \dfrac{K}{N}\right)} + N - 1. \qquad (9.20)$$

(Permission to use excerpts from *Error Correcting Codes*, 2nd ed., by W. W. Peterson and E. J. Weldon, Jr., has been granted by MIT press.)

The BPM codes have one parity digit per block. They have Type $B2$ burst-correcting ability $b = N$. Codes having large values of burst-correcting ability can be constructed by interleaving. BPM interleaved codes require a guard space following the burst period, during which period prior conditions are restored. The guard space is equal to $N - 1$ bits.

Iwadare (1968) devised a class of burst error-correcting codes wherein $K = N - 1$. They are somewhat simpler to implement than are the BPM codes for equal values of b and guard space. These codes also correct burst lengths of $b + N$, and are not phase dependent as are the BPM codes (Peterson and Weldon, 1972, p. 435).

The previously mentioned burst-correcting convolutional codes perform well for burst errors; but poorly for independent random errors.

In the field of real time digital TV the need for correcting both types of errors exists.

A type of code called the self-orthogonal convolutional code (Robinson and Bernstein, 1967) have the ability to correct both independent random errors and burst errors.

A code labeled diffuse self-orthogonal performs somewhere in between the pure burst-correction and the pure self-orthogonal codes. These diffuse types of codes correct t independent errors or less, and burst lengths of b or less. Tong (1978a) and Iwadare (1968b) proposed the diffuse type of self-orthogonal convolutional codes.

The type of code selected here for further discussion is the pure self-orthogonal convolutional code. It is one of the most likely candidates for use in digital TV.

9.4 THE PURE SELF-ORTHOGONAL CONVOLUTIONAL CODE

These codes (Massey, 1963; Robinson and Bernstein, 1967) are the most extensive constructive class of convolutional codes known. Also, they can be decoded simply with majority-logic decoding.

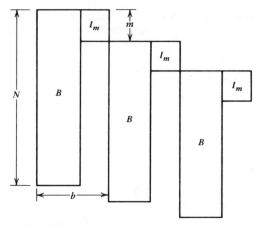

Figure 9.6. Diagram for a semi-infinite parity check matrix.

Unfortunately, for large values of N and K they have relatively low minimum distance (Hamming weights).

Robinson and Bernstein (1967) present a parity check matrix as shown in Fig. 9.6. In this figure I_m is the $m \times m$ identity matrix. The rows of H specify the parity equations that the code satisfies.

The code words are sequences (semi-infinite column vectors) Y such that

$$HY = 0.$$

The resulting code sequences have the form

$$I, I_2 \cdots I_{b-m} C_1 C_2 \cdots C_m I_{b-m+1} \cdots I_{2(b-m)} C_{m+1} \cdots C_{2m} \cdots \qquad (9.21)$$

where the I's are information bits and the C's are parity check bits. Each check digit is included in one parity equation; i.e., the C's depend only on the I's. Any linear recurrent code can be put into this form without changing the maximum distance of the code (Massey, 1963).

As implied in Fig. 9.6, b is the basic block length of the code and m is the number of redundant bits per block. The actual constraint length n_A is

$$n_A = b\left(\frac{N}{m}\right). \qquad (9.22)$$

We assume that m divides N. When $m = N$ the code reduces to a block code with block length b. We call a code with parameters b and m a $(b, b-m)$ code. The $N \times b - m$ matrix $\bar{B}$ and the identity matrix I_m define the parity equations that characterize the code. The blank spaces in Fig. 9.6 represent zeros.

The minimum distance d is defined as the smallest number of positions, in the first n_A set, in which the two code sequences differ in the first $b - m$ information bits. All combinations of t or fewer errors in n_A consecutive bits can be corrected if and only if, $d \geqslant 2t + 1$.

In terms of A_N, b is the minimum number of columns of A_N, including at least one information column from the first block, which add to zero.

The canonical representation of the self-orthogonal convolutional code is a recurrent code for which no two parity equations include the same 2 bits. The rows of H form a set of orthogonal vectors where H has the form given in Fig. 9.6. It can be seen that H has no rectangles, i.e., four 1's in two separate rows of H, which define the corner of a rectangle. It is clear that H has no rectangles and thus describes a self-orthogonal convolutional code if, and only if, the A_N matrix has rectangles.

Robinson and Bernstein show a typical A_N matrix with $b = 2$, $m = 1$, and $t = 2$. They show the following example from Hagelbarger (1962):

$$
A_N = \begin{bmatrix}
1 & 1 & & & & & & & & & & & & \\
0 & 0 & 1 & 1 & & & & & & & & & & \\
1 & 0 & 0 & 0 & 1 & 1 & & & & & & & & \\
0 & 0 & 1 & 0 & 0 & 0 & 1 & 1 & & & & & & \\
0 & 0 & 0 & 0 & 1 & 0 & 0 & 0 & 1 & 1 & & & & \\
1 & 0 & 0 & 0 & 0 & 0 & 1 & 0 & 0 & 0 & 1 & 1 & & \\
1 & 0 & 1 & 0 & 0 & 0 & 0 & 0 & 1 & 0 & 0 & 0 & 1 & 1
\end{bmatrix}. \quad (9.23)
$$

The value of d for this code is found by determining the smallest number of columns of the A_N matrix, including the first column, which sum to zero. Let $C(i)$ denote the ith column of A_N. There are no rectangles in A_N, so we must choose at least four additional columns for each 1 in $C(1)$. The parity columns $C(2)$, $C(6)$, $C(12)$, and $C(14)$, exactly pair the 1's in $C(1)$. Thus $d = 5$ and this code can correct error patterns with $t = 2$ or fewer errors in any of 14 consecutive positions.

If we say that J is the weight of $C(1)$, then $d = J + 1$. There are four 1's in $C(1)$, so $J = 4$ and $d = 5$.

9.4.1 Difference Triangles

In order to maximize the error-correcting capability, n_A is minimized for a given value of J. This is equivalent to minimizing N. For simplicity let the first row of $\overline{B}$ of a $(b, b - 1)$ code be all 1's. Clearly this does not change d and it insures that $N - 1$ is the maximum element in any difference triangle. In a similar manner we set the first m elements in the single $\overline{B}$ to 1 to create a $(b, 1)$ code.

The problem of synthesizing $(b, b - 1)$ and $(b, 1)$ codes is equivalent to synthesizing a set of difference triangles, such that all the elements are distinct. In order to minimize n_A we minimize the maximum element in the set of difference triangles.

Difference triangles having all distinct elements can be formed from a perfect different set, which was first suggested by Hagelbarger (1962).

A perfect difference set is a collection of $q+1$ integers, $(d_1, d_2, \ldots, d_{q+1})$ having the property that their $q^2 + q$ differences $d_i - d_j$, $i \neq j$, $i, j = 1, 2, \ldots, q+1$, are congruent modulo $q^2 + q + 1$ to the integers $1, 2, \ldots, q^2 + q$. For example, the integers $(2, 4, 8)$ form a perfect difference set, since

$$\begin{array}{ll} 2-8 \equiv 1 & 8-2 \equiv 6 \\ 4-2 \equiv 2 & 2-4 \equiv 5 \\ 4-8 \equiv 3 & 8-4 \equiv 4 \pmod{7}. \end{array} \tag{9.24}$$

Singer (1938) has shown that a perfect distance set exists when q is a power of a prime. The elements of a difference set can be reduced modulo $q^2 + q + 1$ and can be reindexed so that the resulting set is in natural order, i.e., $d_i < d_{i+1}$ for $i = 1, 2, \ldots, q$, without changing the property of being a perfect difference set. For convenience let us only consider difference sets in this form. For example, the set $(2, 4, 8)$ is reducible to $(1, 2, 4)$.

A regular array can be formed from a difference set. Difference triangles can be selected from the regular array.

The first row of a regular array $(a_{11}, a_{12}, \ldots, a_{1, q+1})$, is defined by the differences between successive elements of a difference set

$$a_{1, i} = d_{i+1} - d_i, \qquad i = 1, 2, \ldots, q \tag{9.25}$$

except the last element $a_{1, q+1}$, which is defined by

$$a_{1, q+1} = q^2 + q + 1 + d_i - d_{q+1}. \tag{9.26}$$

The remaining elements of a regular array are defined

$$a_{i, j} = \sum_{k=j}^{i+j-1} a_{1, k} \tag{9.27}$$

where

$$a_{1, k} = a_{1, k-q-1} \qquad \text{for } k > q+1, \text{ and } 1 \leqslant i \leqslant q, \ 1 \leqslant j \leqslant q+1.$$

For example, the difference set $(1, 2, 4, 10)$ results in the regular array:

$$\begin{array}{cccc} 1 & 2 & 6 & 4 \\ 3 & 8 & 10 & 5 \\ 9 & 12 & 11 & 7 \end{array} \tag{9.28}$$

Each element in the regular array corresponds to the difference (modulo q^2+q+1) between two members of the difference set:

$$a_{i,j} \equiv d_r - d_j \quad (\text{mod } q^2+q+1) \tag{9.29}$$

where

$$r \equiv i+j \quad (\text{mod } q+1)$$

For example, from the previous array,

$$a_{23} = d_1 - d_3 = 1-4 = 10 (\text{mod } 13). \tag{9.30}$$

From the basic property of a perfect difference set, it can be seen that all the integers from 1 to q^2+q appear only once in the regular array.

Various difference triangles can be constructed from the array by selecting q or fewer consecutive elements. In the example shown $q=3$, so a three element triangle can be formed. The elements $(2,6)$ and $(4,1)$ from the first row yield two triangles

$$\begin{matrix} 2 & 6 & & 4 & 1 \\ & 8 & & & 5 \end{matrix} \tag{9.31}$$

Robinson and Bernstein (1967) give tables of first rows of perfect difference triangles. We use their Table I to form a new self-orthogonal convolutional code. From the table the first row of a triangle which has the elements $(2,5,3,1)$ was chosen. This row was used to find the elements of the corresponding perfect difference set.

From Robinson and Bernstein (1967)

$$a_{i,j} = d_r - d_j \quad (\text{mod } q^2+q+1) \tag{9.32}$$

$$r = i+j \quad (\text{mod } q+1). \tag{9.33}$$

A perfect difference set consists of $(a_{1,1}, a_{1,2}, a_{1,3}, a_{1,4})$.
We let $i=1$, then

$$a_{1,j} = d_{1+j} - d_j. \tag{9.34}$$

Let $j=1$

$$a_{1,1} = d_2 - d_1 \tag{9.35}$$

But

$$a_{1,1} = 2.$$

Let
$$d_1 = 4$$

$$d_2 = a_{1,1} + d_1 = 2 + 4 = 6 \tag{9.36}$$

$$a_{1,2} = d_3 - d_2 \tag{9.37}$$

$$d_3 = a_{1,2} + d_2. \tag{9.38}$$

From the difference triangle $a_{1,2} = 5$

$$d_3 = 5 + 6 = 11. \tag{9.39}$$

Likewise
$$a_{1,4} = 1$$

$$a_{1,4} = q^2 + q + 1 - d_4 + d_1 \tag{9.40}$$

$$1 = 13 - d_4 + 4 = 17 - d_4 \tag{9.41}$$

$$d_4 = 17 - 1 = 16. \tag{9.42}$$

So the perfect difference set is $(4, 6, 11, 16)$ and from it the regular array can be completed:

for $i = 2, j = 1$

$$a_{2,1} = d_3 - d_1 = 11 - 4 = 7, \tag{9.43}$$

for

$$i = 2, j = 2$$

$$a_{2,2} = d_4 - d_2 = 16 - 6 = 10, \tag{9.44}$$

for

$$i = 2, j = 3$$

$$a_{2,3} = d_5 - d_3 = d_1 - d_3$$

$$= 4 - 11 = -7 = +6, \tag{9.45}$$

$$a_{2,4} = d_6 - d_4 + 1 = d_2 - d_4 + 1 = 6 - 16 + 1$$

$$= -9 = 4, \tag{9.46}$$

$$a_{3,1} = d_4 - d_1 = 16 - 4 = 12, \tag{9.47}$$

$$a_{3,2} = d_5 - d_2 = d_1 - d_2 = 4 - 6 = -2 = 11, \tag{9.48}$$

$$a_{3,5} = d_6 - d_3 = d_2 - d_3 = 6 - 11 = -5 = 8, \tag{9.49}$$

$$a_{3,4} = d_7 - d_4 + 1 = d_3 - d_4 + 1 = 11 - 16 + 1$$

$$= -4 = 9. \tag{9.50}$$

So that the array is

$$A = \begin{vmatrix} 2 & 5 & 3 & 1 \\ 7 & 10 & 6 & 4 \\ 12 & 11 & 8 & 9 \end{vmatrix}. \tag{9.51}$$

There are two difference triangles available from A

$$(1) = \begin{matrix} 2 & 5 & 3 \\ 7 & 10 & \\ 12 & & \end{matrix} \tag{9.52}$$

$$(2) = \begin{matrix} 5 & 3 & 1 \\ 10 & 6 & \\ 11 & & \end{matrix} \tag{9.53}$$

Triangle 2 has the smaller maximum number, so we will choose it.
This difference triangle creates the basic orthogonal convolutional code:

$$1\ 0\ 0\ 0\ 0\ 1\ 0\ 0\ 1\ 1\ 0\ 0\ 0\ 0\ 0$$

This gives rise to the code set:

$A_N =$

```
1   1 1
2   0 0 1 1
3   0 0 0 0 1 1
4   0 0 0 0 0 0 1 1
5   0 0 0 0 0 0 0 0 1  1
6   1 0 0 0 0 0 0 0 0  0  1  1
7   0 0 1 0 0 0 0 0 0  0  0  0  1  1
8   0 0 0 0 1 0 0 0 0  0  0  0  0  0  1  1
9   1 0 0 0 0 0 1 0 0  0  0  0  0  0  0  0  1  1
10  1 0 1 0 0 0 0 0 1  0  0  0  0  0  0  0  0  0  1  1
11  0 0 1 0 1 0 0 0 0  0  1  0  0  0  0  0  0  0  0  0  1  1
12  0 0 0 0 1 0 1 0 0  0  0  0  1  0  0  0  0  0  0  0  0  0  1  1
13  0 0 0 0 0 0 1 0 1  0  0  0  0  0  1  0  0  0  0  0  0  0  0  0  1  1
14  0 0 0 0 0 0 0 0 1  0  1  0  0  0  0  0  1  0  0  0  0  0  0  0  0  0  1  1
15  0 0 0 0 0 0 0 0 0  0  1  0  1  0  0  0  0  0  1  0  0  0  0  0  0  0  0  0  1  1
    1 2 3 4 5 6 7 8 9 10 11 12 13 14 15 16 17 18 19 20 21 22 23 24 25 26 27 28 29 30
```

$$(9.54)$$

9.4.2 Feedback Decoding

According to Morrissey (1970) probabilistic feedback decoding may be modeled as a four-state autonomous stochastic machine, in which the states are the same as the original sequential machine, and the transition probabilities between pairs of states are the sums of the probabilities of those input-noise vectors that take the machine from the first state to the second. The state behavior of the four-state autonomous stochastic machine is thus described by the following Markov transition matrix:

$$\pi = \begin{bmatrix} P_{00} & P_{01} & P_{02} & P_{03} \\ P_{10} & P_{11} & P_{12} & P_{13} \\ P_{20} & P_{21} & P_{22} & P_{23} \\ P_{30} & P_{31} & P_{32} & P_{33} \end{bmatrix}. \tag{9.55}$$

Yau (1965) in his Fig. 2 (shown here as Fig. 9.7) shows a transition diagram for the four-state stochastic machine, in which the I's are transitional probabilities. For our particular application we will use this state diagram with modification (Fig. 9.8) to produce one suited for convolutional feedback decoding.

Assign

$$I_1 = (1-p)^2 \tag{9.56}$$

$$I_2 = p(1-p) \tag{9.57}$$

$$I_3 = p(1-p) \tag{9.58}$$

$$I_4 = p^2 \tag{9.59}$$

then

$$I_1 + I_3 = 1 - p \tag{9.60}$$

$$I_2 + I_4 = p \tag{9.61}$$

$$I_1 + I_2 = 1 - p. \tag{9.62}$$

So that the state diagram becomes as shown in Fig. 9.9.

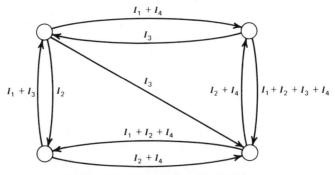

Figure 9.7. A 4-state transition diagram.

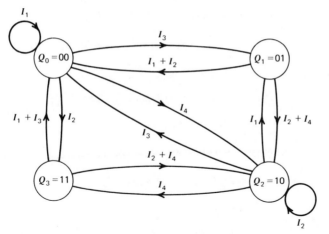

Figure 9.8. Modified transition diagram.

Next, we calculate the probability matrix per Parzen (1964):

$$P_m(i, j) = \sum_{k=1}^{r} P(i, k) P_{m-1}(k, j) \qquad (9.63)$$

from which

$$P_{11} = (1-p)^2, \qquad (9.64)$$

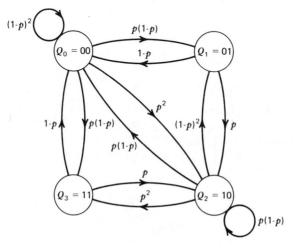

Figure 9.9. Final 4-state transition diagram.

$$P_{12} = P_0(1,1)P(1,2) + P_0(1,2)P(2,2) \qquad (9.65)$$

$$+ P_0(1,3)P(3,2) + P_0(1,4)P(4,2) \qquad (9.66)$$

$$= p(1-p),$$

$$P_{13} = P_0(1,1)P(1,3) + P_0(1,2)P(2,3)$$

$$+ P_0(1,3)P(3,3) = p^2, \qquad (9.67)$$

$$P_{14} = P_0(1,1)P(1,4) + P_0(1,2)P(2,4)$$

$$+ P_0(1,3)P(3,4) + P_0(1,4)P(4,4)$$

$$= p(1-p), \qquad (9.68)$$

$$P_{21} = P_1(2,1)P(1,1) = 1-p, \qquad (9.69)$$

$$P_{22} = P_1(2,2)P(2,2) = 0, \qquad (9.70)$$

$$P_{23} = P_1(2,3)P(3,3) = p, \qquad (9.71)$$

$$P_{24} = P_1(2,4)P(4,4) = 0, \qquad (9.72)$$

$$P_{31} = P_2(3,1)P(1,1) = p(1-p) \qquad (9.73)$$

$$P_{32} = P_2(3,2)P(4,4) = (1-p)^2 \qquad (9.74)$$

$$P_{33} = P_2(3,3)P(3,3) = p(1-p) \qquad (9.75)$$

$$P_{34} = P_2(3,4)P(4,4) = p^2 \qquad (9.76)$$

$$P_{41} = P_3(4,1)P(1,1) = 1-p \qquad (9.77)$$

$$P_{42} = P_3(4,2)P(2,2) = 0, \qquad (9.78)$$

$$P_{43} = P_3(4,3)P(3,3) = p, \qquad (9.79)$$

$$P_{44} = P_3(4,4)P(4,4) = 0. \qquad (9.80)$$

With each state q_i is associated a probability of error P_{qi} equal to the sum of the probabilities of those noise inputs $[e(u+m), \ \varepsilon(u+m)] = [e(u+1), \ \varepsilon(u+1)]$ that cause $e^*(u) \neq e(u)$, given that the system is in state q_i at time $u+m = u+1$. One can find by proper summation that

$$P_{q0} = 0 \qquad (9.81)$$

$$P_{qi} = 2p(1-p) \qquad (9.82)$$

$$P_{q2} = 1 \qquad (9.83)$$

$$P_{q3} = 2p(1-p). \qquad (9.84)$$

The probability of error in feedback decoding (FD) can be written as

$$P_{\text{FD}}(u) = \sum_{i=0}^{3} W_i(u+1)P_{qi}. \tag{9.85}$$

A sufficient condition for P_{FD} to exist is that $\lim u\to\infty W(u) = W = [W_0, W_1, W_2, W_3]$, the steady-state probability of state q_i. If this is the case then

$$P_{\text{FD}} = \lim_{u\to\infty} P_{\text{FD}}(u) = \lim_{u\to\infty} \sum_{i=0}^{3} W_i(u+1)P_{qi}$$

$$= \sum_{i=0}^{3} W_i P_{qi}. \tag{9.86}$$

W does exist for this case (Morrissey, 1970), and may be calculated by solving $W\pi = W$ for W with the added constraint that

$$\sum_{i=0}^{3} W_i = 1. \tag{9.87}$$

The desired solution is (Ash, 1965):

$$W_0 = \frac{1 - 2p + 4p^2 - p^3}{1 - 3p^2 + 2p^3} \tag{9.88}$$

$$W_1 = \frac{p - 3p^3 + 2p^4}{1 - 3p^2 + 2p^3} \tag{9.89}$$

$$W_2 = \frac{3p^2 - 2p^3}{1 - 3p^2 + 2p^3} \tag{9.90}$$

$$W_3 = \frac{p - 3p^2 + 5p^3 - 2p^4}{1 - 3p^2 + 2p^3} \tag{9.91}$$

therefore,

$$P_{\text{FD}} = \sum_{i=0}^{3} W_i P_i = \frac{7p^2 - 12p^3 + 10p^4 - 4p^5}{1 - 3p^2 + 2p^3}. \tag{9.92}$$

In this equation for error probability the symbol p represents the error probability for the BSC assuming the addition of white Gaussian noise to the signal. When coherent reception is assumed in a system that uses PRN sampling and PPM, in conjunction with convolutional coding

$$p = \left[\frac{1}{12\pi M^2 B^2 T^2 (\text{SNR})} \right]^{1/2} \exp\left[-2M^2 B^2 T^2 (\text{SNR}) \right], \qquad (9.93)$$

which includes the PRN/PPM/FM improvement factor. The bit-error rate of the whole PRN/PPM/CONVOLUTIONAL/FM system is written as

$$P_B = \sum_{j=3}^{N} \binom{N}{j} P_{\text{FD}}^j (1 - P_{\text{FD}})^{N-j}. \qquad (9.94)$$

9.4.3 Implementation

In suggesting a method for implementing the canonical self-orthogonal code (CSOC), into a useful communication system, it was decided to use FM as a modulation type. As a preliminary coding step it was decided that PPM encoding will precede the convolutional encoder. The PPM encoding is implemented by use of pseudorandom sampling. In fact the the pulses that sample the video is pseudo random, as in Chapter 6. Ordinarily, PPM sampling pulses are equally spaced and come from an implemented clock. In this case the clock pulses drive a PRN generator, the output of which is used to sample the video and to generate the PPM at a PRN rate. The PRN/PPM pulses constitute the message bit stream to the convolutional encoder.

We assume a video baseband of 4 MHz and a sampling rate of six times this. The Nyquist rate is the well known two, but this assumes equally spaced pulses. It is easy to see that if the pulses are unequally spaced a higher sampling rate is needed. For this case the sampling rate is $4 \times 6 = 24$ MHz. However, we are dealing with TV and must synchronize sampling pulses and TV raster line pulses. In the standard TV system line sync is at a 15,750-Hz rate. The sampling pulses must be a multiple of 15,750 Hz, and a multiplying factor of 1400 was chosen. This gives a basic sampling rate of 22.05 Mbits/sec.

However, due to the fact that a convolutional encoder requires interleaving of message and parity bits, alternate portions of the PRN/PPM message are lost. Therefore, the sampling rate will be set at 44.1 Mbits/sec, in order to compensate for the lost message bits. A rate $R = \frac{1}{2}$ convolutional encoder is assumed, with time slot sharing, to give a transmitted baseband of 44.1 MHz. As mentioned previously this is a broad baseband. In such cases, frequency modulation cannot be done by direct FM means, so that the phase-splitting methods using indirect FM will be used. The transmitter is as shown in Fig. 9.10.

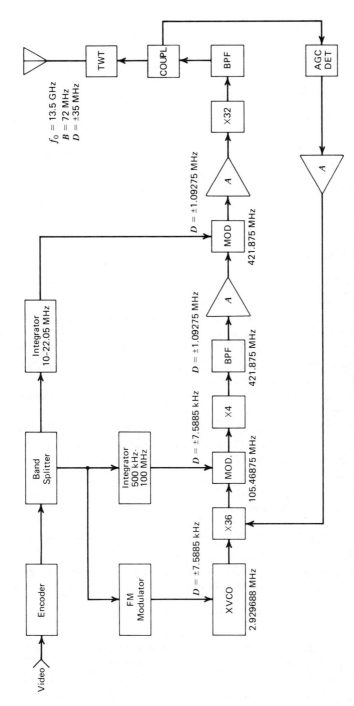

Figure 9.10. Transmitter diagram for a PRN/PPM/CONVOLUTIONAL/FM system.

277

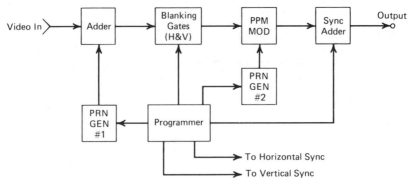

Figure 9.11. Diagram for encoder unit from video input to convolutional encoder.

9.4.4 Encoding

As mentioned previously the encoding is done in three basic steps, PRN, PPM, and convolutional. The first two steps are shown in Fig. 9.11.

Of course, the encoder needs a timing generator or programmer. The programmer creates horizontal and vertical video blanking gates, and drive for the PRN generator 2, which in turn drives the PPM modulator. The PPM output pulses drive the sync adder. The programmer creates the horizontal and vertical sync pulses, which are added to the video.

Also, we again make use of Roberts' modulation (PRN generated noise pulses) to mask the video signal. The PRN generator is driven by the programmer. The two PRN generators 1 and 2 are described in Chapters 5 and 6. The programmer is shown in Fig. 9.12 and is also described in Chapters 5 and 6. The PRN/PPM encoder output furnishes message bits to the convolutional encoder.

9.4.5 The Convolutional Encoder

The encoder, which is shown in Fig. 9.13, is basically a 5-bit feedback shift register having the polynomial, $0 = I + D_2 + D_4 + D_5$. This has three feedback connections, and gives a period of 15 rather than 31. When this PRN generator is initialized by feeding a binary 8, we get the following output:

$$01000[\,111101011001000\,]111101 \qquad (9.95)$$

We use two identical feedback registers, both driven by a 44.1-MHz clock, plus following logic circuitry. This timing is shown in Fig. 9.14. The upper register is preceded by two delay bits to give shift (a); the lower register the normal sift (b). The logic shown generates the 5, 3, 1 self-orthogonal convolutional code described above.

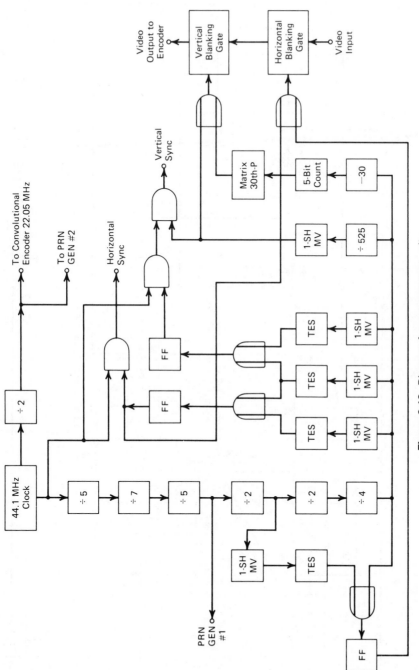

Figure 9.12. Diagram for programmer unit.

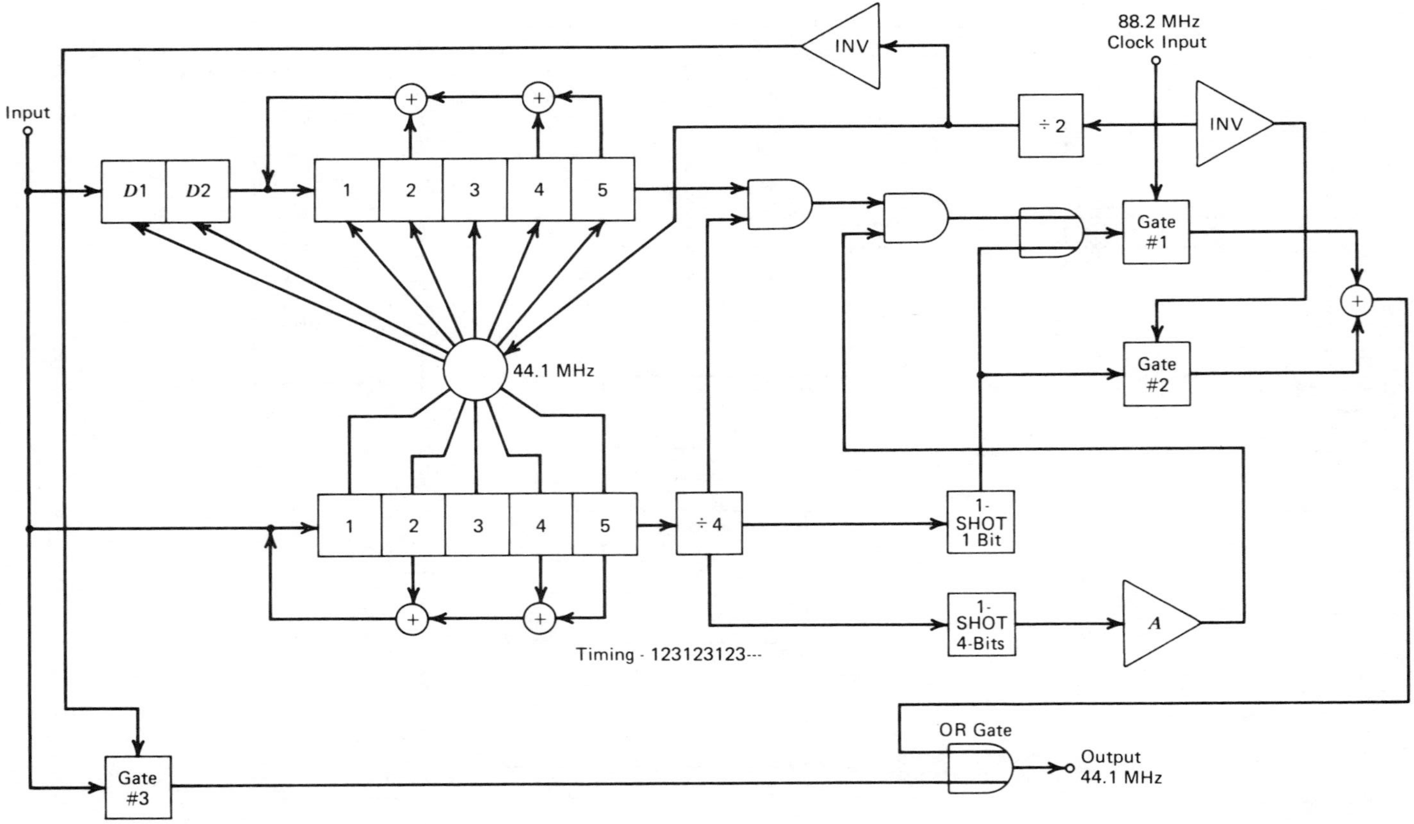

Figure 9.13. Diagram for the self-orthogonal convolutional encoder.

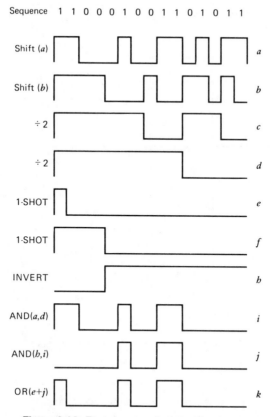

Figure 9.14. Encoder waveform synthesis.

The gates 1 and 2 alternate at a 88.2-MHz rate to form a 30-bit interlaced parity bit stream. The parity bit stream is interlaced with gate 3 to form the convolutional code. The receiver will now be described.

9.4.6 The Receiver

The receiver is the same as described for the PRN/PPM/FM system in Chapter 6.

9.4.7 The Decoder Unit

The decoder unit as shown in Fig. 9.15 uses an integrate-and-dump waveform regeneration circuit followed by the convolutional decoder, and then by PPM detection. A 44.1-MHz pulse generator is divided down to the 15,750-MHz horizontal sync frequency to obtain phase lock. The phase-locked output of

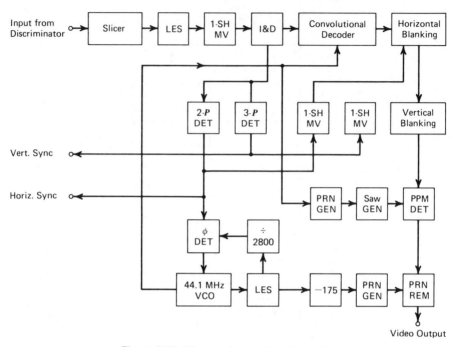

Figure 9.15. Diagram for the decoder unit.

the slaved clock drives the convolutional decoder and the two PRN generator replicas of those in the transmitter. A PRN removal circuit is used to remove the PRN masking the video, just prior to the decoder output.

9.4.8 The Convolutional Decoder Subunit

The convolutional decoder is shown as a section of the general decoder unit of Fig. 9.15. It is described separately and shown in Fig. 9.16.

The input switch commutates alternately between the information and the parity bits. The information bits are sent to a replica of the encoder, which calculates the syndrome of the signal. The parity bits are routed to the memory buffer. The resultant syndrome is sent to the syndrome register, which in turn feeds the threshold detector. Feedback is routed from the detector output, back through the syndrome register.

Errors are corrected by feeding the detector output to an EXCLUSIVE-OR gate; the other input of which is the calculated syndrome.

In greater detail, the decoder estimates one block of the error vector in each decoding cycle. The estimate is subtracted from the corresponding code block to form the estimate of the message block. The first block is decoded using the first N positions of the syndrome S. Next, S is recomputed using the

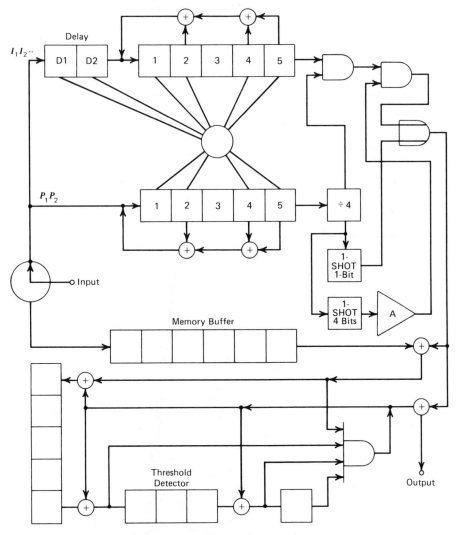

Figure 9.16. Diagram for feedback decoding of the convolutional code.

decoded version of the first block. Positions $m+1$ through $N+m$ of S are used to decode the second block, utilizing the same algorithm as for the first block. In general, an arbitrary block is decoded on the basis of N positions of S, where S is computed using the decoded version of prior blocks. At each step only N positions of S are needed.

The use of the decoded versions of prior blocks to decode, results in a feedback detector, and its operation is referred to as feedback decoding. This decoding method is used for the CSOCs, the implementation of which was described previously.

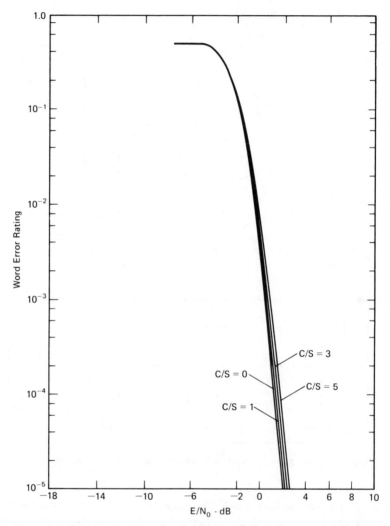

Figure 9.17. Channel-sharing performance for the PRN/PPM/convolutional/FM coding system.

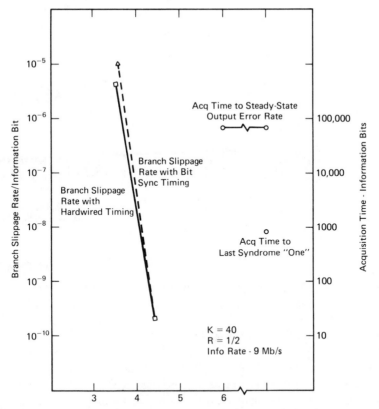

Figure 9.18. Experimental results obtained from a similar convolutional system per Clark and Perkins (1972).

9.4.9 Channel-Sharing Performance

The channel-sharing performance of the PRN/PPM/Convolutional/FM digital TV system is based on $Q = 0.0115$, for $U = 12$, $d = 0.03$, and $\lambda_c = 0.319$. The performance is shown in Fig. 9.17. It is excellent for all the QC/S ratios shown. It shows about 2-dB SNR for a bit-error rate of 10^{-5}. This compares favorably with the performance results obtained by Clark and Perkins (1972), as shown by their Fig. 16, which we reproduce here as Fig. 9.18. They show about 3.5 dB for a 10^{-5} bit-error rate. They used a maximum likelihood convolutional decoder. The added features of PRN/PPM/FM account for the difference in results.

chapter 10

The Reed–Solomon Codes

introduction

The Bose, Chaudhuri, Hocquenghem (BCH) codes were discovered by Hocquenghem (1959) and independently by Bose and Chaudhuri (1960). They constitute a class of cyclic codes which have powerful error-correcting properties and simple decoding algorithms. The most common examples of BCH codes are binary, but the alphabet can equally well be the elements from an arbitrary Galois field, say $GF(q)$. The generator polynomials for these codes are defined in terms of some extension field of $GF(q)$, say $GF(q^m)$. Let α be an element of $GF(q^m)$ of the multiplicative order N, and for arbitrary integers $r \geqslant 0$ and d, $2 \leqslant d \leqslant N$, let $f_r(D), f_{r+1}(D), \ldots, f_{r+d-2}(D)$ be the minimal polynomials of $\alpha^r, \alpha^{r+1}, \ldots, \alpha^{r+d-2}$. For each choice of the preceding parameters (that is, q, m, α, r, and d) there is a BCH code defined; its generator polynomial is defined as

$$g(D) = \mathrm{LCM}\left[\, f_r(D), f_{r+1}(D), \ldots, f_{r+d-2}(D)\,\right]. \tag{10.1}$$

The block length of the code is defined to be N, the multiplicative order of α. Since each of the elements $\alpha_r \cdots \alpha_{r+d-2}$ are roots of $D^N - 1$, each of the polynomials $f_r(D), \ldots, f_{r+d-2(D)}$ divides $D^N - 1$, and thus $g(D)$ divides $D^N - 1$, as required for a cyclic code.

The Reed-Solomon (RS) codes are a subclass of the BCH codes, in which the extension field in which α is defined is the same as the symbol field for the code letters. In this case, the minimal polynomial of α^i is simply $D - \alpha^i$, so we have

$$g(D) = \prod_{i=r}^{r+d-2} (D - \alpha^i). \tag{10.2}$$

The Reed-Solomon code has a minimum distance of the number of parity check digits plus 1. A RS (N, k) code has a code length of N symbols, with k message symbols, and $N - k$ parity check symbols (or $d - 1$). This code has the largest minimum distance possible. They can be designed for very low bit-error rates, dependent upon the numbers of errors it is designed to correct.

10.1 RS CODING EXAMPLE

We use the $(15, 7)$ RS code as an example. It has 8 parity check symbols and a minimum distance of 9. This code can be used in digital TV with excellent results.

The parity symbol generator polynomial for this $(15, 7)$ code is given by

$$g(X) = (X + \alpha)(X + \alpha^2)(X + \alpha^3)(X + \alpha^4)(X + \alpha^5)(X + \alpha^6)(X + \alpha^7)(X + \alpha^8),$$
(10.3)

which can be written as

$$g(X) = \alpha^5 + \alpha^8 X + \alpha^9 X^2 + \alpha X^3 + \alpha^9 X^4 + \alpha^5 X^5 + \alpha^5 X^6 + \alpha^{10} X^7 + X^8. \quad (10.4)$$

This $GF(2^4)$ code consists of the elements $0, 1, \alpha, \alpha^2, \ldots, \alpha^{14}$; 0 and 1 are a subfield $GF(2)$, while $0, 1, \alpha^5$, and α^{10} are a subfield of $GF(2^2)$. When one assumes the irreducible [over $GF(2)$] polynomial $X^4 + X + 1$, the derived code table is as shown in Table 10.1.

Table 10.1

Symbol	Binary Value	Relationship
0	0 0 0 0	0
α^0	1 0 0 0	1
α	0 1 0 0	α
α^2	0 0 1 0	α^2
α^3	0 0 0 1	α^3
α^4	1 1 0 0	$\alpha + 1$
α^5	0 1 1 0	$\alpha^2 + \alpha$
α^6	0 0 1 1	$\alpha^3 + \alpha^2$
α^7	1 1 0 1	$\alpha^3 + \alpha + 1$
α^8	1 0 1 0	$\alpha^2 + 1$
α^9	0 1 0 1	$\alpha^3 + \alpha$
α^{10}	1 1 1 0	$\alpha^2 + \alpha + 1$
α^{11}	0 1 1 1	$\alpha^3 + \alpha^2 + \alpha$
α^{12}	1 1 1 1	$\alpha^3 + \alpha^2 + \alpha + 1$
α^{13}	1 0 1 1	$\alpha^3 + \alpha^2 + 1$
α^{14}	1 0 0 1	$\alpha^3 + 1$

10.2   DECODING ALGORITHMS OF THE MASSEY TYPE

There are several step-by-step decoding algorithms available in the literature. One of the easiest to implement is by Massey (1965). He distinguishes between errors in the information positions, and the errors in the parity check positions by defining

$$e_i(X) = e_r(X)^r + e_{r+1}(X)^{r+1} + \cdots + e_{n-1}(X)^{n-1} \tag{10.5}$$

and

$$e_p(X) = e_0 + e_i(X) + \cdots + e_{r-1}(X)^{r-1}. \tag{10.6}$$

The syndrome is

$$S(X) = S_0 + S_1 X + \cdots + S_{r-1} X^{r-1} \tag{10.7}$$

and is the remainder when $r(X)$ is divided by $g(X)$ and hence can always be formed at the receiver. Since

$$r(X) = f(X) + e(X),$$

and since $g(X)$ divides $f(X)$, it follows that

$$S(X) = Rg[e(X)] \tag{10.8}$$

and

$$S(X) = Rg[e_i(X)] + e_p(X) \tag{10.9}$$

and further

$$S_j = S(\alpha^{m_0 + j - 1}) \qquad j = 1, 2, \ldots. \tag{10.10}$$

Let j be the number of errors to be corrected by the code. A $j \times j$ matrix in S can be formed:

$$M_j = \begin{bmatrix} 1 & 0 & 0 & 0 & \cdots & 0 \\ S_2 & S_1 & 1 & 0 & \cdots & 0 \\ -S_{2j-2} & S_{2j-3} & S_{2j-4} & S_{2j-5} & \cdots & S_{j-1} \end{bmatrix} \tag{10.11}$$

If one removes row 1 and column 1, the $(j-1) \times (j-1)$ matrix is equally valid

$$L_j = \begin{bmatrix} S_1 & 1 & 0 & 0 & \cdots & 0 \\ S_3 & S_2 & S_1 & 1 & \cdots & 0 \\ S_{2j-1} & S_{2j-2} & S_{2j-3} & S_{2j-4} & \cdots & S_j \end{bmatrix}. \tag{10.12}$$

The matrix may also be set up as

$$N_j = \begin{bmatrix} S_1 & S_2 & \cdots & S_j \\ S_2 & S_3 & \cdots & S_{j+1} \\ S_j & S_{j+1} & \cdots & S_{2j-1} \end{bmatrix}. \tag{10.13}$$

The forming of syndromes can best be portrayed by use of an example from p. 287 of Peterson and Weldon (1972).

Suppose that an error equal in value to α^7 occurs in the location corresponding to the location α^3 and an error α^{11} in the position corresponding to α^{10}. Then

$$S_1 = \alpha^7 \alpha^3 \oplus \alpha^{11} \alpha^{10} = \alpha^{10} \oplus \alpha^{21}$$

$$= \alpha^{10} \oplus \alpha^6 (\bmod\ 15) = \alpha^7,$$

$$S_2 = \alpha^7 (\alpha^3)^2 \oplus \alpha^{11} (\alpha^{10})^2 = \alpha^7 \alpha^6 \oplus \alpha^{11} \alpha^{20}$$

$$= \alpha^{13} \oplus \alpha (\bmod\ 15) = \alpha^{12},$$

$$S_3 = \alpha^7 \alpha^9 \oplus \alpha^{11} \alpha^{30} = \alpha^6,$$

$$S_4 = \alpha^7 \alpha^{12} + \alpha^{11} \alpha^{40} = \alpha^{12},$$

$$S_5 = \alpha^7 \alpha^{15} + \alpha^{11} \alpha^{50} = \alpha^{14},$$

$$S_6 = \alpha^7 \alpha^{18} + \alpha^{11} \alpha^{60} = \alpha^{14}. \tag{10.14}$$

The next decoding step is to find the error locations.

Suppose that $V \leqslant t_0$ errors actually occur. These are described by V pairs (Y_i, X_i) for which neither Y_i nor X_i is 0. Then let the equation

$$(X + X_1)(X + X_2) \cdots (X + X_V) = X^V + \sigma_1 X^{V-1} + \cdots + \sigma_{V-1} X + \sigma_V \tag{10.15}$$

from which

$$S_j \sigma_V + S_{j+1} \sigma_{V-1} + \cdots + S_{j+V-1} \sigma_1 + S_{j+V} = 0. \tag{10.16}$$

The decoder is confronted with the problem of determining from (10.16) the σ_i, $1 \leqslant i \leqslant V \leqslant t_0$, given the S_j, $1 \leqslant j \leqslant 2t_0$. This can be accomplished in an iterative manner. At the nth step in the iteration the decoder considers only the first n power sums and attempts to determine a set of l_n values $\sigma_i^{(n)}$, such that the $n - l_n$ equations are

$$S_n + S_{n-1}\sigma_1^{(n)} + \cdots + S_{n-ln}\sigma_{ln}^{(n)} = 0$$

$$S_{n-1} + S_{n-2}\sigma_1^{(n)} + \cdots + S_{n-ln-1}\sigma_{ln}^{(n)} = 0$$

$$\vdots$$

$$S_{ln+1} + S_{ln}\sigma_1^{(n)} + \cdots + S_1\sigma_{ln}^{(n)} = 0 \tag{10.17}$$

when l_n is as small as possible, and the number of equations minimized. [The superscript (n) serves to distinguish the solution at various stages.]

Equation (10.17) can be represented by

$$\sigma(X)^{(n)} = \sigma_0^{(n)} + \sigma_1^{(n)}X + \sigma_2^{(n)}X + \cdots + \sigma_{ln}^{(n)}X^{ln}$$

where

$$\sigma_0^{(n)} = 1. \tag{10.18}$$

Now, suppose that at the nth stage the decoder has determined $\sigma^{(n)}(X)$, with minimal l_n such that (10.18) holds. At the $(n+1)$th stage, the decoder seeks to find the polynomial $\sigma^{(n+1)}(X)$ of the lowest degree such that the equations

$$\sum_{i=0}^{ln+1} S_{j-i}\sigma_i^{(n+)} = 0; \qquad ln+1^{+1} \leqslant j \leqslant n+1 \tag{10.19}$$

hold true.

Define the nth discrepancy d_n as

$$S_{n+1} + S_n\sigma_1^{(n)} + \cdots + S_{n+1-ln}\sigma_{ln}^{(n)} = d_n \tag{10.20}$$

We now return to the example. By definition we set

$$\sigma^{(-1)}(X) = 1, \qquad l_{-1} = 0, \qquad d_{-1} = 1 \tag{10.21}$$

$$\sigma^{(0)}(X) = 1, \qquad l_0 = 0, \qquad d_0 = S_1 = \alpha^7 \neq 0 \tag{10.22}$$

$$\sigma^{(1)}(X) = 1 + S_1 X = 1 + \alpha^7 X, \qquad l = 1 \tag{10.23}$$

$$d_1 = S_2 + S_1\alpha^7 = \alpha^{12} + \alpha^7\alpha^7 = \alpha^{12} + \alpha^{14} = \alpha^5. \tag{10.24}$$

A minimal solution at stage $n+1$ (the next stage) $\sigma^{(n+1)}(X)$, where

$$\sigma^{(n+1)}(X) = \sigma^{(n)}(X) \qquad \text{and} \qquad l_{n+1} - l_n, \qquad d_n = 0 \tag{10.25}$$

$$\sigma^{(n+1)}(X) = \sigma^n(X) - d_n d_m^{-1} X^{(n-m)}\sigma^{(m)}(X) \tag{10.26}$$

and

$$l_{n+1} = \max(l_n, l_m + n - m); \qquad d_n \neq 0. \tag{10.27}$$

So we try making $n = 1$, $m = 0$

$$\sigma^{(2)}(X) = 1 + \alpha^7 X + \left(\frac{\alpha^5}{\alpha^7}\right) X = 1 + (\alpha^7 + \alpha^{13}) X$$

$$= 1 + \alpha^5 X. \tag{10.28}$$

$$l_2 = \max(l_1, 1 - 0 + l_0) = 1 \tag{10.29}$$

$$d_2 = S_3 + S_2 \alpha^5 = \alpha^6 + \alpha^{12} \alpha^5 = \alpha^6 + \alpha^2 = \alpha^3. \tag{10.30}$$

We can now let $n = 2$, $m = 1$

$$\sigma^{(3)}(X) = \sigma^{(2)}(X) + \left(\frac{d_2}{d_1}\right) X \sigma^1(X)$$

$$= (1 + \sigma^5 X) + \left(\frac{\alpha^3}{\alpha^5}\right) X (1 + \alpha^7 X)$$

$$= (1 + \alpha^5 X) + \alpha^{13} X (1 + \alpha^7 X)$$

$$= 1 + \alpha^5 X + \alpha^{13} X + \alpha^5 X^2$$

$$= 1 + \alpha^7 X + \alpha^5 X \tag{10.31}$$

$$l_3 = \max(l_2, l_m + 2 - m) = 2 \tag{10.32}$$

$$d_3 = S_4 + S_3 \alpha^7 + S_2 \alpha^5 = \alpha^{12} + \alpha^6 \alpha^7 + \alpha^{12} \alpha^5$$

$$= \alpha^{12} + \alpha^{13} + \alpha^2 = \alpha^5 \neq 0. \tag{10.33}$$

So we try $n = 3$, $m = 2$, $l_4 = 2$

$$\sigma^{(4)}(X) = (1 + \alpha^5 X + \alpha^{11} X^2) + \left(\frac{d_2}{d_3}\right) + (1 + \alpha^5 X)$$

$$= (1 + \alpha^5 X + \alpha^{11} X^2) + \alpha^{14} X (1 + \alpha^5 X)$$

$$= 1 + X(\alpha^5 + \alpha^{14}) + X^2(\alpha^{11} + \alpha^4)$$

$$= 1 + \alpha^{12} X + \alpha^{13} X^2 \tag{10.34}$$

$$d_4 = d_5 = S_5 + S_4 \sigma_1^{(4)} + S_3 \sigma_2^{(4)}$$

$$= \alpha^{14} + \alpha^{12} \alpha^{12} + \alpha^6 \alpha^{13}$$

$$= \alpha^{14} + \alpha^9 + \alpha^4 = 0. \tag{10.35}$$

Since $d_4 = d_5 = 0$ it is assumed that $\sigma^{(4)}X$ is the solution:

$$\sigma^{(4)}(X) = 1 + \alpha^{12}X + \alpha^{12}X^2. \tag{10.36}$$

Let

$$X = \alpha^3. \tag{10.37}$$

The inverse of $X = \alpha^3$ or $X = \alpha^{15-3} = \alpha^{12}$ so that

$$X = \alpha^{12} \tag{10.38}$$

$$\sigma^{(4)}(X) = 1 + \alpha^{12}\alpha^{12} + \alpha^{13}\alpha^{24} = 1 + \alpha^9 + \alpha^7 = 0. \tag{10.39}$$

The inverse of $X = \alpha^{10}$ is $X = \alpha^5$

$$\sigma^{(4)}(X) = 1 + \alpha^{12}\alpha^5 + \alpha^{13}\alpha^{10} = 1 + \alpha^2 + \alpha^8 = 0. \tag{10.40}$$

Thus it was proved that the error locations were at $X = \alpha^3$ and $X = \alpha^{10}$.

Massey (1965) shows a step-by-step decoder for the RS (15,7) code. This is shown here in Fig. 10.1. The following set of notes describes the operation of this decoder:

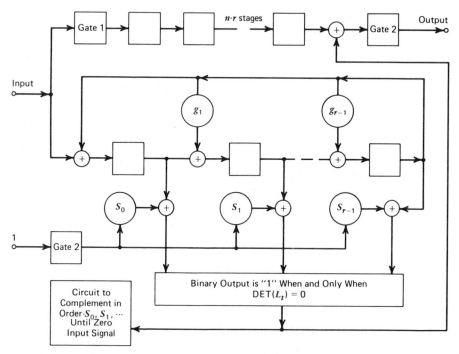

Figure 10.1. The Massey version of a Reed-Solomon decoder.

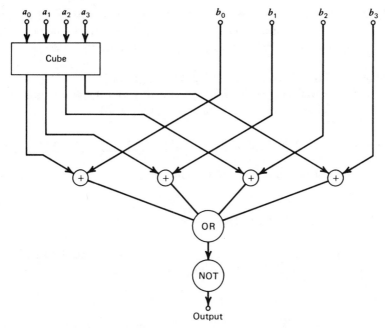

Figure 10.2. The Massey determinant solver unit.

Notes to Fig. 10.1:

Note 1: $(a_0, a_{11}, \ldots, a_{r-1})$ is the syndrome for $e = (0, 0, \ldots, 0, 1)$.

Note 2: Gate 1 is energized while $n - r$ received information digits are read into the upper shift register and is de-energized thereafter. Upper shift register does not shift again until "decode" order is given.

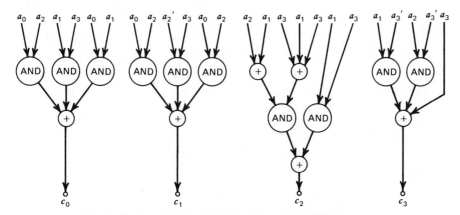

Figure 10.3. Subsection of Massey decoder called the cube box.

> *Note 3:* Complementing circuit begins to function after all n received
> digits are read into the decoder. It ceases to function when it
> receives a "0" input signal at which time the "decode" order is
> given.
>
> *Note 4:* Gates 2 are energized by the decode order.
>
> *Note 5:* Lower shift register, after all n received digits are read in, does
> not shift again until "decode" order is given.

The determinant solving portion of this figure can be implemented as
shown by Massey (1965) and as shown here in Fig. 10.2.

The details of the cube box are shown in Fig. 10.3.

10.3 THE FORNEY ALGORITHM

Forney (1965) makes use of both unknown errors and known location
erasures. He forms eight parity check syndrome equations from errors and
erasures:

$$S_m = \sum_{j=1}^{t} e_j X_j^{rm} + \sum_{k=1}^{S} d_k Y_k^m \qquad m_0 \leqslant m_0 + d - 1. \tag{10.41}$$

He then forms modified syndromes based on the known erasure location
and value:

$$T_n = \sum_{j=1}^{t} E_j X_j^n \qquad 0 \leqslant n \leqslant d - S - 2 \tag{10.42}$$

where

$$E_j = e_j X_j^{m_0} \sigma_d(X_j) \tag{10.43}$$

and

$$\sigma_d(Y_k) = 0. \tag{10.44}$$

The matrix formed is

$$M_t = \begin{bmatrix} T_{2t_0}-2 & T_{2t_0}-3 & \cdots & T_{2t_0}-t-1 \\ T_{2t_0}-3 & T_{2t_0}-4 & \cdots & T_{2t_0}-t-2 \\ \vdots & & & \\ T_{2t_0}-t-1 & T_{2t_0}-t-2 & \cdots & T_{2t_0}-2t \end{bmatrix} \tag{10.45}$$

If $t \leqslant t_0$, then the $t_0 \times t_0$ matrix M has the rank t, where $2t_0 < d - S$, and

$$M = \begin{bmatrix} T_{2t_0}-2 & T_{2t_0}-3 & \cdots & T_{t_0}-1 \\ T_{2t_0}-3 & T_{2t_0}-4 & \cdots & T_{t_0}-2 \\ \vdots & & & \\ T_{t_0}-1 & T_{t_0}-2 & \cdots & T_0 \end{bmatrix}. \tag{10.46}$$

Forney (1965) gives an illustrative example assuming the RS (15,7) code. He assumes errors of value α^4 in position 1 and α in position 4, and erasures of value 1 in position 2 and α^7 in the third position.

$$\left(e_1 = \alpha^4, X_1 = \alpha^{14}, e_2 = \alpha, X_2 = \alpha^{11}, d_1 = 1, Y_1 = \alpha^{13}, d_2 = \alpha^7, Y_2 = \alpha^{12}\right)$$

$$\tag{10.47}$$

The eight parity check equations are formed by using erasures as if they were errors. He gets

$$S_1 = \alpha^{14}, \quad S_2 = \alpha^{13}, \quad S_3 = \alpha^5, \quad S_4 = \alpha^6, \quad S_5 = \alpha^9,$$

$$S_6 = \alpha^{13}, \quad S_7 = \alpha^{10}, \quad S_8 = \alpha^4. \tag{10.48}$$

Next, modified syndromes are formed assuming

$$\sigma_{d0} = 1 \tag{10.49}$$

$$\sigma_{d1} = Y_1 + Y_2 = \alpha^{13} + \alpha^{12} = \alpha \tag{10.50}$$

$$\sigma_{d2} = Y_1 Y_2 = \alpha^{13}\alpha^{12} = \alpha^{10}, \tag{10.51}$$

then

$$T_0 = S_3 + \sigma_{d1}S_2 + \sigma_{d2}S_1$$

$$= \alpha^5 + \alpha\alpha^{13} + \alpha^{10}\alpha^{14} = \alpha^8 \tag{10.52}$$

$$T_1 = S_4 + \sigma_{d1}S_3 + \sigma_{d2}S_2 = \alpha^8 \tag{10.53}$$

$$T_2 = 0, \quad T_3 = \alpha^3, \quad T_4 = \alpha^{13}, \quad T_5 = \alpha^3. \tag{10.54}$$

The matrix equations become

$$T_5 = T_4\sigma_{e1} + T_3\sigma_{e2} + T_2\sigma_{e3} \tag{10.55}$$

$$T_4 = T_3\sigma_{e1} + T_2\sigma_{e2} + T_1\sigma_{e3} \tag{10.56}$$

$$T_3 = T_2\sigma_{e1} + T_1\sigma_{e2} + T_0\sigma_{e3} \tag{10.57}$$

or

$$\alpha^3 = \alpha^{13}\sigma_{e1} + \alpha^3\sigma_{e2} \tag{10.58}$$

$$\alpha^{13} = \alpha^3\sigma_{e1} \qquad + \alpha^8\sigma_{e3} \tag{10.59}$$

$$\alpha^3 = \qquad + \alpha^8\sigma_{e2} + \alpha^8\sigma_{e3}. \tag{10.60}$$

This is a Galois field matrix of three equations and three unknowns. Solve for σ_{e1}, σ_{e2}, and σ_{e3}. A solution is found by using the Gauss-Jordan triangulation to give

$$\alpha^5 = \sigma_{e1} + \alpha^5\sigma_{e2}$$

$$\alpha^{10} = \qquad \sigma_{e2} + \sigma_{e3} \tag{10.61}$$

$$0 = 0.$$

This is easily solved to give $\sigma_{e1} = \alpha^{10}$, $\sigma_{e2} = \alpha^{10}$, and $\sigma_{e3} = 0$.
We now use the error locater polynomial

$$\sigma_e(X) = X^2 + \sigma_{e1}X + \sigma_{e2} = X^2 + \alpha^{10}X + \alpha^{10} \tag{10.62}$$

when $X = \alpha^{14}$ and $X = \alpha^{11}$

$$\sigma_e(X) = 0. \tag{10.63}$$

This proves that the errors were located at $X_1 = \alpha^{14}$ and $X_2 = \alpha^{11}$, or actually at the inverse of these values

$$X_1 = \alpha^{15-14} = \alpha \text{ (position 1)} \tag{10.64}$$

$$X_2 = \alpha^{15-11} = \alpha^4 \text{ (position 4)}. \tag{10.65}$$

This example used by Forney was for two errors and two erasures. There is no theoretical reason for limiting $t = 2$, even though some say this. The Forney system should be able to correct three errors with one erasure. The proof of this statement follows.

10.4 AN EXTENSION OF THE FORNEY SYSTEM (A NEW ALGORITHM)

A computer program can be set up using Galois field arithmetic. A Fortran program can be written for the Gauss-Jordan triangulation, using Galois field arithmetic.

We first assume that $\alpha = 2$ and work with powers of 2, using modulo-15 arithmetic and EXCLUSIVE-OR processing in place of addition and subtraction, since this is modulo-2 for both. Such a program has been written and run on the Univac 1108 computer. The results prove conclusively that three errors can be corrected consistently using (15, 7) RS coding. The computer program is shown in the Appendix along with the results. This computer program constitutes what is known in the industry as software, or as a software algorithm. The use of Galois field arithmetic in a computer program constitutes a new algorithm for decoding a RS (15, 7) code.

While this computer programming might be said to be of academic interest only, this is not the case. A special computer having a capability of handling the written Fortran program can be implemented and used, as we now term it, as software.

Such an approach is outlined in the section on implementation of the (15, 7) RS decoding algorithm.

10.5 IMPLEMENTATION OF THE RS (15, 7) CODING MODEM

RS (15, 7) coding can be used for digital TV. Some D/A conversion is needed between video analog signals and the RS encoder. PPM coding is chosen. This is shown in Fig. 10.4. The video signal is sampled at a rate of 12.6 MHz and PPM modulated. This PPM signal is fed into the RS encoder. This encoder also provides for Roberts' modulation of the video signal, as is done in the previous systems. Provision for video blanking gates at the horizontal and vertical sync positions is made prior to PPM modulation. The TV sync is added after PPM modulation. The programmer is the heart of the unit, for it contains the clock and the synthesis of video sync signals. This programmer is detailed in Fig. 10.5. It need not be described in detail here.

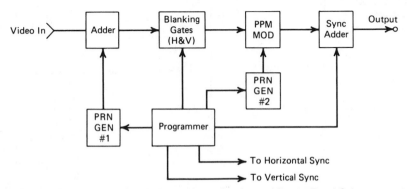

Figure 10.4. Diagram for the PPM encoder unit—from video to Reed-Solomon encoder.

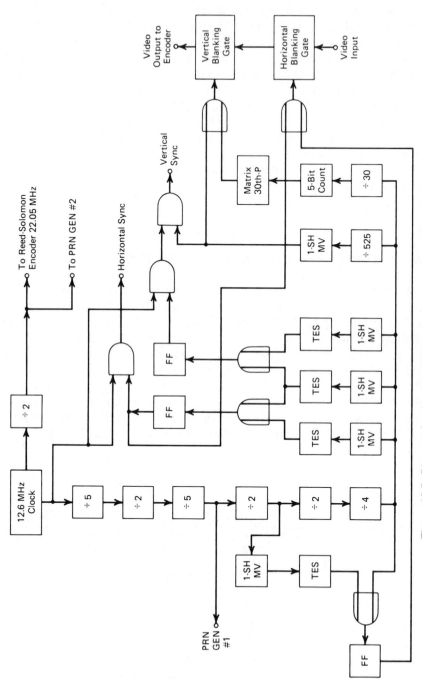

Figure 10.5. Diagram for the programmer unit for the Reed-Solomon system.

The sampling pulse must be a multiple of 15,750, and a multiplying factor of 800 is chosen to give a 12.6-MHz clock and sampling frequency.

10.5.1   The RS Encoder

The implementation of the encoder is shown in Fig. 10.6. The message symbols are formed in the lower register, which consists of seven groups of 4-bit registers, using feedforward multiplicative elements. The multiplication constants are formed by using (10.3). The seven groups of message symbols are fed into the feedback paths of the upper parity check generator shift register, and also out to the modulator unit. During this period the output switch is in number 2 position and the gate is open. When all the message symbol groups have been fed into the upper registers, the output switch goes to position 1, and the gate is closed. Then the eight groups of parity symbols are sent to the modulator unit. Thus we have seven message symbol groups followed by eight parity symbol groups, sent to the modulator. The process repeats itself. The drive for the RS encoder is the PPM waveforms.

The signals from the system previously described are fed to an FM modulator unit, prior to amplification and transmission.

10.5.2   The Demodulator

The output of the FM detector is fed to the decoder unit. The decoder unit uses pulse regeneration and bit synchronization. This is followed by the RS decoder as shown in Fig. 10.7. Integrate and dump is used to aid in removing noise components.

The RS output is fed to the PPM demodulator and then to the unit that removes the Roberts' modulation from the video.

Provision is also made for TV sync separation and phase lock for synchronization. Much of this was described previously in detail for the other systems.

10.5.3   The RS Decoder

This is shown in Fig. 10.8. The output of the syndrome generator is buffered. The buffered output is sent to an 8-bit accumulator. The four left-hand positions in the accumulator are reserved for each of 15 RS four-tuple symbols. The first symbol is fed into the accumulator simultaneously with the parallel output from a 4-bit counter. The 4-bit counter gives symbol location.

The combined group of eight, consisting of four syndrome bits and four locator bits are fed in parallel into an interface accumulator, and hence into an 8-bit accumulator in the special Galois field computer.

The interface logic is used to interrupt the computer processing to see whether or not it is ready to receive the first 8 bits. Interrogation continues until the computer is ready to process. When the computer is ready to accept

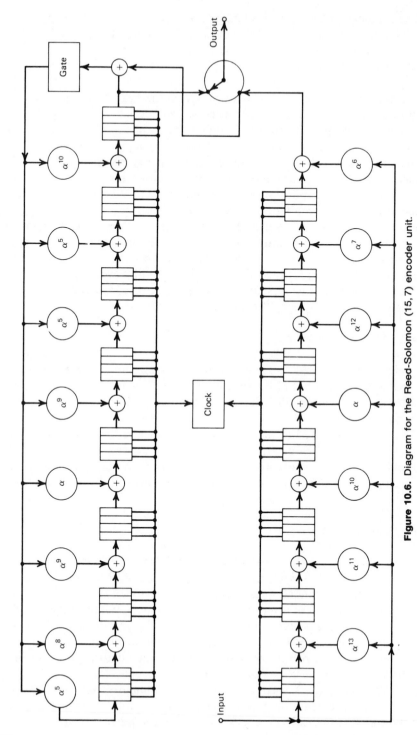

Figure 10.6. Diagram for the Reed–Solomon (15,7) encoder unit.

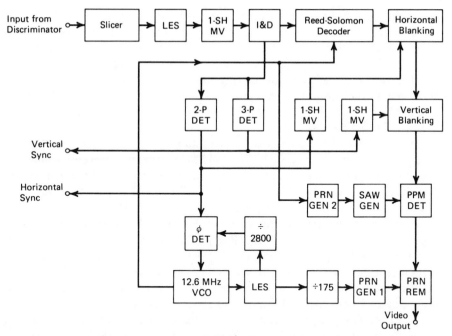

Figure 10.7. Diagram for the Reed-Solomon decoder unit.

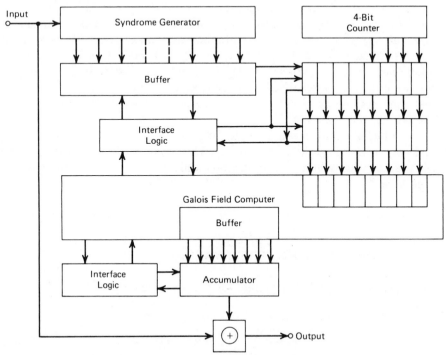

Figure 10.8. Diagram for the Reed-Solomon (15, 7) decoder unit.

the 8 bits, an enabling pulse is sent back to the accumulator controls and the buffer unit. The first 8 bits are loaded into the computer and the second 8 bits are readied for transmittal. This process is repeated until all 15 four-tuple symbols of the first words have entered the computer for processing.

It takes a finite time for solving the error location and error value equations in the computer. This process is long enough to prevent the use of this decoding method for real time usage. However, most digital TV systems are used in real time.

It is not beyond the state of the art to build this special computer so that it functions at very high speed. The whole computer processing time for decoding one word must be done within the time frame of the last word bit. The duration of each word bit is 79.4 nsec. The processing time for one word must be done within a 40-nsec period.

A special purpose computer designed to take advantage of parallel shift register operation could be built. The present state of the art allows logic circuitry to be built at operating rates of 500 Mbits/sec. This rate should be sufficient to enable the use of real time digital TV computation.

An output buffer in the computer could be hardwired to an external accumulator having parallel inputs and serial outputs.

The decoding process is completed by EXCLUSIVE-OR gating the computer output with the input word in synchronism. The noise produced syndrome pulses would then cancel the errors in the received message word. This, of course, assumes that no more than three errors have occurred in a word.

Perhaps a simpler solution exists in the use of two syndrome calculators and buffers, with two computer interface accumulators. While word B is being corrected, word A is shifted out of and word C shifted into this additional syndrome buffer unit. At the completion of decoding, the roles of the two units are interchanged. A much slower computer would be needed for real time communications. This alternate method would require only one special purpose computer.

Another alternate method would be to store a complete frame of video and play it back into the empty frames. This would work at the expense of a slight amount of picture quality degradation. Contour interpolation could be used here.

10.6 CHANNEL-SHARING PERFORMANCE

The equation for symbol error rate for the RS system is

$$P_S = \sum_{j=t+1}^{N} \binom{N}{j} p^j (1-p)^{N-j}. \tag{10.66}$$

Since we have the SNR augmented by the use of PRN/PPM, there is an improvement factor, so that

$$P = \frac{\exp\left[-6.66\,(\text{SNR})\right]}{\left[13.33\pi\,(\text{SNR})\right]^{1/2}}.$$ (10.67)

Numerical assessment of performance was obtained by letting $N = 15$, $J = 4$, and $Q = 0.172$. The channel-sharing performance of the $(15, 7)$ RS codes, in the PRN/PPM/RS/FM format, is shown in Fig. 10.9. It is very good.

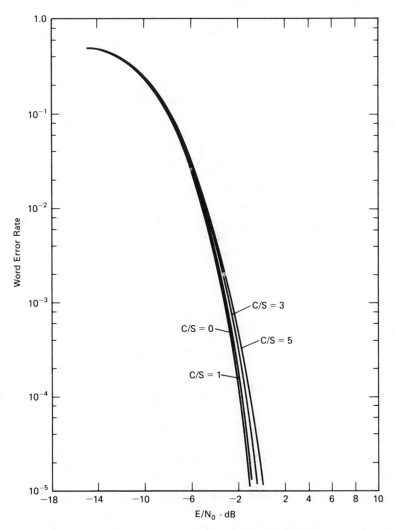

Figure 10.9. Channel-sharing performance of the Reed-Solomon $(15, 7)$ code.

10.7 APPENDIX A

Assume an error syndrome as shown:

$$R = 000\alpha^7 000\alpha^2 000\alpha^{10}000$$

$$e_1 = \alpha^7 \qquad\qquad e_2 = \alpha^2 \qquad\qquad e_3 = \alpha^{10} \qquad\qquad d_1 = \alpha^3$$

$$X_1 = \alpha^{11} \qquad\qquad X_2 = \alpha^7 \qquad\qquad X_3 = \alpha^3 \qquad\qquad Y_1 = \alpha^5.$$

$$(10A.1)$$

This shows three errors and one erasure:

$$S_1 = \alpha^7\alpha^{11} + \alpha^2\alpha^7 + \alpha^{10}\alpha^3 + \alpha^3\alpha^5 = \alpha^3 + \alpha^9 + \alpha^{13} + \alpha^8 = \alpha^9$$

$$(10A.2)$$

$$S_2 = \alpha^7\alpha^{22} + \alpha^2\alpha^{14} + \alpha^{10}\alpha^6 + \alpha^3\alpha^{10} = \alpha^{14} + \alpha + \alpha + \alpha^{13} = \alpha^2$$

$$(10A.3)$$

$$S_3 = \alpha^7\alpha^{33} + \alpha^2\alpha^{21} + \alpha^{10}\alpha^9 + \alpha^3\alpha^{15}$$

$$= \alpha^{10} + \alpha^8 + \alpha^4 + \alpha^3 = \alpha^{14} \qquad\qquad (10A.4)$$

$$S_4 = \alpha^7\alpha^{44} + \alpha^2\alpha^{28} + \alpha^{10}\alpha^{12} + \alpha^3\alpha^{20}$$

$$= \alpha^6 + 1 + \alpha^7 + \alpha^8 = \alpha^4 \qquad\qquad (10A.5)$$

$$S_5 = \alpha^7\alpha^{55} + \alpha^2\alpha^{35} + \alpha^{10}\alpha^{15} + \alpha^3\alpha^{25}$$

$$= \alpha^2 + \alpha^7 + \alpha^{10} + \alpha^{13} = \alpha^8. \qquad\qquad (10A.6)$$

$$S_6 = \alpha^7\alpha^{66} + \alpha^2\alpha^{42} + \alpha^{10}\alpha^{18} + \alpha^3\alpha^{30}$$

$$= \alpha^{13} + \alpha^{14} + \alpha^{13} + \alpha^3 = \alpha^0 = 1, \qquad\qquad (10A.7)$$

$$S_7 = \alpha^7\alpha^{77} + \alpha^2\alpha^{49} + \alpha^{10}\alpha^{21} + \alpha^3\alpha^{35}$$

$$= \alpha^9 + \alpha^6 + \alpha + \alpha^8 = \alpha^0 = 1, \qquad\qquad (10A.8)$$

$$S_8 = \alpha^7\alpha^{88} + \alpha^2\alpha^{56} + \alpha^{10}\alpha^{24} + \alpha^3\alpha^{40}$$

$$= \alpha^5 + \alpha^{13} + \alpha^4 + \alpha^{13} = \alpha^8. \qquad\qquad (10A.9)$$

Since there is only one erasure

$$\sigma_{d0} = 1, \tag{10A.10}$$

$$\sigma_{d1} = \alpha^5, \tag{10A.11}$$

$$\sigma_{d2} = 0, \tag{10A.12}$$

$$T_0 = S_3 + \alpha^5 S_2 = \alpha^{14} + \alpha^5 \alpha^2 = \alpha^{14} + \alpha^7 = \alpha, \tag{10A.13}$$

$$T_1 = S_4 + \alpha^5 S_3 = \alpha^4 + \alpha^5 \alpha^{14} = \alpha^4 + \alpha^4 = 0, \tag{10A.14}$$

$$T_2 = S_5 + \alpha^5 S_4 = \alpha^8 + \alpha^5 \alpha^4 = \alpha^8 + \alpha^9 = \alpha^{12}, \tag{10A.15}$$

$$T_3 = S_6 + \alpha^5 S_5 = \alpha^0 + \alpha^5 \alpha^8 = \alpha^0 + \alpha^{13} = \alpha^6, \tag{10A.16}$$

$$T_4 = S_7 + \alpha^5 S_6 = \alpha^0 + \alpha^5 \alpha^0 = \alpha^0 + \alpha^5 = \alpha^{10} \tag{10A.17}$$

$$T_5 = S_8 + \alpha^5 S_7 = \alpha^8 + \alpha^5 \alpha^0 = \alpha^8 + \alpha^5 = \alpha^4. \tag{10A.18}$$

As shown above the matrix to be solved is

$$T_5 = T_4 \sigma_{e1} + T_3 \sigma_{e2} + T_2 \sigma_{e3} \tag{10A.19}$$

$$T_4 = T_3 \sigma_{e1} + T_2 \sigma_{e2} + T_1 \sigma_{e3} \tag{10A.20}$$

$$T_3 = T_2 \sigma_{e1} + T_1 \sigma_{e2} + T_0 \sigma_{e3} \tag{10A.21}$$

or

$$\alpha^4 = \alpha^{10} \sigma_{e1} + \alpha^6 \sigma_{e2} + \alpha^{12} \sigma_{e3} \tag{10A.22}$$

$$\alpha^{10} = \alpha^6 \sigma_{e1} + \alpha^{12} \sigma_{e2} + 0 \tag{10A.23}$$

$$\alpha^6 = \alpha^{12} \sigma_{e1} + 0 + \alpha. \tag{10A.24}$$

When we let $\alpha = 2$

$$16 = 1024 \sigma_{e1} + 64 \sigma_{e2} + 4096 \sigma_{e3} \tag{10A.25}$$

$$1024 = 64 \sigma_{e1} + 4096 \sigma_{e2} + 0 \tag{10A.26}$$

$$64 = 4096 \sigma_{e1} + 0 + 2 \sigma_{e3}. \tag{10A.27}$$

We solve for σ_{e1}, σ_{e2}, and σ_{e3}. The answers are in the powers of 2, and are easily changed back to give powers of α. In the Fortran program, which is listed in Table 10A-1, the symbol (0000) is set equal to -1. In the computer program we also evaluate the expression

$$\sigma_e(X) = X^3 + \sigma_{e1} X^2 + \sigma_{e2} X + \sigma_{e3} \tag{10A.28}$$

```
 1    C     FORTRAN PROGRAM FOR SIMULATING THE DECODING OF A REED-SOLOMON (15,7) CODE.
 2    C     GALOIS FIELD ARITHMETIC IS USED. THE SIMULATED INPUT CONSISTS OF 8 COLUMNS AND 16 ROWS.
 3    C     THE FIRST FOUR COLUMNS REPRESENT REED-SOLOMON SYNDROME ERRORS IN VALUE; THE LAST 4 THE
 4    C     ERROR LOCATIONS IN BINARY SYMBOLS. THE SYMBOLS USED FOR THIS DEMONSTRATION ARE LISTED
 5    C     BELOW.
 6    C                                         00000000      0
 7    C                                         00000001      1
 8    C                                         00000010      2
 9    C                                         00000011      3
10    C                                         11010100      4
11    C                                         00000101      5
12    C                                         00000110      6
13    C                                         00000111      7
14    C                                         00101000      8
15    C                                         00001001      9
16    C                                         00011010      10
17    C                                         00001011      11
18    C                                         11101100      12
19    C                                         00001101      13
20    C                                         00001110      14
21    C                                         00001111      15
22    C     THE REED-SOLOMON (15,7) CODE SYMBOLS ARE:
23    C                                         0000         -1
24    C                                         1000          1
25    C                                         0100          ALPHA
26    C                                         0010          ALPHA**2
27    C                                         0001          ALPHA**3
28    C                                         1100          ALPHA**4
29    C                                         0110          ALPHA**5
30    C                                         0011          ALPHA**6
31    C                                         1101          ALPHA**7
32    C                                         1010          ALPHA**8
33    C                                         0101          ALPHA**9
34    C                                         1110          ALPHA**10
35    C                                         0111          ALPHA**11
36    C                                              1111       ALPHA**12
37    C                                              1011       ALPHA**13
38    C                                              1001       ALPHA**14
39          DIMENSION A(10,10),B(10),NN(10),F(16,4),FF(16,4),ZC(5),ZNS(5),
40         *NSE(5)
41          DIMENSION KQB(16),KAC(16),KRB(16),KSB(16)
42          DIMENSION K5(16)
43          DIMENSION FA(16,8),KPB(16),KEB(16)
44          DIMENSION NSYW(8)
45          DIMENSION ABC(16)
46          DIMENSION MP(6),MF(15),IJBER(16)
47          DIMENSION MFK(16),IFACE(10)
48          DIMENSION ME(15)
49          DIMENSION MB(5),ZA(5),ZZA(5),MD(5),IARY(16)
50          DATA IARY/-1,3,2,6,1,9,5,11,0,14,8,13,4,7,10,12/
51          COMMON/XXX/IOUT,F,FF,FA
52          INTEGER FA,A,B,F,FF
53        1 READ(5,100) N
54      100 FORMAT(I2)
55          READ(5,201) ((FA(KAN,KAY),KAY=1,8),KAN=1,16)
56      201 FORMAT(8I1)
57          READ(5,101) ((F(L,M),M=1,4),L=1,16)
58      101 FORMAT(4I1)
59          READ(5,102) ((FF(LL,MM),MM=1,4),LL=1,16)
60      102 FORMAT(4I1)
61    C     SPLIT THE 8 x 16 MATRIX INTO TWO 4 x 16 MATRICES, IN ORDER TO SEPARATE
62    C     ERROR VALUES FROM ERROR LOCATIONS
63          DO 55 JAY=1,16
64          JAR=8
65          DO 55 JAM=1,4
66          KPB(JAY)=KPB(JAY)+(FA(JAY,JAM)*JAR)
67          KEB(JAY)=KEB(JAY)+(FA(JAY,JAM+4)*JAR)
68          JAR=JAR/2
69       55 CONTINUE
70    C     FORM BINARY NUMBERS FROM THE MATRIX ROWS.
71          DO 56 JAY=1,16
72          IOU=KPB(JAY)+1
```

```
73              KQB(JAY)=IARY(IOU)
74           56 CONTINUE
75      C       FORM 8 PARITY CHECKS.
76              DO 58 JAX=1,8
77              JMU=-1
78              DO 57 JAY=1,16
79              K5(JAY)=15-KEB(JAY)
80              KAC(JAY)=(15-KEB(JAY))*JAX
81              KRB(JAY)=KAC(JAY)+KQB(JAY)
82              IF(KQB(JAY).EQ.-1) KRB(JAY)=-1
83              ABC(JAY)=FLOAT(KRB(JAY))/15.
84              KSB(JAY)=KRB(JAY)-(INT(ABC(JAY))*15)
85              CALL IEXOR(JMU,KSB(JAY))
86              JMU=IOUT
87           57 CONTINUE
88      C       FORM 5 MODIFIED SYNDROMES FROM ERASURE VALUES PER FORNEY (1965).
89              MY1=K5(11)
90              LAC=MY1+NSYN(2)
91              ABK=FLOAT(LAC)/15.
92              LAD=LAC-(INT(ABK)*15)
93              CALL IEXOR(LAD,NSYN(3))
94              LT0=IOUT
95              LAE=MY1+NSYN(3)
96              ABL=FLOAT(LAE)/15.
97              LAF=LAE-(INT(ABL)*15)
98              CALL IEXOR(LAF,NSYN(4))
99              LT1=IOUT
100             LAG=MY1+NSYN(4)
101             ABM=FLOAT(LAG)/15.
102             LAH=LAG-(INT(ABM)*15)
103             CALL IEXOR(LAH,NSYN(5))
104             LT2=IOUT
105             LAI=MY1+NSYN(5)
106             ABN=FLOAT(LAI)/15.
107             LAJ=LAI-(INT(ABN)*15)
108             CALL IEXOR(LAJ,NSYN(6))
109             LT3=IOUT
110             LAK=MY1+NSYN(6)
111             ABO=FLOAT(LAK)/15.
112             LAM=LAK-(INT(ABO)*15)
113             CALL IEXOR(LAM,NSYN(7))
114             LT4=IOUT
115             LAN=MY1+NSYN(7)
116             ABP=FLOAT(LAN)/15.
117             LAO=LAN-(INT(ABP)*15)
118             CALL IEXOR(LAO,NSYN(8))
119             LT5=IOUT
120     C       FORM ERROR LOCATION MATRIX (3 EQUATIONS, 3 UNKNOWNS)
121     C       BY ASSUMING ALPHA=2.
122             A(1,1)=2**LT4
123             A(1,2)=2**LT3
124             A(1,3)=2**LT2
125             A(2,1)=2**LT3
126             A(2,2)=2**LT2
127             A(2,3)=2**LT1
128             A(3,1)=2**LT2
129             A(3,2)=2**LT1
130             A(3,3)=2**LT0
131             B(1)=2**LT5
132             B(2)=2**LT4
133             B(3)=2**LT3
134     C       FORM GAUSS-JORDAN TRIANGULATION MATRIX.
135             NS=N-1
136             DO 10 K=1,NS
137             M2=A(J,K)*32768/A(K,K)
138             IF(A(K,K).EQ.0) M2=A(K,K)*32768/A(J,K)
139             DO  7 I=K,N
140             AA=ALOG10(2.)
141             X=FLOAT(A(J,I)*32768)
142             YES=A(K,I)*M2
143             Y=FLOAT(YES)
```

```
144            IF(A(K,K).EQ.0)  GO TO 81
145            GO TO 82
146         81 X=FLOAT(A(K,I)*32768)
147            YES=A(J,I)*M2
148            Y=FLOAT(YES)
149         82 CONTINUE
150            IF(X.GE.32768.) X=X/32768.
151            IF(Y.GE.32768.) Y=Y/32768.
152            IF(X.GT.0.) GO TO 40
153            X=.1
154            AA=1.
155         40 XX=ALOG10(X)/AA
156            AA=ALOG10(2.)
157            IF(Y.GT.0.) GO TO 41
158            Y=.1
159            AA=1.
160         41 CONTINUE
161            YY=ALOG10(Y)/AA
162            INPUT1=INT(XX)
163            INPUT2=INT(YY)
164            CALL IEXOR(INPUT1,INPUT2)
165            NALPH=2
166            A(J,I)=32768.*(NALPH**IOUT)
167            IF(A(J,I).GE.32768.) A(J,I)=A(J,I)/32768.
168            IF(IOUT.EQ.-1) GO TO 21
169            GO TO 22
170         21 A(J,I)=0.
171         22 AA=ALOG10(2)
172            IF(A(K,K).EQ.0.) GO TO  83
173            GO TO 87
174         83 A(K,I)=32768.*(NALPH**IOUT)
175            IF(A(K,I).GE.32768.) A(K,I)=A(K,I)/32768.
176            IF(IOUT.EQ.-1) GO TO 88
177            GO TO 89
178         88 A(K,I)=0.
179         89 AA=ALOG10(2)
180            A(J,I)=A(K,I)
181         87 CONTINUE
182          7 CONTINUE
183            AA=ALOG10(2)
184            W=FLOAT(B(J)*32768)
185            ZES=B(K)*M2
186            Z=FLOAT(ZES)
187            IF(A(K,K).EQ.0.) GO TO 77
188            GO TO 78
189         77 W=FLOAT(B(K)*32768)
190            ZES=B(J)*M2
191            Z=FLOAT(ZES)
192         78 CONTINUE
193            IF(W.GE.32768.) W=W/32768.
194            IF(Z.GE.32768.) Z=Z/32768.
195            W=.1
196            AA=1.
197         42 WW=ALOG10(W)/AA
198            AA=ALOG10(2.)
199            IF(Z.GT.0.) GO TO 43
200            Z=.1
201            AA=1.
202         43 CONTINUE
203            ZZ=ALOG10(Z)/AA
204            INPUT1=INT(WW)
205            INPUT2=INT(ZZ)
206            CALL IEXOR(INPUT1,INPUT2)
207            NALPH=2
208            B(J)=32768.*(NALPH**IOUT)
209            IF(B(J).GE.32768.) B(J)=B(J)/32768.
210            IF(IOUT.EQ.-1) GO TO 11
211            GO TO 12
212         11 B(J)=0.
213         12 AA=ALOG10(2.)
214            IF(A(K,K).EQ.0.) GO TO 79
```

```
215              GO TO  84
216           79 B(K)=32768.*(NALPH**IOUT)
217              IF(B(K).GE.32768.) B(K)=B(K)/32768.
218              IF(IOUT.GT.-1) GO TO 85
219              GO TO 86
220           85 B(K)=0.
221           86 AA=ALOG10(2.)
222              B(J)=B(K)
223           84 CONTINUE
224           10 CONTINUE
225    C         SOLVE GAUSS-JORDAN MATRIX FOR VALUES OF UNKNOWNS.
226              AA=ALOG10(2.)
227              NN(N)=B(N)*32768./A(N,N)
228              IF(NN(N).GE.32768.) NN(N)=NN(N)/32768
229              DO 44 LA=1,NS
230              K=N-LA
231              MJ=-1
232              NSK=N-K
233              DO 45  MK=1,NSK
234              AA=ALOG10(2.)
235              I=N-MK+1
236              MB(I)=A(K,I)*NN(I)
237              AA=ALOG10(2.)
238              ZA(I)=FLOAT(MB(I))
239              IF(ZA(I).GE.32768.) ZA(I)=ZA(I)/32768.
240              IF(ZA(I).GT.0.) GO TO 46
241              ZA(I)=.1
242              AA=1.
243           46 CONTINUE
244              ZZA(I)=ALOG10(ZA(I))/AA
245              AA=ALOG10(2.)
246              MD(I)=INT(ZZA(I))
247              CALL IEXOR(MD(I),MJ)
248              MJ=IOUT
249           45 CONTINUE
250              LB=MJ
251              ZD=FLOAT(B(K))
252              IF(ZD.GE.0.) GO TO 48
253              ZD=.1
254              AA=1.
255           48 RD=ALOG10(ZD)/AA
256              MH=INT(RD)
257              CALL IEXOR(LB,MH)
258              MI=(2**IOUT)*32768
259              NN(K)=MI/A(K,K)
260              IF(NN(K).GE.32768) NN(K)=NN(K)/32768
261              IF(IOUT.EQ.-1)  GO TO 49
262              GO TO 50
263           49 NN(K)=0
264           50 AA=ALOG10(2.)
265           44 CONTINUE
266              DO 64 MA=1,5
267              AA=ALOG10(2.)
268              ZC(MA)=FLOAT(NN(MA))
269              IF(ZC(MA).GT.0.) GO TO  65
270              ZC(MA)=.1
271              AA=1.
272           65 ZNS(MA)=ALOG10(ZC(MA))/AA
273              NSF(MA)=INT(ZNS(MA))
274           64 NSE(MA)=INT(ZNS(MA))
275    C         SOLVE THE ERROR LOCATION POLYNOMIALS FOR ERROR LOCATIONS.
276              IBJ=1
277              DO 90 JB=1,15
278              JB=JB-1
279              MU=-1
280              JF=N+1
281              DO  91 JE=1,JF
282              JH=JE-1
283              JI=JF-JE
284              MO=NSE(JH)+(JI*JBE)
285              IF(JH.EQ.0) M)=JI*JBE
286              AC=FLOAT(MO)/15.
287              MP(JE)=MO-INT(AC)*15)
```

```
288              IF(NSE(JH).EQ.-1) MP(JE)=-1
289              CALL IEXOR(MU,MP(JE))
290              MU=IOUT
291         91 CONTINUE
292     C        THE ERROR SPOTS WILL SHOW UP AS -1 NUMBERS.
293     C        PACK ERROR LOCATIONS AT LOWER END OF OF ACCUMULATOR.
294              MF(JBE)=MU
295              IF(MF(JBE).EQ.-1) IJBER(IJB)=JBE
296              IF(MF(JBE).EQ.-1) IJBER(16)=IJB
297              IF(MF(JBE).EQ.-1) IJB=IJB+1
298         90 CONTINUE
299     C        SOLVE FOR ERROR VALUES PER FORNEY.
300              MY1=5
301              CALL IEXOR(IJBER(2),IJBER(1))
302              MS=IOUT
303              CALL IEXOR(MS,MY1)
304              MSIG1=IOUT
305              MT=IJBER(1)+MY1
306              AE=FLOAT(MT)/15
307              IB=MT-(INT(AE)*15)
308              MW=IJBER(2)+MY1
309              AF=FLOAT(MW)/15.
310              IC=MW-(INT(AF)*15)
311              MX=IJBER(2)+IJBER(1)
312              AG=FLOAT(MX)/15.
313              ID=MX-(INT(AG)*15)
314              CALL IEXOR(IB,IC)
315              MY=IOUT
316              CALL IEXOR(MY,ID)
317              MSIG2=IOUT
318              IA=IJBER(2)+IJBER(1)+MY1
319              AH=FLOAT(IA)/15.
320              MSIG3=IA-(INT(AH)*15)
321              IE=IJBER3)*8
322              AI=FLOAT(IE)/15.
323              IF=IE-(INT(AI)*15)
324              IG=IJBER(3)*7
325              AJ=FLOAT(IJ)/15.
326              IH=IG-(INT(AJ)*15)
327              KB=MSIG1+IH
328              BI=FLOAT(KB)/15.
329              KC=KB-(INT(BI)*15)
330              II=IJBER(3)*6
331              AK=FLOAT(II)/15.
332              IJ=II-(INT(AK)*15)
333              KD=MSIG2+IJ
334              BJ=FLOAT(KD)/15.
335              KE=KD-(INT(BJ)*15)
336              IK=IJBER(3)*5
337              AL=FLOAT(IK)/15.
338              IL=IK-(INT(AL)*15)
339              KF=MSIG3+IL
340              BK=FLOAT(KF)/15.
341              KG=KF-(INT(BK)*15)
342              CALL IEXOR(IF,KC)
343              IM=IOUT
344              CALL IEXOR(IM,KE)
345              IN=IOUT
346              CALL IEXOR(IH,KG)
347              IDEN=IOUT
348              IO=MSIG1+NSYN(7)
349              AM=FLOAT(IO)/15.
350              IP=IO-(INT(AM)*15)
351              IQ=MSIG2+NSYN(6)
352              AN=FLOAT(IQ)/15.
353              IR=IQ-(INT(AN)*15)
354              IS=MSIG3+NSYN(5)
355              AO=FLOAT(IS)/15.
356              IT=IS-(INT(AO)*15)
357              CALL IEXOR(NSYN(8),IP)
358              IU=IOUT
359              CALL IEXOR(IU,IR)
360              IV=IOUT
```

```
361            CALL IEXOR(IV,IT)
362            INUM=IOUT
363            IF(INUM.LT.IDEN) INUM=INUM+15
364            ME(1)=INUM-IDEN
365            IW=ME(1)+IF
366            AP=FLOAT(IW)/15.
367            IX=IW-(INT(AP)*15)
368            CALL IEXOR(IX,NSYN(8))
369            MS7P=IOUT
370            JJ=MF(1)+IJ
371            AR=FLOAT(J)/15.
372            JK=JJ-(INT(AR)*15)
373            CALL IEXOR(JK,NSYN(6))
374            MS6P=IOUT
375            JL=MF(1)+IL
376            AS=FLOAT(JL)/15.
377            JM=JL-(INT(AS)*15)
378            CALL IEXOR(JM,NSYN(5))
379            MS5P=IOUT
380            CALL IEXOR(IJBER(1),MY1)
381            MSIG1P=IOUT
382            MSIG2P=IB
383            JN=MS7P+MSIG1P
384            BA=FLOAT(JN)/15.
385            JO=JN-(INT(BA)*15)
386            JP=MS6P+MSIG2P
387            BB=FLOAT(JP)/15.
388            JQ=JP-(INT(BB)*15)
389            CALL IEXOR(MS8P,JO)
390            JR=IOUT
391            CALL IEXOR(JR,JQ)
392            JNUM=IOUT
393            JS=IJBER(2)*8
394            BC=FLOAT(JS)*15)
395            JT=JS-(INT(BC)*15)
396            JU=IJBER(2)*7
397             BD=FLOAT(JU)/15.
398             JV=JU-(INT(BD)*15)
399             JW=JV+MSIG1P
400             BE=FLOAT(JW)/15.
401             JX=JW-(INT(BE)*15)
402             JY=IJBER(2)*6
403             BF=FLOAT(JY)/15.
404             JZ=JY-(INT(BF)*15)
405             KH=JZ+MSIG2P
406             BG=FLOAT(KH)/15.
407             KJ=KH-(INT(BG)*15)
408             CALL IEXOR(JT,JX)
409             KI=IOUT
410            CALL IEXOR(KI,KJ)
411            JDEN=IOUT
412            IF(JNUM.LT.JDEN) JNUM=JNUM+15
413            ME(2)=JNUM-JDEN
414            NAN=IJBER(2)*8
415            XA=FLOAT(NAN)/15.
416            NBN=NAN-(INT(XA)*15)
417            NCN=ME(2)+NBN
418            XB=FLOAT(NCN)/15.
419            NDN=NCN-(INT(XB)*15)
420            CALL IEXOR(NDN,MS8P)
421            MS3PP=IOUT
422            NEN=IJBER(2)*7
423            XC=FLOAT(NEN)/15.
424            NFN=NEN-(INT(XC)*15)
425            NGN=ME(2)+NFN
426            XD=FLOAT(NGN)*15)
427            NHN=NGN-(INT(XD)*15)
428            CALL IEXOR(MS7P,NHN)
429            MS7PP=IOUT
430            NIN=IJBER(2)*6
431            XE=FLOAT(NIN)/15.
432            NJN=NIN-(INT(XE)*15)
```

```
433                   NKN=ME(2)+NJN
434                   XF=FLOAT(NKN)/15.
435                   NLN=NKN-(INT(XF)*15)
436                   CALL IEXOR(MS6P,NLN)
437                   MS6PP=IOUT
438                   MSGPP=5
439                   NMN=MSGPP+MS7PP
440                   XG=FLOAT(NMN)*15)
441                   NNN=NMN-(INT(XG)*15)
442                   CALL IEXOR(MS8PP,NNN)
443                   KNUM=IOUT
444                   NON=IJBER(1)*7
445                   XH=FLOAT(NON)/15.
446                   NPN=NON-(INT(XH)*15)
447                   NQN=MSGPP+NPN
448                   XI=FLOAT(NQN)/15.
449                   NRN=NQN-(INT(XI)*15)
450                   NSN=IJBER(1)*8
451                   XJ=FLOAT(NSN)/15.
452                   NTN=NSN-(INT(XJ)*15)
453                   CALL IEXOR(NTN,NRN)
454                   KDEN=IOUT
455                   IF(KNUM.LT.KDEN) KNUM=KNUM+15
456                   ME(3)=KNUM-KDEN
457                   IBER1=15-IJBER(3)
458                   IBER2=15-IJBER(2)
459                   IBER3=15-IJBER(1)
460                   WRITE(6,9)
461                   WRITE(6,205) (KEB(JAY),JAY=1,16)
462               205 FORMAT(4X,16(4X,I2)/)
463                   WRITE(6,6)A(1,1),A(1,2),A(1,3),B(1)
464                   WRITE(6,6)A(2,1),A(2,2),A(2,3),B(2)
465                   WRITE(6,6)A(3,1),A(3,2),A(3,3),B(3)
466                   WRITE(6,214)
467                   WRITE(6,8) (NSE(MA), MA=1,15)
468                   WRITE(6,108)
469               WRITE(6,212) (MF(JBE), JBE=1,15)
470           212 FORMAT(5X,15(4X,I2)///)
471                   WRITE(6,206)
472                   WRITE(6,80) IBER1,IBER2,IBER3
473                   WRITE(6,207)
474                   WRITE(6,80) ME(1),ME(2),ME(3)
475     C             RECREATE THE SYNDROME INTO A "1","O" STRING 60 BITS LONG, AND
476     C             FILL A 60 BIT ACCUMULATOR, HARD WIRE ACCUMULATOR OUTPUT TO
477     C             HARDWARE ERROR CORRECTION DEVICES.
478                   DO 110 IAN=1,15
479                   MF(IAN)=-1
480           110 CONTINUE
481                   KRT=IJBER(16)
482                   DO 111 IAN=1, KRT
483                   JIM=15-IJBER(IAN)
484                   LBC=(KRT+1)-IAN
485                   MF(JIM)=ME(LBC)
486           111 CONTINUE
487                   WRITE(6,208)
488                   WRITE(6,212) (MF(KRT),KRT=1,15)
489                   ISTART=32
490                   IEND=35
491                   LF=1
492                   LK=0
493                   DO 51 LC=1,15
494                   IF(MF(LC).EQ.-1)   MFK(LC)=0
495                   IF(MF(LC).EQ.0)    MFK(LC)=8
496                   IF(MF(LC).EQ.1)    MFK(LC)=4
497                   IF(MF(LC).EQ.2)    MFK(LC)=2
498                   IF(MF(LC).EQ.3)    MFK(LC)=1
499                   IF(MF(LC).EQ.4)    MFK(LC)=12
500                   IF(MF(LC).EQ.5)    MFK(LC)=6
501                   IF(MF(LC).EQ.6)    MFK(LC)=3
502                   IF(MF(LC).EQ.7)    MFK(LC)=13
503                   IF(MF(LC).EQ.8)    MFK(LC)=10
504                   IF(MF(LC).EQ.9)    MFK(LC)=5
```

```
505              IF(MF(LC).EQ.10)     MFK(LC)=14
506              IF(MF(LC).EQ.11)     MFK(LC)=7
507              IF(MF(LC).EQ.12)     MFK(LC)=15
508              IF(MF(LC).EQ.13)     MFK(LC)=11
509              IF(MF(LC).EQ.14)     MFK(LC)=9
510              DO 52 LD=ISTART,IEND
511              IF(FLD(LD,1,MFK(LC)).EQ.0)   INAME='0'
512              IF(FLD(LD,1,MFK(LC)).EQ.1)   INAME='1'
513          52 CONTINUE
514          51 CONTINUE
515              WRITE(6,209)
516              WRITE(6,53) (IFACE(I),I=1,10)
517          53 FORMAT(20X,10A6)
518           6 FORMAT(34X,4(6X,I5)///)
519           9 FORMAT(1H1,T45,'THE FORMED SYNDROMES'/)
520           4 FORMAT(T44,'A1',T55,'A2',T66,'A3',T78,'B'/)
521         213 FORMAT(T45,'GAUSS-JORDAN TRIANGULATION'/)
522         214 FORMAT(T45,'SOLUTION TO MATRIX'/)
523         108 FORMAT(T28,'INVERSE ERROR LOCATIONS,ERRORS ARE AT -1 POINTS'/)
524         206 FORMAT(T38,'ERROR LOCATIONS'/)
525         207 FORMAT(T38,'ERROR VALUES'/)
526         209 FORMAT(T30,'COMPUTER OUTPUT STREAM TO ERROR CORRECTOR'/)
527         208 FORMAT(T38,'ERROR LOCATION SYNDROME'/)
528           8 FORMAT(34X,3(6X,I5)///)
529          80 FORMAT(34X,3(4X,I2)//)
530              STOP
531              END

             END OF COMPILATION:                    NO DIAGNOSTICS

 1              SUBROUTINE IEXOR(LAA,LAB)
 2         C    SUBROUTINE FOR EXCLUSIVE-OR CALCULATIONS
 3              INTEGER F,FF
 4              COMMON/XXX/KTK,F,FF
 5              DIMENSION F(16,4),FF(16,4)
 6              K=LAA+2
 7              KK=LAB+2
 8              KKK1=F(K,1)
 9              KKK2=F(K,2)
10              KKK3=F(K,3)
11              KKK4=F(K,4)
12              KK1=FF(KK,1)
13              KK2=FF(KK,2)
14              KK3=FF(KK,3)
15              KK4=FF(KK,4)
16              IF(KKK1.EQ.0)   KAK=0
17              IF(KKK1.EQ.1)   KAK=8
18              IF(KKK2.EQ.0)   KBK=0
19              IF(KKK2.EQ.1)   KBK=4
20              IF(KKK3.EQ.0)   KCK=0
21              IF(KKK3.EQ.1)   KCK=2
22              IF(KKK4.EQ.0)   KDK=0
23              IF(KKK4.EQ.1)   KDK=1
24              KEK=KAK+KBK+KCK+KDK
25              IF(KK1.EQ.0)    KFK=0
26              IF(KK1.EQ.1)    KFK=8
27              IF(KK2.EQ.0)    KGK=0
28              IF(KK2.EQ.1)    KGK=4
29              IF(KK3.EQ.0)    KHK=0
30              IF(KK3.EQ.1)    KHK=2
31              IF(KK4.EQ.0)    KOK=0
32              IF(KK4.EQ.1)    KOK=1
33              KPK=KFK+KGK+KHK+KOK
34              KSK=XOR(KEK,KPK)
35              IF(KSK.EQ.0)    KTK=-1
36              IF(KSK.EQ.8)    KTK=0
37              IF(KSK.EQ.4)    KTK=1
38              IF(KSK.EQ.2)    KTK=2
39              IF(KSK.EQ.1)    KTK=3
40              IF(KSK.EQ.12)   KTK=4
41              IF(KSK.EQ.6)    KTK=5
42              IF(KSK.EQ.3)    KTK=6
43              IF(KSK.EQ.13)   KTK=7
```

```
44               IF(KSK.EQ.10)  KTK=8
45               IF(KSK.EQ.5)   KTK=9
46               IF(KSK.EQ.14)  KTK=10
47               IF(KSK.EQ.7)   KTK=11
48               IF(KSK.EQ.15)  KTK=12
49               IF(KSK.EQ.11)  KTK=13
50               IF(KSK.EQ.9)   KTK=14
51               RETURN
52               END
```

END OF COMPILATION   NO DIAGNOSTICS

RESULTS:

THE FORMED SYNDROMES

0	1	2	3	4	5	6	7	8	9	10	11	12	13	14	15
-1	-1	-1	-1	7	-1	-1	-1	2	-1	3	-1	10	-1	-1	-1

GAUSS-JORDAN TRIANGULATION

A1	A2	A3	B
1024	64	4096	16
0	128	256	32
0	0	1	64

SOLUTION TO MATRIX

13   5   6

INVERSE ERROR LOCATIONS; ERRORS ARE AT -1 POINTS.

14	12	-1	8	12	13	-1	1	3	2	-1	12	2	11	0

ERROR LOCATIONS

4   8   12

ERROR VALUES

7   2   10

ERROR LOCATION SYNDROME

-1	-1	-1	7	-1	-1	-1	2	-1	-1	-1	10	-1	-1	-1

COMPUTER OUTPUT STREAM TO ERROR CORRECTOR

000000000000110100000000000000010000000000111000000000000

by letting X take on values from $\alpha^0 \rightarrow \alpha^{14}$. Three of these 15 values will be zero, which shows up on the computer output as -1. By inspection one can see that the computer correctly located the three error location positions, which were at positions 11, 7, and 3, the inverses of which are 4, 8, and 12.

The computer was also programmed to give error values, by using the three following Forney equations:

$$e_1 = \frac{S_8 + \sigma_1 S_7 + \sigma_2 S_6 + \sigma_3 S_5}{X_1^8 + \sigma_1 X_1^7 + \sigma_2 X_1^6 + \sigma_3 X_1^5} \tag{10A.29}$$

where

$$\sigma_1 = X_2 + X_3 + Y_1 \tag{10A.30}$$

$$\sigma_2 = X_3 Y_1 + X_2 Y_1 + X_2 X_3 \tag{10A.31}$$

$$\sigma_3 = X_2 X_3 Y_1. \tag{10A.32}$$

Having found e_1 we modify syndromes:

$$S_8^1 = S_8 + e_1 X_1^8 \tag{10A.33}$$

$$S_7^1 = S_7 + e_1 X_1^7 \tag{10A.34}$$

$$S_6^1 = S_6 + e_1 X_1^6 \tag{10A.35}$$

$$\sigma_1^1 = Y_1 + X_3 \text{ for } X_2 = 0 \tag{10A.36}$$

$$\sigma_2^1 = X_3 Y_1 \text{ for } X_2 = 0 \tag{10A.37}$$

$$e_2 = \frac{S_8^1 + \sigma_1^1 S_7^1 + \sigma_2^1 S_6^1 + \sigma_3^1 S_5^1}{X_2^8 + X_2^7 \sigma_1^1 + X_2^6 \sigma_2^1 + X_2^5 \sigma_3^1}. \tag{10A.38}$$

Having found e_2 and e_1 we modify again

$$S_8^{11} = S_8^1 + e_2 X_2^8 \tag{10A.39}$$

$$S_7^{11} = S_7^1 + e_2 X_2^7 \tag{10A.40}$$

$$S_6^{11} = S_6^1 + e_2 X_2^6 \tag{10A.41}$$

$$\sigma_1^{11} = X_3 \quad (Y_1 = 0) \tag{10A.42}$$

$$\sigma_2^{11} = 0 \quad (Y_1 = 0) \tag{10A.43}$$

$$\sigma_3^{11} = 0 \quad (Y_1 = 0) \tag{10A.44}$$

$$e_3 = \frac{S_8^{11} + S_7^{11} \sigma_1^{11}}{Y_1^8 + Y_1^7 \sigma_1^{11}}. \tag{10A.45}$$

The computer run shows the three error values as α^7, α^2, and α^{10}.

chapter 11

Summary, Results, and Conclusions

summary

There are many modulation techniques available for use with digital TV communications. The three basic methods are PAM, PDM, and PCM. The best performance criterion for these modulation systems is the mean-square-error. These three systems are compared as to performance in Chapter 2. The comparison curves shown at the end of that chapter show a clear advantage to the PCM/FM modulation system. This is why PCM has become the work horse for digital TV. Since a major portion of the book deals with PCM based systems, Chapter 3 is used to detail a basic system.

The major problem in digital TV communications is bandwidth reduction. There have been many proposals for reducing the transmission bandwidth requirements. Therefore, Chapter 4 is devoted to this subject. An attempt was made in this chapter to thoroughly cover every aspect of this lengthy subject. There are, no doubt, some bandwidth reduction techniques that have been omitted. The state-of-the-art changes so rapidly that by the time this book appears in print, there will probably be some new methods for reducing bandwidth.

However, this book is primarily concerned with the channel transmission aspects of digital TV. Bandwidth reduction is very important, but there are many applications in which these reduced bandwidths must be transmitted from one geographical location to another. The main thrust of this book is concerned with applications of techniques that will efficiently encode and communicate reduced bandwidth signals from point to point.

Six of the chapters in this book are concerned with the communications end of digital TV.

11.1 COMBINATIONS OF PULSE MODULATION SYSTEMS

In Chapters 5, 6, and 7 pulse modulation systems using a new sampling technique are described. This new sampling technique is termed "spread spectrum sampling." It could also be called "pseudorandom sampling." In this method video sampling occurs at a pseudorandom rate, rather than at a fixed rate. In order to use this sampling the whole modem must be geared to pseudorandomness.

In Chapter 5, spread spectrum sampling is applied along with video to the pulse-position modulator, which in turn is used to drive the PCM quantizer. The quantizer output is frequency modulated. The process is reversed at the receiver. The resulting signal is termed PAM/PCM/PRN/FM and has a transmission bandwidth of 44.1 MHz.

The system implementation is described in detail. It includes the use of Robert's modulation, a form of pseudorandom dither for a 2:1 bandwidth reduction in the form of 3-bits/word of quantization. However, because of the spread spectrum portion of the signal the bandwidth saved by using dither is lost.

The channel sharing capabilities of the system is assessed by running a curve on a digital computer of word error rate versus SNR. Four such curves are shown in Fig. 5.9. The results are based on the ratio of unwanted users power to users power, as described in Chapter 5. When there is only one user, or a condition of no interference, the performance is about 2 dB better than for standard PCM/FM.

However, because of the relatively large pulse duty factor the system has poor channel-sharing capabilities.

For comparison interest a DPCM/FM system is checked for word error rate versus SNR ratio. For no interference conditions the results are about 3.5 dB better than for the PAM/PCM/PRN/FM system described. However, DPCM too, has poor channel sharing capabilities. This is shown in Fig. 5.10.

In Chapter 6, a PRN/PPM/FM system is described that makes use of this new spread spectrum sampling technique. Again, the PPM signal is formed from pseudorandom driving pulses.

However, the transmission bandwidth required for this system is 22.05 MHz. By use of a special technique the pulse duty factor was made quite low, so that the system has excellent channel-sharing capabilities. Its overall performance is several decibels better than for the two systems analyzed in Chapter 5. This performance is shown in Fig. 6.11.

Chapter 7 describes two multi-channel PRN/PPM/FM systems. One system uses 13 series-parallel simultaneous channels; the other 25-channels. Both of these systems have system bit-error rate performance that excels that of single channel PRN/PPM/FM. This is shown in Fig. 7.39.

11.1.1 Bi-Phase Transorthogonal Coding

In Chapter 8 a Bi-Phase Transorthogonal system is described. The details of implementation are discussed, and the performance assessed. This is a block type code using a 7×7 bit matrix of ones and zeros.

A method for implementation is described in detail, and its word error rate performance is assessed. This is shown in Fig. 8.16. The results are good but are poorer than for PRN/PPM/FM.

11.1.2 Error-Correction Codes

Chapters 9 and 10 dealt with two types of error-correction codes; the Convolutional and the RS codes. As mentioned in Chapter 4 there is a real need for error-correction codes in digital TV communication systems.

In Chapter 9 spread spectrum sampling is again used to produce PRN/PPM. This time it was combined with convolutional coding and the whole system was frequency-modulated. This produces a combination coding system called PRN/PPM/CONVOLUTIONAL/FM. The implementation of this system is described in detail and mathematical analysis is made of the system performance. The computer run gave a word error rate performance that is excellent. It is an excellent system to use for channel sharing. These results are shown in Fig. 9.17. For no interference, a word error rate of 10^{-5} is found for 1.5 dB of signal to noise. This can be compared with the results obtained by Merrill and Thompson (1972), who show a 2.6-dB SNR for 10^{-5} bit-error rate.

In Chapter 10 spread spectrum sampling was again used to produce PRN/PPM. This time it was combined with RS (15-7) coding and FM to get PRN/PPM/RS/FM. The implementation of this system is described in detail, which included a new decoding algorithm.

The system is mathematically analyzed and its word error rate versus SNR is assessed. The results are shown in Fig. 10.9. Its performance was a little worse than that of the three PRN/PPM/FM systems, and the PRN/PPM/convolutional/FM system.

11.2 RESULTS

Performance comparison can be made for the eight systems described in Chapters 5–10. There are three classes of communication systems described in these chapters; pulse modulation, block coding, and error-correction coding. The PRN/PPM/FM systems are definitely pulse-modulation systems, while PCM and DPCM systems are border line cases between pulse modulation and coding. The bi-phase transorthogonal codes are in a class of nonerror detection and correction block type codes. Convolutional codes and RS codes are error-correcting codes.

The bandwidths for each of these codes was chosen based on optimum performance of the system. The chosen transmission bandwidths are

PAM/PCM/PRN/FM	44.1 MHz
DPCM/FM	22.05 MHz
PRN/PPM/FM	22.05 MHz
PRN/PPM/FM (13-channel)	6.65 MHz
PRN/PPM/FM (25-channel)	2.64 MHz
Bi-phase Transorthogonal	22.05 MHz
PRN/PPM/convolutional/FM	44.1 MHz
PRN/PPM/RS/FM	12.6 MHz

SNR is the most important parameter in assessing word error rate. The channel SNRs should be modified to reflect the FM improvement factor $3M^2$, where M is the modulation index; the PPM/FM improvement factor of $(4/9)BT$, where BT is the time bandwidth product; and the PRN improvement factor BT.

In the PCM systems a quantization factor of $\log_2 L$ must be used as is explained in Chapter 5.

All of the preceding factors are used in the word error rate assessments for all systems except the PRN/PPM/FM systems, which assessed bit-error rates. Word error rates are meaningless in PRN/PPM/FM systems. Some have termed such factors to be "processing gain." In other words, the processing gain of the entire communication system is included in the assessment of error probability.

All of these systems are assessed on performance in channel-sharing systems. In videophone channel-sharing systems, a common buffer is used for controlling to some extent the orthogonality between various user signals. However, in military use this is not always possible to do. Therefore, the channel sharing portrayed here does not use a common buffer, but relies on the orthogonality of the codes chosen. The cross-correlation coefficients for each system is carefully assessed and used in the computations.

Fig. 11.1 compares the performance of all eight systems for no interference conditions, or on the assumption that only one of the 12 possible users is active.

The performance of each system was adjusted for bandwidth ratio; the reference bandwidth being 22.05 MHz. Because of their low bandwidth, the multi-channel PRN/PPM/FM systems are stand outs in performance. At 10^{-5} error rate the PRN/PPM/RS/FM system is about 1.5 dB better than the PRN/PPM/FM system and the PRN/PPM/convolutional/FM system. The performances of the DPCM/FM system and the bi-phase transorthogonal/FM system are about equal.

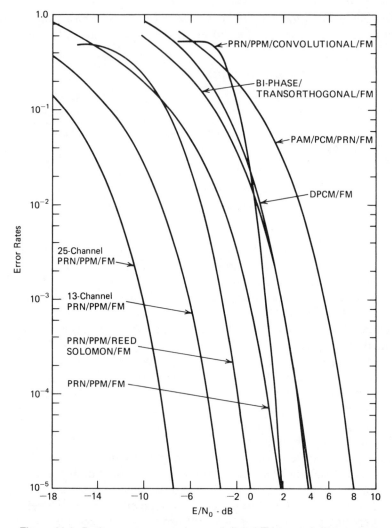

Figure 11.1. Performance comparisons of digital TV transmission systems.

At this point it should be mentioned that the particular RS code chosen here is the $(15,7)$ version. There are better RS codes than this; for example, the concatenated RS $(16,6)$ and the concatenated RS $(32,20)$.

It might be mentioned at this point, that Blasbalg *et al.* (1964) have been taking advantage of PRN/PPM/FM systems for years, especially in voice communication systems called random access discrete address (RADA) systems. It is basically a superior system and especially useful in channel-sharing systems. This is due to the selection of PRN codes having low cross correlation and to the use of low duty factor pulses. These two factors aid the orthogonality facet of these channel-sharing systems.

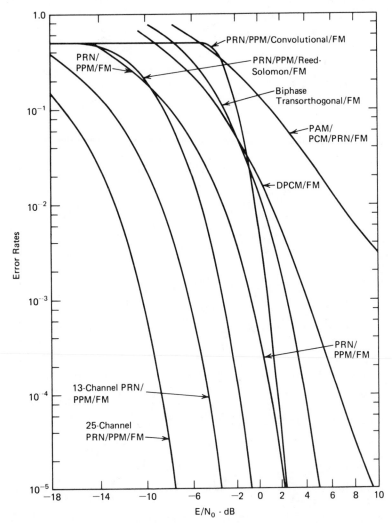

Figure 11.2. Channel-sharing performance comparisons of digital TV transmission systems for clutter-to-signal ratios of unity.

In comparing performance in channel sharing the ratio of unwanted user power to wanted user power is varied in steps 1, 3, and 5. This accounts for geographical geometry. The unwanted user might be closer to you than the wanted user, and therefore would come in with greater signal strength.

In Fig. 11.2, equal user signal powers are assumed. It was also assumed that all 12 users were simultaneously "on the air." In assuming power ratios complete orthogonality was assumed to exist. This ratio was then modified by a factor that takes into account the correct cross-correlation coefficient.

In Fig. 11.2 again, the multi-channel PRN/PPM/FM systems are standouts. All of the other codes have retained their relative performance

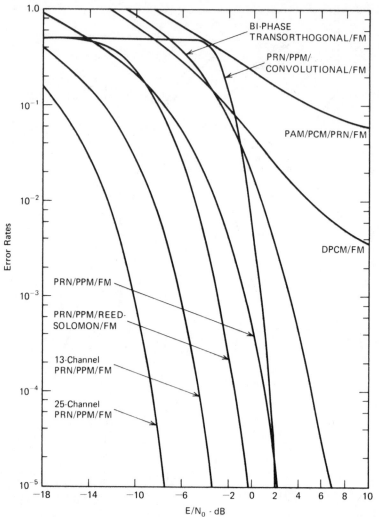

Figure 11.3. Channel-sharing performance comparisons of digital TV transmission systems for clutter-to-signal ratios of 3:1.

positions with respect to each other. Performance degradation shows, however, especially in the cases of the PAM/PCM/PRN/FM and DPCM/FM systems.

In Fig. 11.3, it was assumed that the C/S ratio was 3:1. Again, we have the same two standouts. The bi-phase transorthogonal/FM system is noticeably degraded in performance, but is still quite usable. However, the DPCM/FM and the PAM/PCM/PRN/FM systems have deteriorated to the point of nonutility.

In Fig. 11.4, it was assumed that the C/S ratio was 5:1. The relative performance of the various systems remains about the same, but degradation to most systems is quite noticeable.

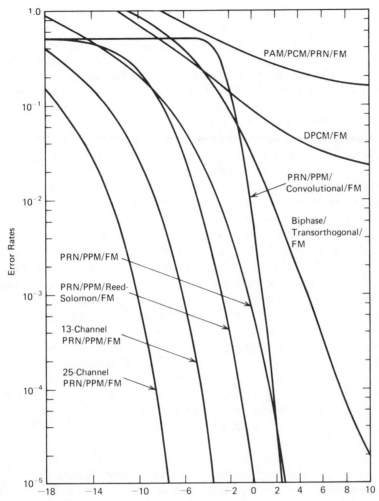

Figure 11.4. Channel-sharing performance comparisons of digital TV transmission systems for clutter-to-signal ratios of 5:1.

CONCLUSIONS

From the results shown it is easy to see that there are two standout systems for digital TV use, and especially for channel-sharing situations; the multi-channel PRN/PPM/FM systems. Following the multi-channel systems, in order of performance for error rates of 10^{-5}, are PRN/PPM/RS/FM, single-channel PRN/PPM/FM, PRN/PPM/convolutional/FM, bi-phase transorthogonal/FM, DPCM/FM, and PAM/PCM/PRN/FM. The single channel PRN/PPM/FM system is the easiest implemented and has good performance.

However, it should be mentioned that the multi-channel PRN/PPM/FM systems are a little beyond the state of the art, at the present time, as mentioned in Chapter 7. When present state-of-the-art conditions are used to compare system performance, the cost economy choice is single channel PRN/PPM/FM. If cost is not a consideration, the present state-of-the-art choice is PRN/PPM/REED-SOLOMON/FM. However, the state of the art can change very rapidly these days. Soon, we might be able to select the multi-channel PRN/PPM/FM systems.

It should be obvious, moreover, that the entire field of digital TV communication systems could not be covered here. It is possible, or even likely, that other stand-out systems exist or will soon exist.

There are many proposals for methods to reduce bandwidth in digital TV systems. Some form of sacrifice is incurred in all types of bandwidth reduction. However, some of these ideas are practical and worth applying.

Systems that produce reductions of 2:1 seem to have fewer penalties. Dither plus PCM or DPCM is one such method. Good results are obtained by the use of dither and the system implementation is not too complex. The significant bit selection systems are also worth pursuing, for a 2:1 reduction.

Dither can be combined with DPCM for a 3:1 bandwidth reduction, and quite successfully. However, DPCM systems are affected strongly by channel noise and need the aid of error correction methods.

The use of bit-plane encoding can give a 6:1 bandwidth reduction, at the cost of added implementation complexity. Hybrid systems combining interframe, and intraframe coding, with conditional replenishment look promising. However, the problem of distortion from objects in motion, and from fast panning, have not been completely solved as yet.

Hybrid systems that combine interframe systems with DPCM or with transform coding look promising.

Slant transform coding seems to be practical for real time digital TV. The other forms of transform coding; Karhunen-Loeve, fast Fourier, Walsh-Hadamard, and Haar are applicable to nonreal time digital TV. Transform coding can give excellent picture quality at greatly reduced bandwidths, but the system implementation complexity is high.

However, the use of the FFT with other types of transforms can speed up the computational ability of many of the other transform systems. For example, the DCT when combined with the FFT has been implemented to give excellent picture quality for real-time digital TV. The number theoretic transforms make use of a modified radix-2 FFT, and attain excellent picture quality. They could be made useful in the field of real time digital TV by the use of modulo type logic or computers.

The sine transform, which is called by some a "fast Karhunen-Loeve transform, could be combined with the FFT to give very excellent picture quality for real time digital TV use. It seems certain, as the state-of-the-art advances, that the transform systems will become more practical for use with real-time digital TV.

Bibliography and References

1. DIGITAL TELEVISION COMMUNICATION SYSTEMS

Aaron, M. R., J. S. Fleishman, R. A. McDonald, and E. N. Protonotarios (1969) "Response of Delta Modulation to Gaussian Signals," B. S. T. J., **48**, pp. 1167–1196, May/June.

Abate, J. E. (1967) "Linear and Adaptive Delta Modulation," *Proc. IEEE*, pp. 298–308, Mar.

Abbot, R. P., (1971), "A Differential Pulse Code Modulation Coder for Videotelephony Using Four Bits per Sample," *IEEE Trans Communi. Technology*, pp. 907–912, Dec.

Aeronutronic, Inc. (1959), "Performance Comparison of PAM/FM, PDM/FM, and PCM/FM Systems," Report no. U-743, (AD-234960), Vols. I, II, and III, Dec.

Agarwal, R. C., and C. S. Burrus, (1974), "Fast Convolution Using Fermat Number transforms with Applications to Digital Filtering"-*IEEE Trans. Acoust., Speech, and Signal Processing,* **22**, pp. 87–97, Apr.

Agarwal, R. C., and C. S. Burrus, (1975), "Number Theoretic Transforms to Implement Fast Digital Convolution" *Proc. IEEE*, **22**, pp. 550–560, Apr.

Agarwal, V. K., and T. J. Stephens, Jr., (1970), "On-Board Data Processor (Picture Bandwidth Compression)," Data Systems Lab., IRAD Report No. R 733.3–32, Feb.

Agarwal, J. P. and J. B. O'Neal, Jr., (1973), "Low Bit Rate DPCM Monochrome Television Signals," *IEEE Trans. Communications,* pp. 706–714, June.

Agrawal, J. P. (1972), "Comparison of Information Theory and Subjective Performance Measures for Television," Report Airforce Office Sci. Res. Contr.-F 44620-69-C-0033, June.

Ahiezer, N. I., and M. Krein, (1962), "Some Questions in the Theory of Moments," - *Ameri. Math. Soc.*, Providence, RI.

Ahmed, N., *et al.* (1973), "On Notation and Definition of Terms Related to a Class of Complete Orthogonal Functions," - *IEE Trans. Electromag. Compat.*, **15**, pp. 75–80, May.

Ahmed, N., T. Natarajan, and K. R. Rao, (1974), "Discrete Cosine Transform," *IEEE Trans. on Computat.*, pp. 90–93, Jan.

Ahmed, N., and K. R. Rao, (1975), "Orthogonal Transforms for Digital Signal Processing," Springer, New York/Berlin.

Aho, A. V., J. E. Hopcroft, and J. D. Ullman, (1974), *The Design and Analysis of Computer Algorithms,* Addison-Wesley, p. 101.

Albert, A. (1972), *Regression and the Moore-Penrose Pseudoinverse* Academic Press, New York.

Amello, G. F., W. J. Bertram, and M. F. Tompsett, (1971), "Charge-Coupled Imaging Devices! Design Considerations" - *IEEE Trans. Elect. Devices*, **18**, pp. 986–992, Nov.

Anderson, G. B., and T. S. Huang, (1971), "Piecewise Fourier Transformation for Picture Bandwidth Compression," *IEEE Trans. Commun. Tech.*, pp. 133–140, Apr.

Anderson, G. B., and T. S. Huang, (1972), "Correction to Piecewise Fourier Transformation for Picture Bandwidth Compression," *IEEE Trans. Commun. Tech.*, pp. 488–492, June.

Anderson, G. B. (1967), "Images and Fourier Transform," S. M. Thesis, Dept. Elec. Engr., M. I. T., Cambridge, May.

Anderson, G. B. and T. S. Huang, (1969), "Errors in Frequency-Domain Image Processing," in *Spring Joint Comput. Conf.*, pp. 173–185.

Ando, H. (1959), "Quantizing System of Higher Components of TV Signals," *J. Inst. Elect. Commun. Eng.* (Japan), **42**, p. 5.

Andrews, H. C. (1970), "Walsh Functions in Image Processing, Feature Selection and Pattern Recognition," *Proc. Symp. Walsh Functions*, Washington, DC., pp. 26–32.

Andrews, H. C., and W. K. Pratt, (1968), "Fourier Transform Coding of Images," *Conf. Rec. Hawaii Int. Conf. Syst. Sci.*, pp. 677–679, Jan.

Andrews, H. C. (1970), "Computer Techniques in Image Processing,"-Academic Press, New York.

Andrews, H. C., And W. K. Pratt, (1969), "Transform Image Coding," *Proc. Symp. Comput. Proc. Commun.*, pp. 63–84, Apr.

Andrews, H. C., and W. K. Pratt, (1970), "Digital Image Processing," *Appl. Walsh Functions Symp.*, pp. 183–194, Apr.

Andrews, H. C., and K. L. Caspari, (1971), "A Generalized Technique for Spectral Analysis," *IEEE Trans. Comput.*, pp. 16–25, Jan.

Andrews, H. C., and K. L. Caspari, (1971), "Degrees of Freedom and Computational Requirements in Matrix Manipulation," *IEEE Trans. Comput.*, Feb.

Andrews, H. C., A. G. Tescher, and R. P. Kruger, (1972), "Image Processing by Digital Computer," *IEEE Spectrum*, pp. 20–32, July.

Andrews, H. C., and C. L. Patterson, (1974), "Outer Product Expansions and Their Uses in Digital Image Processing," Aerospace Corp., El Segundo, CA, Report ATR-74 (8139)-2, Jan.

Andrews, H. C., and C. L. Patterson, (1976), "Singular Value Decompositions and Digital Image Processing," *IEEE Trans. Acoust. Speech, Signal Processing*, **24**, no. 1, pp. 26–53, Feb.

Andrews, H. C., and C. L. Patterson, (1974), "Singular Value Decomposition (SVD) Image Coding," *IEEE Trans. Commun.*, pp. 425–432, Apr.

Andrews, H. C., and W. K. Pratt, (1968), "TV Bandwidth Reduction by Encoding Spatial Frequencies," *J. Soc. Motion Picture and TV Eng.*, **77**, no. 12, pp. 1279–1281, Dec.

Andrews, H. C., J. M. Davies, and G. R. Schwarz, (1967), "Adaptive Data Compression," *Proc. IEEE*, **55**, pp. 267–277, Mar.

Andrews, H. C. (1968), "A High Speed Algorithm for the Computer Generation of Fourier Transforms," *IEEE Trans. Comput.*, p. 373, Apr.

Andrews, H. C. (1971), "Some Unitary Transformations in Pattern Recognition and Image Processing," *IFIP Congress* (Yugoslavia), Aug.

Andrews, H. C. (1970), "Fourier and Hadamard Image Transform Channel Error Tolerance," *Proc. UMR Mervin J. Kelly Commun. Conf.*, Rolla, MO, pp. 10-4-1–10-4-6, Oct.

Andrews, H. C. (1968), "Fourier Coding of Images," Univ. South California, Los Angeles, Report 271, June.

Angel, E., and A. K. Jain, (1972), "A Nearest Neighbors Approach to Multidimensional Filtering," *Conf. Decision and Control*, New Orleans, LA, Dec.

Angel, E., and A. K. Jain, (1973), "A Dimensionality Reducing Model for Distributed Filtering," *IEEE Trans. Automatic Controls*, **18**, pp. 59–62, Feb.

Angel, E., and R. Bellman, (1972), "Dynamic Programming and Partial Differential Equations," Academic, New York.

Anglin, H. (Ed.) (1965), "Demonstration of Feasibility of Air-to-Ground Photo Transmission in Near Real Time," Tech. Report No. SEG-TR-65-42,. Systems Eng. Group Res. Tech. Div. Airforce Syst. Command, Wright-Patterson Airforce Base, OH, Nov.

Apple, G. G. and P. A. Wintz, (1969), "*Experimental PCM System Employing Karhunen-Loeve Sampling,*" *Int. Symp. Information Theory,* Ellenville, NY, Jan.

Arazi, B. (1974), "Hadamard Transforms of Magnified, Expanded, and Periodic Signals," *IEEE Trans. Commun.*, pp. 848–850, June.

Arazi, B. (1974), "Two-Dimensional Digital Processing of One-Dimensional Signal," *IEEE Trans. Acoust., Speech, Signal Proc.,* 22, pp. 81–86, Apr.

Arnstein, W., H. W. Mergler, and B. Singer, (1964), "Digital Linear Interpolation and Binary Rate Multiplier," *Contr. Eng.,* II, pp. 79–83, June.

Arps, R. B. (1971), "The Statistical Dependence of Run-Lengths in Printed Matter," *Codierung, Nachrichtentech, Fachberichte*, 40, pp. 218–226.

Arps, R. B. (1969), "Entropy of Printed Matter at the Threshold of Legibility," Radioscience Lab., Stanford Univ., Stanford, CA, Sci. Report 31, June.

Arps, R. B., R. L. Erdman, A. S. Neal, and C. E. Schlaepfer, (1969), "Character Legibility Versus Resolution in Image Processing of Printed Matter," *IEEE Trans. Man-Mach. Syst.*, pp. 66–71, Sept.

Arps, R. B. (1970), "A Model for Legibility Dependence on Spatial Resolution," *SID IDEA Symp.*, pp. 8–81.

Ausherman, D. A., S. J. Dwyer, and G. S. Lodwick, (1972), "Extraction of Connected Edges from Knee Radiographs," *IEEE Trans. Comput.*, pp. 753–757, July.

Averback, E., and A. S. Coriell, (1961), "Short Term Memory in Vision," *B. S. T. J.*, 40, pp. 309–328, July.

Baker, H. D. (1963), "Initial Stages of Dark and Light Adaptation," *J. Opt. Soc. Am.*, 53, pp. 98–103, Jan.

Balder, J. C., and C. Kramer, (1962), "Video Transmission by Delta Modulation Using Tunnel Diodes," *Proc. IRE*, 50 pp. 428–431, Apr.

Barlow, H. B. (1964), "Dark Adpatation: A New Hypothesis," *Vision Res.*, 4, pp. 47–58, Jan./Feb.

Barlow, H. B. (1961), "*The Coding of Sensory Messages—Current Problems in Animal Behavior,*" Cambridge University Press, London.

Barrett, E. B., and R. N. Devich, (1976), "Linear Programming Compensation for Space-Variant Image Degradation," *Proc. SPIE/OSA Conf. Image Processing*, J. C. Urbach, (Ed.), Pacific Grove, CA, pp. 152–158.

Barstow, J. B., and H. N. Christopher, (1954), "The Measurement of Random Monochrome Video Interference," *Trans. AIEE, 73, (Commun. Electron.)*, 73, pp. 735–741, Jan.

Barstow, J. B., and H. N. Christopher, (1962), "The Measurement of Random Video Interference to Monochrome and Color Television Pictures," *Trans. AIEE (Commun. Electron.)*, 81, pp. 313–320, Nov.

Baumunk, J. F. and S. H. Roth, (1960), "Pictorial Data Transmission From Space Vehicle," *Elec. Eng.*, 79, pp. 134–138, Feb.

Bauch, H. H., H. Haberle, H. G. Mushmann, H. Ohnsorge, G. A. Wengenroth, and H. J. Woite, (1974), "Picture Coding," *IEEE Trans. Commun.*, pp. 1158–1167, Sept.

Bayless, J. W., S. J. Campanella, and A. J. Goldberg, (1973), "Digital Voice Communications," *IEEE Spectrum*, 10, pp. 28–34, Oct.

Beaudette, C. G. (1971), "An Efficient Facsimile System for Weather Graphics," *MIT Symp. Bandwidth Compression Proc.*, Gordon & Breach, pp. 217–229.

Becker, H. D., and J. G. Lawton, (1958/1961), "Theoretical Comparison of Binaty Data Transmission Systems," Cornell Aeronautical Lab., Buffalo, NY, Report PAM-621-3841B3, May/(Mar. 1961).

Beddoes, M. P. (1961) "Experiments with a Slope Feedback-Coder for Television Compression," *IRE Trans. Broadcasting,* pp. 12–28, Mar.

Beddoes, M. P. (1957), "A Variable Velocity Magnetic Scanning Process,"-IEE Monograph, No. 241R, pt. C, June.

Beddoes, M. P. (1963), "Two Channel Method for Compressing Bandwidth of Television Signals," *Proc. IEE,* **110**, pp. 369-378, Feb.

Bedfor, L. H., and O. S. Puckles, (1934), "A Velocity Modulation Television System," *J. IEE*, **75**, p. 63.

Bell, D. A. (1953), "*Economy of Bandwidth in Television,*" *J. Brit. IRE*, **13**, pp. 447–470.

Bell, D. A., and E. A. Howson, (1960), "Reduction of Television Bandwidth by Frequency Interlace," *J. Brit. IRE*, **29**, pp. 127–136, Feb.

Bell, D. A., and G. E. D. Swann, (1956), "Spectrum of Television Signals," *Wireless Eng.*, **33**, pp. 253–256, Nov.

Bellman, R. (1960), *Introduction to Matrix Analysis*, McGraw Hill, New York.

Bender, E. C., and R. D. Howson, (1948), "The Picturephone System-Wideband Data Service,"- *B. S. T. J.*, **27**, pp. 667–681, Feb.

Bennett, W. R. (1948), "Spectra of Quantized Signals," *B. S. T. J.*, **27**, pp. 446–472, July.

Benson, R. W., and I. J. Hirsh, (1953), "*Some Variables in Audio Spectrometry*"- *J. Acoust. Soc. Am.*, **25**, pp. 499–505, May.

Bially, T. (1969), "Space-Filling Curves: Their Generation and Their Application to Bandwidth Reduction," *IEEE Trans. Inform. Theory*, pp. 658–664, Nov.

Biberman, L. M. (1966), *Reticles in Electro-Optic Devices*, Pergamon, New York.

Biberman, L. M., and S. Nudelman, (1971), *Photoelectronic Imaging Devices*, Vols. I and II, Plenum Press, New York.

Biderman, K. (1970), "Image Encoding in Modulated Gratings from 1899-1970," *Opt. Acta.*, **17**, pp. 631–635, Aug.

Biernson, G., and A. W. Snyder, (1965), "Electromagnetic Effects in the Cone of the Human Retina," *Electron. Lett.*, pp. 89–90, June.

Billings, A. R. (1958), "A Coder for Halving the Bandwidth of Signals," *Proc. IEE*, **105**, pt. B, pp. 182–184, Mar.

Bisignani, W. T., G. P. Richards, and J. W. Whelan, (1966), "The Improved Gray Scale and the Coarse-Fine PCM Systems, Two New Digital TV Bandwidth Reduction Techniques," *Proc. IEEE*, **54**, pp. 376–390, Mar.

Blachman, N. M. (1972), "*Some Comments Concerning Walsh Functions,*" *IEEE Trans. Inform. Theory*, **18**, p. 427, May.

Blackman, R. B., and J. W. Tukey, (1959), *The Measurement of Power Spectra from the Point of View of Communications Engineering*, Dover, New York.

Blackwell, H. R. (1963), "Neural Theories of Simple Visual Discrimination," *J. Opt. Soc. Amer.*, **53**, pp. 129–160, Jan.

Bobilin, R. T., "*A Study of Prefilters and Sampling Rate for Intraframe Picturephone Coders*" - Unpublished memorandum.

Blasbalg, H., and R. Van Blerkom, (1962), "Message Compression," *IRE Trans. Space Electron. Telemetry*, pp. 228–238, Sept.

Bodycomb, J. V., and A. H. Haddad, (1970), "Some Properties of a Predictive Quantizing System," *IEEE Trans. Commun. Tech.*, pp. 682–684, Oct.

Boothroyd, W. P., and E. M. Creamer, (1949), "Dot Systems of Color TV," *Electronics*, pp. 88–92, Dec; (1950), pp. 96–99, Jan.

Bostelmann, G. (1974), "*A Simple High Quality DPCM Coder for Videotelephony Using 8 M Bits per Second*," *Nachrichtentech, Z.*, **27**, pp. 115, 117, Mar.

Bosworth, R. H., and J. C. Candy, (1969), "A Companded One-Bit Coder for Television Transmission," *B. S. T. J.*, **48**, May.

Bowyer, D. E. (1971), "Rate Calculations for Compressed Hadamard Transformed Data," Teledyne Brown Eng., Huntsville, AL, Summary Report MDS-Int-1338, June.

Bowyer, D. E. (1971), "Walsh Functions, Hadamard Matrices and Data Compression," Walsh Function Symp.

Boynton, R. M. (1961), Some Temporal Factors in Vision, and Sensory Communications, MIT Press, pp. 739–756.

Bradley, S. D. (1959), "Optimizing a Scheme for Run Length Coding," *Proc. IEEE*, **57**, pp. 108–109, July.

Bradley, S. D. (1970), "Data Compression for Image Storage and Transmission," *Proc. SID*, **11**, 4th Quarter, pp. 147–150.

Brainard, R. C. (1967), "Subjective Evaluation of PCM Noise—Feedback for Television," *Proc. IEEE*, **55**, pp. 346–353, Mar.

Brainard, R. C., and J. C. Candy (1967), "Direct-Feedback Coders: Design and Performance," *Proc. IEEE*, pp. 776–786, May.

Brainard, R. C., F. W. Kammerer, and E. G. Kimme, (1962), "Estimation of the Subjective Effects of Noise in Low-Resolution Television Systems," *IRE Trans. Inform. Theory*, **17**, pp. 99–106, Feb.

Brainard, R. C., R. Sadacca, L. J. Lopez, and G. N. Ornstein, (1966), "Development and Evaluation of a Catalog Technique for Measuring Image Quality," TR 1150, U. S. Army Personnel Res. Office, Aug.

Brainard, R. C., F. W. Mounts, and B. Prasada, (1967), "Low-Resolution TV: Subjective Effects of Frame Repetition and Picture Replenishment," *B. S. T. J.*, **57**, pp. 261–271, Jan.

Brainard, R. C., C. C. Cutler, and D. E. Rowlinson, "Picturephone Communications with Delay," private communications.

Brainard, R. C. (1967), "Low Resolution TV: Subjective Effects of Nosie Added to a Signal," *B. S. T. J.*, **46**, pp. 233–260, Jan.

Brault, J. W., and O. R. White, (1971), "The Analysis and Restoration of Astronomical Data via the Fast Fourier Transform," *Astron. Astrophys.*, **13**, pp. 169–189.

Brigham, E. O. (1974), *The Fast Fourier Transform*, Prentice-Hall, Englewood Cliffs, NJ.

Brindley, G. S. (1960), "*Physiology of the Retian and Visual Pathway*" Arnold, London, p. 144 ff.

Broadbent, D. E. (1958), *Perception and Communications*, Pergamon Press, London, p. 299.

Broadbent, D. E. (1961), "*Human Perception and Animal Learning*," in Current Problems in Animal Behavior, Cambridge Univ. Press, London.

Brofferio, S., and F. Rocca, (1970), "Simulation of Methods for Redundancy Reduction of the Video Signal," *Nat. Telemetry Conf.*

Broste, N. A. (1970), "Digital Generation of Random Sequences with Specified Autocorrelation and Probability Density Functions," U. S. Army Missile Command, Redstone Arsenal, AL, Report RE-TR-70-5, Mar.

Brown, E. F. (1969), "A Sliding-Scale Direct-Feedback PCM Coder for Television," - *B. S. T. J.*, **48**, pp. 1537–1553, May-June.

Brown, E. F. (1963), "A New Crispener Circuit for Television Images," *J. SMPTE*, **72**, pp. 849–853, Nov.

Brown, J. L. Jr., (1960), "Mean-Square Truncation Error in Series Expansions of Random Functions," *J. SIAM,* **8**, pp. 18–32, Mar.

Brown, H. E. (1971), "The Picturephone System: Transmission Plan," *B. S. T. J.,* pp. 351–391, Feb.

Brown, E. F. (1967), "Low-Resolution TV: Subjective Comparison of Interlaced and Non-Interlaced Pictures," *B. S. T. J.,* **46**, pp. 199–232, Jan.

Brown, C. G. (1970), "*Signal Processing Techniques Using Walsh Functions,*" Applications of Walsh Functions Symp., Apr.

Brown, R. D. (1977), "A Recursive Algorithm for Sequency-Ordered Fast Walsh Transform," *IEEE Trans. Comput.,* **26**, pp. 819–822, Aug.

Bruce, R. A. (1964), "Optimum Pre-Emphasis and De-Emphasis Networks for Transmission of Television PCM," *IEEE Trans. Commun. Tech.,* pp. 91–96, Sept.

Bruce, J. D. (1965), "Optimum Quantization," MIT. Res. Lab. Electronics, Tech. Report 429, Mar.

Bruch, W. (1961), "Transcoders PAL-NTSC: The Translation of a PAL Signal into an NTSC Signal and Visa Versa NTSC PAL," *Telefunken Zeitung,* **37**, pp. 115–135.

Budrikis, Z. L., and A. J. Seyler, (1965), "Detail Perception After Scene Changes in Television Image Presentation," *IEEE Trans. Inform. Theory,* pp. 31–43, Jan.

Budrikis, Z. L. (1961), "Visual Thresholds and the Visability of Random Noise in TV," *Proc. IRE* (Australia), **22**, pp. 751–759, Dec.

Budrikis, Z. L. "Visual Smoothing with Application to Television," Ph.D. dissertation, Univ. of Western Australia, Nedlands, Australia.

Budrikis Z. L. (1972), "Visual Fidelity Criterion and Modeling," *Proc. IEEE,* **60**, pp. 771–779, July.

Burkhardt, D. A. (1966), "Brightness and Increment Threshold," *J. Opt. Soc. Amer.,* **56**, pp. 979–981, July.

Burt, D. J. (Ed.), (1975), "Development of a CCD Image Sensor—Phase I,"-Report No. 16194C, ESTEC Contract No. 2256/74AK (England), Nov.

Butin, H. (1978), "On an Ordering of Walsh Functions," *IEEE Trans. Comput.,* **27**, pp. 87–90, Jan.

Butin, H. (1972), "A Compact Definition of Walsh Functions," *IEEE Trans. Comput,* **21**, pp. 590–592, June.

Cabrey, R. L. (1960), "Video Transmission Over Telephone Cable Pairs by PCM," *Proc. IRE,* **48**, pp. 1546–1561, Sept.

Cahn, L. *et al* (1957), "Experiments in Processing Pictorial Information with a Digital Computer," *Proc. Eastern Joint Comput. Conf.,* pp. 221–229.

Campbell, F. W., and J. G. Robson, (1968), "Application of Fourier Analysis to the Visibility of Grating," *J. Physiol.* (London), **197**, pp. 551–566.

Campbell, S. G. (1971), "Facsimile Transmission Systems," *Int. Conf. Commun.,* pp. 7.12–7.16.

Candy, J. C., M. A. Franke, B. G. Haskell, and F. W. Mounts, (1971), "Transmitting Television as Clusters of Frame-to-Frame Differences," *B. S. T. J.,* **50**, July-Aug.

Candy, J. C., and R. H. Bosworth, (1972), "Methods for Designing Differential Quantizers Based on Subjective Evaluations of Edge Busyness," *B. S. T. J.,* **51**, Sept.

Candy, J. C., and F. W. Mounts, (1971), "Redundancy Reduction System with Data Editing," U. S. Patent No. 3, 571, 807, Mar.

Cannon, T. M. (1974), "Digital Image Deburring by Nonlinear Homomorphic Filtering," Ph.D. dissertation, Univ. Utah, Dept. Computer Sci., Aug.

Capon, J. (1959), "A Probabilistic Model for Run-Length Coding of Pictures," *IRE Trans. Inform. Theory,* pp. 157–163, Dec.

Carl, J. W., and R. V. Smartwood, (1973), "A Hybrid Walsh Transform Computer," *IEEE Trans. Comput.*, **22**, pp. 669–672, July.

Chambers, J. P. (1973), "The Use of Coding Techniques to Reduce the Tape Consumption of Digital TV Recording," Conf. on Video and Data Recording, Birmingham, England, July.

Chan, D., and R. W. Donaldson, (1971), "Subjective Evaluation of Pre-and Post Filtering in PAM, PCM, and DPCM Voice Communication Systems," *IEEE Trans. Commun. Tech.*, pp. 601–612.

Chan, D. and R. W. Donaldson, (1971), "Optimum Pre-and Post Filtering of Sampled Signals, with Application to Pulse Modulation and Data Compression Systems," *IEEE Trans. Commun. Tech.*, pp. 141–157, Apr.

Chang, K., and R. W. Donaldson, (1972), "Analysis of Optimization and Sensitivity of DPCM Systems Operating on Noisy Communication Channels," *IEEE Trans. Commun. Tech.*, pp. 338–350, June.

Chen, W. H. (1973), "Slant Transform Image Coding," Univ. Southern California Image Processing Inst., Los Angeles, USCEE Report 441, May.

Chen, W. H., and W. K. Pratt, (1973), "Color Image Coding with the Slant Transform," *Proc. Symp. Walsh Functions*, Apr.

Cheng, D. K., and J. J. Liu, (1977), "An Algorithm for Sequency Ordering of Hadamard Functions," *IEEE Trans. Comput.*, **26**, pp. 308–309, Mar.

Cherry, E., *et al.* (1963), "An Experimental Study of the Possible Bandwidth Compression of Visual Image Signals," *Proc. IEEE*, pp. 1507–1517, Nov.

Cherry, E. C. (1962), "The Bandwidth Compression of TV and Facsimile," *Television Soc. J.*, **10**, pp. 40–49, Feb.

Cherry, E. C., and G. G. Gouriet, (1953), "Some Possibilities for the Compression of TV Signals by Recording," *Proc. IEE*, **100**, pp. 9–18, Jan.

Cherry, E. C. (1957), *On Human Communications*, MIT Press, Cambridge, MA.

Cherry, E. C., M. H. Kubba, D. E. Pearson, and M. P. Barton, (1963), "An Experimental Study of the Possible Bandwidth Compression of Visual Image Signals," *Proc. IEEE*, **51**, pp. 1507–1517, Nov.

Cherry, E. C., B. Prasada, and D. G. Holloway, (1958), "Open Loop System of TV Compression," Provisional Patent (British) No. 32376.

Chow, M. C. (1971), "Variable Length Redundancy Removal Coders for Differentially Coded Video Telephone Signals," *IEEE Trans. Commun. Tech.*, pp. 923–926, Dec.

Chu, D. C. (1974), "Spectrum Shaping for Computer Generated Holograms," Ph. D. dissertation, Dept. Elec. Eng., Stanford Univ. Stanford, CA.

Claire, E. J., S. M. Farber, and R. R. Green, (1971), "Practical Techniques for Transform Data Compression/Image Coding," *Proc. Applications of Walsh Functions Symp.* (Washington DC) pp. 2–6, Apr.

Claire, E. J. (1970), "Bandwidth Reduction in Image Tannsmission," Walsh Functions Symp., pp. 39.9–39.14.

Clinger, A. H., and D. R. Ziemer, (1961), "Reducing Channel Capacity Requirements in Digital Imagery Transmission: A Study Report," *7th Conf. Rec. Nat. Commun. Symp.*, pp. 75–85.

Cohen, F. (1972), "A Switched Quantizer for Nonlinear Coding of Video Signals," *Nachrichtentech Z.*, **25**, pp. 551–559, Dec.

Colas-Baudelarie, P. (1973), "Digital Picture Processing and Psychophysics," Dept. Rept. UTEC-C Sc-74-025, Mar.

Cole, E. R. (1973), "The Removal of Unknown Image Blurs by Homorphic Filtering," Univ. Utah, Dept. Comput. Sci., Report UTEC-C-Sc-029, June.

Connor, D. J., R. C. Brainard, and J. O. Limb, (1972), "Intraframe Coding for Picture Transmission," *Proc. IEEE*, **60**, pp. 779–791, June.

Connor, D. J. and J. O. Limb, "Properties of Frame Difference Signals Generated by Moving Images," to be published.

Connor, D. J., J. O. Limb, R. F. W. Pease, and W. G. Scholes, "Detecting the Moving Area in Noisy Video Signals," to be published.

Connor, D. J., B. G. Haskell, and F. W. Mounts, (1973), "A Frame-to-Frame Picturephone Coder for Signals Containing Differential Quantizing Noise," *B. S. T. J.*, **52**, pp. 35–51, Jan.

Connor, D. J., R. F. W. Pease, and W. G. Scholes, (1972), "Television Coding Using Two-Dimensional Spatial Prediction," *B. S. T. J.*, **50**, no. 3, pp. 1049–1061, Mar.

Cooley, J. W., and J. W. Tukey, (1956), "An Algorithm for the Machine Calculation of Complex Fourier Series," *Math. Computation*, **19**, pp. 297–301.

Cooley, J. W., P. A. W. Lewis, and P. D. Welch, (1969), "The Finite Fourier Transform," *IEEE Trans. Audio Electroacoust.*, **17**, pp. 77–85, June.

Corrington, M. S. (1978), "Implementation of Fast Cosine Transforms Using Real Arithmetic," *Proc. IEEE Nat. Aerospace and Electron. Conf.*, Dayton, OH, pp. 350–357.

Crawford, B. H. (1947), "Visual Adaptation in Relation to Brief Conditioning Stimuli," *Proc. Royal Soc.* (London), **134B**, pp. 283–302.

Crawther, W. R., C. M Rader, (1966), "Efficient Coding of Vocoder Channel Signals Using Linear Transformation," *Proc. IEEE*, **54**, pp. 1591–1595, Nov.

Cunningham, J. E. (1963), "Temporal Filtering of Motion Pictures," Sc. D. dissertation, Dept. Elec. Eng., MIT, Cambridge, MA.

Cunningham, J. E. (1959), "A Study of Some Methods of Picture Coding," M. S. Thesis, MIT, Cambridge, MA.

Curry, H. B. and I. J. Schoenberg, (1966), "On Polyan Frequency Functions IV. The Fundamental Spline Functions and Their Limits," *J. Anal. Math.*, **17**, pp. 219–235.

Cutler, C. C. (1960), "Transmission System Employing Quantization," Patent 2927962, Mar.

Cutler, C. C. (1952), "Differential Quantization of Communication Signals," Patent 2605361, July.

Davenport, W. B., and W. L. Root, (1958), *An Introduction to the Theory of Random Signals and Noise*, McGraw-Hill, New York.

Davisson, L. D. (1972), "Rate-Distortion Theory and Application," *Proc. IEEE*, **60**, pp. 800–808, July.

De Boor, C. (1969), "On Uniform Approximation by Splines," *J. Approx. Theory*, **1**, pp. 219–235.

De Jager, F. (1952), "Delta Modulation A Method of PCM Transmission Using the 1-Unit Code," Philips Res. Report, **7**, pp. 442–466.

De Lange, H. (1959), "Theoretical and Experimental Characteristics of Random Noise in TV," *J. British IRE*, **19**, June.

De Palma, J. J. and E. M. Lowry, (1962), "Sine Wave Response of the Visual System," *J Opt. Sco. Amer.*, **52**, Mar.

Deryugin, N. J. (1951), "The Power Spectrum and the Autocorrelation of the TV Signal," *Telecommunications*, **7**, pp. 1–12.

Deryugin, N. J. (1958), "Theoretical Possibilities of Compressing Bandwidth of TV Channel," *Telecommunications*, **8**, pp. 805–810.

Deutsch, R. (1965), *Estimation Theory*, Prentice-Hall, Englewood Cliffs, NJ.

Deutsch, S. (1971), "Visual Displays Using Pseudorandom Dot Scan," *Proc. Soc. Information Display*, **12**, pp. 131–146.

Deutsch, S. (1962), "Narrow-Band TV Uses Pseudorandom Scan," *Electronics*, **35**, pp. 49–51, Apr.

Deutsch, S. (1956), "The Possibilities of Reduced TV Bandwidth," *IRE Trans. Broadcast Television Receivers,* **2**, pp. 69–82, Oct.

Deutsch, S. (1965), "Pseudorandom Dot Scan TV Systems," *IEEE Trans. Broadcasting*, pp. 11–21, July.

Devereux, V. G., and D. J. Meares, (1972), "PCM of Video Signals: Subjective Effects of Random Digit Errors," BBC Res. Report 14.

Devereux, V. G. (1973), "Digital Video: Differential Coding of PAL Signals Based on Differences Between Samples One Subcarrier Period Apart," BBC Res. Report 7.

Doba, S. Jr., and J. W. Richey, (1960), *"Clampers in Video Transmission"* Trans. *AIEE*, **69**, pp. 477–487, Dec.

Dodd, P. D., and F. B. Wood, (1966), "Image Information, Classification, and Coding," *IEEE Int. Conf. Rec.*, pp. 60–71.

Dome, R. B. (1950), *"Frequency Interlace Color TV."* Electronics, **23**, pp. 1323–1331, Sept.

Donaldson, R. W., and R. J. Douville, (1969), "Analysis, Subjective Evaluation, Optimization and Comparison of the Performance Capabilities of PCM, DPCM, ΔM, AM, and PM Voice Communication," *IEEE Trans. Commun. Tech*, **17**, pp. 421–431, Aug.

Donaldson, R. W., and D. Chan, (1969), "Analysis and Subjective Evaluation of the DPCM Voice Communications Channel," *IEEE Trans. Commun. Tech.*, **17**, pp. 10–19, Feb.

Donchin, E. (1966), "A Multivariate Approach to the Analysis of Average Evoked Potentials," *IEEE Trans. Biomed. Eng.*, **13**, pp. 131–139, July.

Draper, N. R., and H. Smith, (1966), "Applied Regression Analysis," Wiley, New York.

Drury, R. W., (Ed.) (1970), Electronic Imaging Systems Symp., - Washington DC., *SPSE Publ.*

Dubois, E., and A. N. Venetsanopoulos, (1978), "The Discrete Fourier Transform Over Finite Rings with Application to Fast Convolution," *IEEE Trans. Comput.*, **27**, pp. 586–593, July.

Edward, J. A., and M. M. Fitelson, (1973), "Notes on Maximum Entropy Processing," *IEEE Trans. Inform. Theory*, **19**, pp. 232–234, Mar.

Ekstrom, M. P. (1978), "A Numerical Algorithm for Identifying Spread Functions of Shift-Invariant Imaging Systems," *IEEE Trans. Comput.*, **22**, pp. 322–328, Apr.

Ekstrom, M. P. (1972), "Computational Aspects of Solving Planar Integral Equations by Regularization," *Proc. Sixth Asilomar Conf. on Circuits and Syst.*, Pacific Grove, CA, pp. 489–493, Nov.

Ekstrom, M. P. (1974), "An Iterative Improvement Approach to the Numerical Solution of Vector Toeplitz Systems," - *IEEE Trans. Comput.*, **23**, pp. 320–325, March.

Elias, P. (1955), *"Predictive Coding,"* IRE Trans. Inform. Theory, pp. 16–33, Mar.

Elias, P. (1970), *"Bounds on Performance of Optimum Quantizers,"* IEEE Trans. Inform. Theory, pp. 172–184, Mar.

Eliot, A. R., and Y. Y. Shum, (1972), "A Parallel Array Hardware Implementation of the Fast Hadamard and Walsh Functions," Symp. Applications of Walsh Functions, Mar.

Enomoto, H., and K. Shibata, (1965), "Features of Hadamard Transformed TV Signal," *Nat. Conf. IECE of Japan*, p. 881.

Enomoto, H., and K. Shibata, (1969), "TV Signal Coding Method by Orthogonal Transformations," *Joint Conf. IECE* (Japan), p. 1430.

Enomoto, H., and K. Shibata, (1969), "Experiments on TV PCM System by Orthogonal Transformation," - *Joint Conf. IECE* (Japan), p. 2219.

Enomoto, H., and K. Shibata, (1970), "Orthogonal Transform Coding System for TV Signals," *J. Inst. TV Eng.* (Japan), **24**, pp. 99–108, Feb.

Enomoto, H., and K. Shibata, (1971),"Orthogonal Transform Coding for TV Signals," *IEEE Trans. Electromag. Compat.*, **13**, pp. 11–17, Aug.

Essman, J. E. (1972), "Theory and Application of DPCM and Comparison of DPCM with other Time Predictive Techniques," Ph. D. dissertation, Purdue Univ., Lafayette, IN, June.

Essman, J. E., and P. A. Wintz, (1968), "Redundancy Reduction of Pictorial Information," Midwestern Simulation Council Meeting on Adv. Math. Methods in Simulation, Oakland Univ.

Essman, J. E., and P. A. Wintz, (1973), "The Effects on Channel Errors in DPCM Systems and Comparison with PCM Systems," *IEEE Trans. Commun.*, pp. 867–877, Aug.

Fano, R. M. (1961), *Transmission of Information,* Wiley, New York, pp. 170–172.

Fano, R. M. (1961), *Transmission of Information, A Statistical Theory of Communication,* MIT Press, Cambridge, MA.

Fatechand, R. (1959), "Theoretical and Experimental Characteristics of Random Noise in TV," *J. British IRE*, pp. 335–344, June.

Fine, T. (1964), "Properties of an Optimal Digital System and Applications," *IEEE Trans. Inform. Theory,* **10**, pp. 287–296, Oct.

Fink, D. G. (1957), *Television Engineering Handbook*, McGraw-Hill, New York.

Fino, B. J., and V. R. Algazi, (1974), "The Slant Haar Transform," *Proc. IEEE,* **62**, pp. 653–654.

Fino, B. J., and V. R. Algazi, (1976), "Unified Matrix Treatment of the Fast Walsh-Hadamard Transform," *IEEE Trans. Comput.*, **25**, pp. 1142–1146, Nov.

Fitts, P. M. (1956), "Stimulate Correlates of Visual Pattern Recognition: A Probability Approach," *J. Experim. Psych.*, **51**, pp. 1–11.

Fleischer, P. E. (1964), "Sufficient Conditions for Achieving Minimum Distortion in a Quantizer," *IEEE Int. Conv. Rec.*, **12**, pp. 104–111.

Fox, L. (1954), "Practical Solution of Linear Equations and Inversion of Matrices," *Appl. Math. Ser. Nat. Bur. Stand.*, **39**, pp. 1–54.

Foy, W. H. Jr. (1964), "*Entropy of Simple Line Drawings,*" *IEEE Trans. Inform. Theory*, **10**, pp. 165–167, Apr.

Frankovitch, J. M., and H. P. Peterson, (1957), "A Functional Description of the Lincoln TX-2 Computer," *Proc. Western Joint Comput. Conf.,* pp. 146–155, Feb.

Franks, L. E. (1964), "Power Spectral Density of Random Facsimile Signals," *Proc. IEEE*, **52**, pp. 431–432, Apr.

Franks, L. E. (1966), "A Model for the Random Video Process," *B. S. T. J.*, **45**, pp. 609–630, Apr.

Fredendall, G. L., K. H. Powers, and H. Staras, (1957–1958), "Some Relations Between TV Picture Redundancy and Bandwidth Requirements," *Proc. Conf. on Physical Problems of Color Television, Acta Electronica*, **2**, pp. 378–383.

Fredendall, G. L., and W. L. Behrend, (1960), "Picture Quality Procedure for Evaluation Subjective Effects of Interference," *Proc. IRE*, **48**, pp. 1030–1034, June.

Freeny, S. L., R. B. Kieburtz, K. V. Mina, and S. K. Tewksury, (1971), "System Analysis of a TDM-FDM Translator/Digital A-Type Channel Bank," *IEEE Trans. Commun. Tech.*, pp. 1050–1069, Dec 19.

Frei, W., and P. A. Jaeger, (1973), "Some Basic Considerations for the Source Coding, of Color Pictures," *Int. Conf. Commun.*, Seattle, WA.

Frieden, B. R. (1971), "Restoring with Maximum Likelihood," Univ. Arizona, Optical Sci. Center, Tucson, AZ, Tech. Report 67, Feb.

Frieden, B. R. (1974), "2-D Restoration by Decision Allocation of Pseudograms," *Proc. Spring Meeting Optical Soc. Amer.*, Washington, DC, Apr.

Frieden, B. R. (1967), "Bandlimited Reconstruction of Optical Objects and Spectra," *J. Opt. Sci. Amer.*, pp. 1013–1019, Aug.

Frieden, B. R. (1976), "Maximum Entropy Restoration of Garrymede," *Proc. SPIE/OSA Conf. Image Processing*, J. C. Urbach, (Ed.), Pacific Grove, CA, **74**, pp. 160–165, Feb.

Fu, K. S. (1973), *Sequential Methods in Pattern Recognition and Machine Learning,* Academic Press, New York.

Fukanga, K. (1973), *An Introduction to Statistical Pattern Recognition*, Academic Press, New York.

Fukinuki, T. (1974), "Optimization of DPCM for TV Signals with Consideration of Visual Property," - *IEEE Trans. Commun.*, pp. 821–826, June.

Fukinuki, T., *et al.* (1971), "Degradation of TV Signals by Transmission Errors in DPCM" (in Japanese), *Rec. Tech. Group Commun. Syst., Inst. Electron. Commun. Eng. (Japan)*, 113, Mar.

Fukushima, K., and H. Ando, (1964), "TV Band Compression by Multimode Interpolation," *J. Inst. Elec. Commun. Eng.* (Japan), 47, pp. 55–64.

Fultz, K. E., and D. B. Pennick, (1965), "The T. I. Carrier System," -, *B. S. T. J.*, 44, p. 1405, Sept.

Gabor, D., and P. C. J. Hill, (1961), "TV Band Compression by Contour Interpolation," *Proc. Inst. Elec. Eng.*, 108, pt. B, pp. 303–315.

Gabor, D. (1959), "Television Compression by Contour Interpolation," no. 2, Nuovo Cimento, suppl., p. 467.

Gallager, R. G. (1968), *Information Theory and Reliable Communications*," Wiley, New York, pp. 442–502.

Gardenhire, L. W. (1971), "A Method for Synchronizing Statistical Encoding," Pat. Appl. Rad. Inc., May.

Gattis, J. L., and P. A. Wintz, (1971), "Automated Techniques for Data Analysis and Transmission," Purdue Univ., Tech. Report TR- EE-71-37, Aug.

Geddes, W. K. E. (1963), "Picture Processing by Quantization of the Time Derivative," BBC Eng. Div. Report T-114.

Geddes, W. K. E. and G. F. Newell, (1962), "Test of Three Systems of Bandwidth Compression of TV Signals," *Proc. IEE*, 109, pt. 5, pp. 311–324, July.

Gentleman, W. M. (1968), "Matrix Manipulation and Fast Fourier Transformations," *B. S. T. J.*, 47, pp. 1099–1103, July-Aug.

Gerrard, A. A., and J. E. Thompson, (1973), "An Experimental DPCM Encoder for Viewphone Signals," *IERE J.*, 43, Mar.

Gibson, J. J. (1950), *The Perception of the Visual World*, Houghton Mifflin, Boston, MA.

Gicca, F. A. (1962), "Digital Spacecraft TV," *Space Aeronautics*, 38, pp. 73–78, Dec.

Gilbert, E. N., and E. F. Moore, (1959), "Variable Length Binary Encodings," *B. S. T. J.*, 38, pp. 933–966, July.

Gish, H. (1967), "*Optimum Quantization of Random Sequences*," Defense Supply Agency, Report AD 656042, May.

Gish, H., and J. N. Pierce, (1960), "Asymptotically Efficient Quantizing," *IEEE Trans. Inform. Theory*, 14, pp. 676–683, Sept.

Goblick, T. J., and J. L. Hoslinger, (1967), "Analog Source Digitation: A Comparison of Theory and Practice," *IEEE Trans. Inform. Theory*, 13, pp. 323–326, Apr.

Goldfarb, D., and L. Lapidus, (1968), "Conjugate Gradient Method for Nonlinear Programming Problems with Linear Constraints," *Ind. Eng. Chem. Fundamentals*, 7, pp. 142–151.

Golding, L. S. and R. K. Garlow, (1971), "Frequency Interleaved Sampling of a Color TV Signal," *IEEE Trans. Commun. Tech.*, 19, pp. 972–979, Dec.

Golomb, S. W., I. S. Reed, and T. K. Truong, "Integer Convolutions Over the Finite Field GF $(3 \times 2^n + 1)$," *SIAM J. Appl. Math.*, 32, Mar.

Golub, G. H., and C. Reinsch, (1970), "Singular Value Decomposition and Least Squares Solutions," *Numer. Math.*, 14, pp. 403–420.

Golub, G. H. (1969), "Matrix Decomposition and Statistical Calculations," in *Statistical Computation*, R. C. Milton and J. A. Nelder, (Eds.), Academic, New York, pp. 365–367.

Golub, G. H., and M. A. Saunders, (1970), "Linear Least Squares and Quadratic Programming," in *Integer and Nonlinear Programming*, J. Abadic, (Ed.), Amsterdam, The Netherlands: North-Holland, p. 229.

Good, I. J. (1958), *"The Interaction Algorithm and Practical Fourier Analysis,"* J. Royal Statistical Sci., Series B, **20**, pp. 361–372.

Goodall, W. W. (1961), "Television by PCM," *B. S. T. J.*, **30**, pp. 33–49, Jan.

Goodman, J. W. (1968), *Introduction to Fourier Optics*, McGraw-Hill, New York.

Goodman, D. J. (1969), "The Application of Delta Modualtion to Analog-to-PCM Encoding," *B. S. T. J.*, **48**, pp. 321–343, Feb.

Goodman, L. M. (1971), "Binary Linear Transformation for Redundancy Reduction," *Proc. IEEE*, **55**, pp. 467–468, Mar.

Goodman, L. M., and P. R. Drouliet, (1966), "Asymptotically Optimum Pre-Emphasis and De-Emphasis Networks for Sampling and Quantizing," *Proc. IEEE*, **54**, pp. 795–796, May.

Gorog, J., *et al.* (1971), "An Experimental Low-Cost Graphic Information Distribution System Terminal," *SID Digital Tech. Papers*, pp. 14–15, May.

Gouriet, G. G. (1952), "A Method of Measuring TV Picture Detail and its Applications," *Electron. Eng.*, p. 306.

Gouriet, G. G. (1957), "Bandwidth Compression of a TV Signal," *Proc. IEE*, Paper No. 2357R, (**104B**, p. 265), May.

Gouriet, G. G. (1956), "Bandwidth Reduction in Relation to TV," *Brit. Commun. Electron.*, **3**, pp. 424–429.

Gouriet, G. G. (1952), "Dot Interlaced TV," *Electron. Eng.*, **24**, pp. 166–171, Apr.

Graham, R. E. (1957-1958), "Communication Theory Applied to TV Coding," *Acta Electronica*, **2**, pp. 334–343.

Graham, R. E. (1958), "Predictive Quantizing of TV Signals," *IRE WESCON Conv. Rec.*, pt. 4, pp. 147–157.

Graham, R. E., and J. L. Kelly, Jr., (1958), "A Computer Simulation Chain for Research on Picture Coding," *IRE WESCON Conv. Rec.*, pt. 4, pp. 41–46.

Graham, R. E. (1958), "Subjective Experiments in Visual Communication," *IRE Nat. Conv. Rec.*, pt. 4, pp. 100–106.

Graham, D. N. (1967), "Image Transmission by Two-Dimensional Contour Coding," *Proc. IEEE*, **55**, pp. 336–346, Mar.

Gregory, R. L. (1956), "An Experimental Treatment of Vision as an Information Source and Noisy Channel," Symp. Inform. Theory, London, Butterworth Sci. Publ., 1956.

Grenander, V., and G. Szego, (1958), Toeplitz Forms and Their Applications, Univ. of California Press, Berkeley and Los Angeles.

Greville, T. N. E. (1967), "Spline Functions Interpolation, and Numerical Quadrature," *Math. Methods for Digital Comput.*, **2**, A Ralston, and H. S. Wilf, (Eds.), Wiley, New York.

Greville, T. N. E. (1969), *Theory and Applications of Spline Functions,* Academic, New York.

Guillemin, E. A. (1957), *Synthesis of Passive Networks*, Wiley, New York, p. 279.

Guzman, A. (1968), *Decomposition of a Visual Scene into Three Dimensional Bodies, Fall Joint Comput. Conf., AFIPS Proc.*, Washington DC, Spartan Press, pp. 291–304.

Haantjes, J., and K. Teer, (1954), "Multiplex TV Transmission: Subcarrier and Dot Interlace Systems," *Wireless Eng.*, **31**, pt. II, pp. 266–273, Oct.

Haar, A. (1955), "Zur Theorie der Orthogonalen Funktionen-System," Inaugural dissertation, *Math. Annalen*, **5**, pp. 17–31.

Habibi, A. "Performance of Zero-Memory Quantizers Using Rate Distortion Criteria," to be published.

Habibi A. (1971), "Comparison at Nth Order DPCM Encoder with Linear Transformations and Block Quantization Techniques," *IEEE Trans. Commun. Tech.*, pp. 948–956, Dec.

Habibi, A. (1974), *"Hybrid Coding of Pictorial Data,"* *IEEE Trans. Comun.*, pp. 614–624, May.

Habibi, A. and P. A. Wintz, (1971), *"Image Coding by Linear Transformation and Block Quantization,"* *IEEE Trans. Commun. Tech.*, pp. 50–62, Feb.

Habibi, A., and P. A. Wintz, (1969), "Optimum Linear Transformations for Encoding Two-Dimensional Data," Purdue Univ., Tech. Report TR-EE-69-15, Lafayette, IN, May.

Habibi, A., and P. A. Wintz, (1970), "Linear Transformations for Encoding Two-Dimensional Sources," School of Elec. Eng., Purdue Univ., Lafayette, IN, Tech. Report TR-EE-70-2, June.

Habibi, A. (1972), "Two-Dimensional Bayesian Estimate of Images," *Proc. IEEE*, **60**, pp. 878–883, July.

Habibi, A., and R. S. Hershell, (1974), "A Unified Representation of Differential Pulse Code Modualtion (DPCM) and Transform Coding Systems," *IEEE Trans. Commun.*, pp. 692–696, May.

Habibi, A., and P. A. Wintz, (1970), "Picture Coding by Linear Transformation," IEEE Int. Symp. Inform. Theory, The Netherlands, June 1970.

Habibi, A. (1972) "Data Modulation and DPCM Coding of Color Signals," *Proc. Int. Telemetry Conf.*, **8**, Oct.

Habibi, A. (1975), Study of On-Board Compression of Earth Resources Data," Tech. Report, Contract No. NAS 2-8394, TRW No. 26566, TRW Syst. Group, Redondo Beach, CA, Sept.

Hall, E. L. (1974), "Almost Uniform Distribution for Computer Image Enhancement," *IEEE Trans. Comput.*, pp. 207–208, Feb.

Hall, E. L. *et al* (1971), "A Survey of Preprocessing and Feature Extraction Techniques for Radiographic Images," *IEEE Trans. Comput.*, pp. 1032–1044, Sept.

Hallbert, B. (1960), *Photogrammetry*, McGraw-Hill, New York.

Han, K. S., G. J. Berzins and S. T. Donaldson, (1973), "Pulsed X-Ray Imaging by Incoherent Holographic Techniques," *Proc. Electro-Opt. Syst. Des. Conf.*, Sept.

Hanson, R. J. (1971), "A Numerical Method for Solving Fredholm Integral Equations of the First Kind Using Singular Values," *SIAM J. Numer. Anal.*, **8**, pp. 616–622, Sept.

Happ, W. W. (1969), "Coding Schemes for Run-Length Information Based on Poisson Distribution," *Nat. Telemetering Conf. Rec.*, pp. 257–261, Apr.

Harmuth, H. F. (1968), "A Generalized Concept of Frequency and Some Applications," *IEEE Trans. Inform. Theory*, **14**, pp. 375–382, May.

Harmuth, H. F. (1969), "Applications of Walsh Functions in Communications," *IEEE Spectrum*, **6**, pp. 82–91, Nov.

Harmuth, H. F. (1970), Transmission of Information by Orthogonal Functions, Springer-Verlag, New York, pp. 22–23.

Harris, J. L. (1966), "Diffraction and Resolving Power," *J. Opt. Soc. Amer.*, **56**, pp. 569–574, May.

Harrison, C. W. (1952), "Experiments with Linear Prediction in TV," *B. S. T. J.*, pp. 764–784, July.

Hartline, H. K. (1964), "The Response of Single Optic Nerve Fibers of the Vertebrate Eye to Illumination of the Retina," *Basic Readings in Neuropsychology*, Tokyo, John Weatherhill, Inc.

Haskell, B. G. (1972), "Buffer and Channel Sharing by Several Interframe Picturephone Coders," *B. S. T. J.*, **51**, pp. 261–289, Jan.

Haskell, B. G., J. O. Limb, and R. F. W. Pease, (1973), "Frame-to-Frame Redundancy Reduction System Which Transmits and Intraframe Coded Signal," U. S. Pat. No. 3,767,817, Oct.

Haskell, B. G. (1974), "Frame-to-Frame Coding of TV Pictures Using Two-Dimensional Fourier Transforms," *IEEE Trans. Inform. Theory*, pp. 119–120, Jan.

Haskell, B. G., and J. O. Limb (1972), "Prediction Video Encoding Using Measured Subject Velocity," U. S. Patent No. 3,632,865, Jan.

Haskell, B. G., F. W. Mounts, and J. C. Candy, "Interframe Coding of Videotelephone Pictures," *Proc. IEEE,* **60**, pp. 792–800, July.

Hayes, J. F., and R. Bobilin, "Efficient Waveform Encoding," School of Elec. Eng., Purdue Univ., Lafayette, IN, Tech. Report TR-EE-69-4, Feb.

Hayes, J. F., A. Habibi, and P. A. Wintz, (1970), "Rate Distortion Function for a Gaussian Source Model of Images," *IEEE Trans. Inform. Theory*, pp. 507–509, July.

Haynes, H. E., and D. T. Hogers, (1958), "Stop-Go Scanning Saves Spectrum Space," *Electronics*, Sept.

Helstrom, C. W. (1967), "Image Restoration by the Method of Least Squares," *J. Opt. Soc. Amer.*, **57**, pp. 297–303, Mar.

Henderson, K. N. (1964), "Some Notes on Walsh Functions," *IEEE Trans. Electromag. Compat.*, **13**, pp. 50–52, Feb.

Herbst, N., and P. M. Will, (1969), "Design of a Experimental Laboratory for Pattern Recognition and Signal Processing," IBM Res. Rept. RC-2619, Sept.

Hershel, R. S. (1971), "*Unified Approach to Restoring Degraded Images in the Presence of Noise,*" Univ. of Arizona, Opt. Sci. Center, Tucson, AZ, Tech. Report 72, Dec.

Hestenes, M. R. (1966), "*Calculus of Variations and Optimal Control Theory,*" Wiley-Interscience, New York.

Hestenes, M. R. (1975), "*Optimization Theory,*" Wiley-Interscience, New York.

Hestenes, M. R. (1975), "Pseudoinverses and Conjugate Gradients," *Commun. Ass. Comput. Mach.*, **18**, pp. 40–43, Jan.

Hestenes, M. R. (1975), "*The Conjugate Gradient Method for Solving Linear Systems,*" Proc. *Symp. Applied Math. and Numerical Analysis.*

Hestenes, M. R., and E. Stiefel, (1952), "Methods of Conjugate Gradients for Solving Linear Systems," *J. Res. Nat. Bur. Standards*, **49**, pp. 409–436.

Hochman, D., and D. R. Weber (1970), "DACOM Facsimile Data Compression Techniques," *Int. Conf. Commun.*. pp. 20.14–20.21.

Hochman, D., H. Katzman, and D. R. Weber, (1967), "Application of Redundancy Reduction to TV Bandwidth Compresion," *Proc. IEEE*, **55**, pp. 263–267, Mar.

Hoerl, A. E., and R. W. Kennard, (1970), "Ridge Regression: Biased Estimation for Non-Orthogonal Problems," *Technometrics*, **12**, pp. 55–68, Feb.

Holway, A. H., and E. G. Boring, (1941), "Determinants of Apparent Visual Size with Distance Variant," - *Amer. J. Psychology,* **51**, pp. 21–37.

Honnel, M. A., and M. D. Prince, (1951), "Television Image Reproduction by Use of Velocity-Modulation Principles," *Proc. IRE*, **39**, pp. 265–268, Mar.

Horner, J. L. (1969), "Optical Spatial Filtering with the Least-Mean-Square Error Filter," *J. Opt. Soc. Amer.*, **59**, pp. 297–303, May.

Horner, J. L. (1970), "Optical Restoration of Images Blurred by Atmospheric Turbulence Using Optimal Filter Theory," *Appl. Opt.*, **9**, pp. 167–171, Jan.

Hotelling, H. (1933), "Analysis of a Complex of Statistical Variables into Principle Components," *J. Educ. Psychology,* **24**, pp. 417–441, 498, 520.

Hou, H. S., and H. C. Andrews, (1977), "Least Squares Image Restoration Using Spline Functions," *IEEE Trans. Comput.*, **26**, pp. 856–873, Sept.

Huang, T. S., and J. W. Woods, (1969), "Picture Bandwidth Compression by Block Quantization," Int. Symp. Info. Theory, Ellenville, NY.

Huang, T. S., and M. T. Chikhaoui, (1967), "The Effect of BSC on PCM Picture Quality," *IEEE Trans. Inform. Theory*, **13**, pp. 270–273, Apr.

Huang, T. S., O. J. Tretiak, B. Prasada, and Y. Yamaguchi, (1967), "Design Considerations on PCM Transmission of Low Resolution Monochrome Still Pictures," *Proc. IEEE*, **55**, pp. 331–335, Mar.

Huang, T. S. (1965), "The Subjective Effect of Two-Dimensional Pictorial Noise," *IEEE Trans. Inform. Theory*, pp. 43–53, Jan.

Huang, T. S., and O. J. Tretiak, (1968), "A Pseudorandom Multiplex System for Facsimile Transmission," *IEEE Trans. Commun. Tech.*, pp. 436–438, June.

Huang, T. S. (1972), "Run-Length Coding and its Extensions," in *Picture Bandwidth Compression*, Gordon & Breach, New York, pp. 231–266.

Huang, T. S. (1965), "PCM Picture Transmission," *IEEE Spectrum*, **2**, pp. 57–63, Dec.

Huang, T. S., and F. W. Scoville, (1965), "The Subjective Effect of Spatial and Brightness Quantization in PCM Picture Transmission," *Proc. NEREM Conv. Rec.*, Nov.

Huang, T. S. (1966), "Digital Picture Coding," *Proc. Nat. Electron. Conf.*, pp. 793–797.

Huang, T. S., and P. M. Narendra, (1975), "*Image Restoration by Singular Value Decomposition,*" *Appl. Opt.*, **14**, no. 9, pp. 2213–2216, Sept.

Huang, T. S. (1976), "*Restoring Images with Shift Varying Degradations,*" *Proc. SPIE/OSA Conf. on Image Processing*, J. C. Urbach, (Ed.), Pacific Grove, CA, **74**, pp. 149–151, Feb.

Huang, T. S., G. J. Yang, and G. Y. Tang, (1978), "A Fast Two-Dimensional Median Filtering Algorithm," *Proc. IEEE Computer Soc. Conf. on Pattern Recognition and Image Processing*, Chicago, IL, pp. 128–131, May/June.

Huang, T. S., D. S. Baker, and S. P. Berger, (1975), "Iterative Image Restoration," *Appl. Optics*, **14**, pp. 1165–1168, May.

Huang, T. S., and P. T. Hartmen, (1967), "Subjective Effect of Additive White Pictorial Noise," MIT Res. Lab. of Elect., Q. P. R., No. 85, pp. 317–319, Apr.

Huang, T. S. (1963), "Two-Dimensional Power Density Spectrum of TV Random Noise," Quarterly Progress Report, No. 69, MIT Res. Lab., Cambridge, MA, pp. 143–149, Apr.

Huang, J. J. Y., and P. M. Schultheis, (1963), "Block Quantization of Correlated Gaussian Random Noise," *IEEE Trans. Commun.*, **11**, pp. 289–296, Sept.

Hunt, B. R., D. H. Janney, and R. K. Zeigler, (1970), "An Introduction to Restoration and Enhancement of Radiographic Images," Los Alamos Sci. Lab., Los Alamos, NM, Report LA-4305.

Hunt, B. R., D. H. Janney, and R. K. Zeigler, (1973), "Radiographic Image Enhancement by Digital Computers," *Mater. Eval. J. ASNT.*, **31**, pp. 1–5.

Hunt, B. R. (1973), "The Application of Constrained Least Squares Estimation to Image Restoration by Digital Computer," *IEEE Trans. Comput.*, **22**, pp. 805–812, Sept.

Hunt, B. R. (1972), "A Theorem on the Difficulty of Numerical Deconvolution," *IEEE Trans. Audio Electroacoust.*, **20**, pp. 94–95, Mar.

Hunt, B. R., and H. C. Andrews, (1973), "Comparison of Different Filter Structures for Image Restoration," *Proc. 6th Annu. Hawaii Int. Conf. Syst. Sci.*, Jan.

Hunt, B. R. (1972), "Data Structures and Computational Organization in Digital Image Enhancement," *Proc. IEEE*, **60**, pp. 884–887, July.

Hunt, B. R. "Bayesian Methods in Digital Image Restoration," to be published.

Hunt, B. R., and D. H. Janney, (1974), "Digital Image Processing at Los Alamos Scientific Laboratory," *Computer*, **7**, pp. 57–62, May.

Hunt, B. R. (1975), "Digital Image Processing," *Proc. IEEE*, **63**, pp. 693–708, Apr.

Hunt, R. W. G. (1967), *The Reproduction of Color*, Wiley, New York.

Iijima, Y., T. Oshima, and T. Ishiguro, (1972), "Spectrum of Granular Noise of Double Integration ΔM," *IECE Commun. Group Meeting Rec.,* Paper CS72-13, May.

Inose, H. and Y. Yasuda, (1963), "A Unity Bit Coding Method by Negative Feedback," *Proc. IEEE,* **51,** pp. 1524–1535, Nov.

Inoue, S. and K. Shibata, (1968), "Quantizing Noise of Orthogonal Transform TV PCM System," *Nat. Conf. IECE (Japan),* p. 1310.

Ishiguro, T. and H. Kaneko, (1971), "A Nonlinear DPCM Code Based on ΔM/DPCM Code Conversion with a Digital Filter," *IEEE Commun. Conf. Rec.,* pp. 1.27–1.32.

Iwersen, J. E. (1969), "Calculated Quantizing Noise of Single Integration Delta-Modulation Coders," *B. S. T. J.,* **48,** pp. 2359–2389, Sept.

Jain, A. K., and E. Angel, (1974), "Image Restoration, Modeling, and Reduction of Dimensionality," *IEEE Trans. Comput.,* **23,** pp. 470–476, May.

Jain, A. K. (1976), "Fast Inversion of Banded Toeplitz Matrices by Circular Decomposition," Tech. Report AJ-76-001, Dept. of Elec. Eng., SUNY, Buffalo, NY, Jan.

Jain, A. K. (1972), "Linear and Nonlinear Interpolation for Two-Dimensional Image Enhancement," Conf. Decision and Control, Dec.

Jain, A. K. (1974), "A Fast Karhunen-Loeve Transform for Finite Discrete Images," *Proc. of Nat. Elec. Conf.,* Chicago, IL, pp. 323–328, Oct.

Jain, A. K., S. H. Wang, and Y. Z. Liao, (1976), "Fast Karhunen-Loeve Transform Data Compression Studies," *Proc. Nat. Elec. Conf.,* pp. 6.5-1–6.5-5.

Jain, A. K. (1976), "Computer Program for Fast Karhunen-Loeve Algorithm," NASA Contract NAS 8-31434, Final Report, Dept. of Elec. Eng., SUNY, Buffalo, NY, Feb.

Jain, A. K. (1977), "A Fast Karhunen-Loeve Transform for Digital Restoration of Images Degraded by White and Colored Noise," *IEEE Trans. Comput.,* **26,** pp. 560–571, June.

Kaul, P. (1970), "Differential PCM Encoding of TV Signals Using a Digital Loop," *Int. Conf. Commun., Conf. Rec.,* Paper 70-CP-202, pp. 2-16–2-23.

Kay, N. D., C. F. Knapp, and W. F. Schreiber, (1958), "Synthetic Highs—An Experimental TV Bandwidth Reduction System," *84th SMPTE Conv.,* pp. 1–18, Oct.

Kazumoto, I., Yukihiko II Jima, Tatsuo Ishiguro, Haruo Kaneko, and Seiichero Shigaki— "*Interframe Coding for 4-MHz Color TV Signals,*" *IEEE Trans. Commun.,* **23,** Dec. 1975, pp. 1461–1465.

Kekre, H. B., and J. K. Solanki, (1971), "*Modified Slant and Modified Haar Transforms for Picture Bandwidth Compression,*" *Proc. Midwest Symp. Circuits & Systems,* pp. 718–721.

Kell, R. D. (1929), "Improvements Relating to Electronic Picture Transmission Systems," British Patent No. 344811, Apr.

Kelly, D. H. (1966), "Vision Research with Harmonically Related Pure Stimuli," ITEK Corp., Palo Alto, CA, VIDYA Report 221, May.

Kelly, R. D. (1962), "Information Capacity of a Single Retinal Channel," *IRE Trans. Inform. Theory,* **8,** pp. 221–226, Apr.

Kelly, D. H. (1961), "Visual Response to Time-Dependent Stimuli," *J. Opt. Soc. Amer.,* **51,** pp. 422–429, Apr.

Kelly, D. H. (1966), "Frequency Doubling in Visual Response," *J. Opt. Soc. A.,* **56,** pp. 1628–1633, Nov.

Kendall, M. G. (1943), *The Advanced Theory of Statistics,* Charles Griffin, London, 1943.

Kennedy, J. D., S. J. Clark, and W. A. Parkyn, (1970), "Digital Imagery Processing Techniques," McDonnell Douglas Astronautics Co. Report No. MDC-GO 402, Jan.

Kennedy, J. D. (1970), "Walsh Function Imagery Analysis," *Proc. Walsh Functions Symp.,* Washington DC, pp. 7–10.

Kent, W. H., (1973), "Charge Distribution in Buried-Channel Devices," *B. S. T. J.,* **52,** pp. 1009–1023.558, July/Aug.

Kharekevich, A. A. (1958), "Comparison of a Few Possible Methods of Transmitting Simple Drawings," *Telecommunications* (USSR), (transl. *Elektrosvyas*), (London), No. 5, pp. 527–540, May.

Kim, C. E., and M. G. Strintzis, (1978), "High Speed Two-Dimensional Convolution," *Proc. IEEE Computer Soc. Conf. Pattern Recognition and Image Processing*, pp. 123–127, May/June.

Kimme, E. G., "Methods of Optimal System Design for a PCM Video System Employing Quantization Noise Feedback," unpublished.

Kimme, E. G., and F. F. Kuo, (1963), "Synthesis of Optimal Filters for a Feedback Quantization System," *IEEE Trans. Circuit Theory*, **10**, pp. 405–413, Sept.

Kitsopoulos, S. C. and E. R. Kretzmer, (1961), "Computer Simulation of a TV Coding Scheme," *Proc. IRE*, **49**, pp. 1076–1077, June.

Klie, J., "Video Telephone System with Increased Resolution for Facsimile Transmission," to be published.

Knapp, C. F., and W. F. Screiber, (1958), "TV Bandwidth Reduction by Digital Coding," *IRE Nat. Conv. Rec.*, pt. 4, pp. 88–90.

Knauer, S. C. (1976), "Real-Time Video Compression Algorithm for Hadamard Transform Processing," *IEEE Trans. Electromagn. Compat.*, **8**, pp. 28–36, Feb.

Knauer, S. C. (1976), "Criteria for Building 3-D Vector Sets in Interlaced Video Systems," *Proc. Nat. Telecommun. Conf.*, Dallas, TX, pp. 44:5-1–44:5-5, Nov. 29-Dec. 1.

Knight, J. M. Jr. (1962), "Maximum Acceptable Bit Error Rates for PCM Analog and Digital TV Systems," *Nat. Telemetering Conf. Proc.*, Paper 10-3, pp. 1–9.

Koch, W. E. (1956), "Video Transmission Over Channels Having Narrow Frequency Bands," *IRE Wescon Conf.*

Koschman, A. (1954), "On the Filtering of Non-Stationary Time Series," *Nat. Electronics Conf. Proc.*, p. 126.

Kramer, H. P., and M. V. Mathews, (1956), "A Linear Coding for Transmitting a Set of Correlated Signals," *IRE Trans. Inform. Theory*, **2**, pp. 41–46, Sept.

Kretzmer, E. R. (1960), "Reduced Bandwidth Transmission Systems," U. S. Patent No. 2,949,505, Aug.

Kretzmer, E. R. (1956), "Reduced Alphabet Representation of TV Signals," *IRE Nat. Conv. Rec.*, pt. 4, pp. 140–147.

Kretzmer, E. R. (1952), "Statistics of TV Signals," *B. S. T. J.*, **31**, pp. 751–763, July.

Kruger, R. P., W. B. Thompson, and A. F. Turner, (1974), "Computer Diagnosis of Pneumoconiosis," *IEEE Trans. Syst. Man Cybern.*, pp. 40–49, Jan.

Kubba, M. H. (1963), "Automatic Picture Detail Detection in the Presence of Random Noise," *Proc. IEEE*, **51**, pp. 1518–1523, Nov.

Kummerow, T. (1972), "Statistics for Efficient Linear and Nonlinear Picture Coding," *Proc. Int. Telemetering Conf.*, **8**, pp. 149–161, Oct.

Kunz, H. O. (1979), "On the Equivalence Between One-Dimensional Discrete Walsh-Hadamard and Multidimensional Discrete Fourier Transforms," *IEEE Trans. Comput.*, **28**, pp. 267–268, Mar.

Kuo, B. C. (1970), *Discrete Data Control Systems*, Prentice-Hall, Englewood Cliffs, NJ.

Kuroyanagi, N. (1972), "Space-Division Digital Networks for Video Signals," *IEEE Trans. Commun.*, **20**, pp. 275–281, June.

Kurtenbach, A. J., and P. A. Wintz, (1969), "Quantizing for Noisy Channels," *IEEE Trans. Commun. Tech.*, **17**, pp. 291–302, Apr.

Kurtenbach, A. J., and P. A. Wintz, (1968), "Data Compression for Second Order Processes," *Proc. Nat. Telemetry Conf.*

Lackey, R. B., and D. Meltzer, (1971), "A Simplified Definition of Walsh Functions," *IEEE Trans. Comput.*, **20**, pp. 211–213, Feb.

Laemmel, A. E. (1951), "Coding Processing for Bandwidth Reduction in Picture Transmission," - Microwave Res. Inst., Polytechnic Inst., Brooklyn, NY, Rep. R-246-51, Aug.

Laemmel, A. E. (1967), "Dimension Reducing Mapping for Signal Compression," Microwave Res. Inst., Polytechnic Inst., Brooklyn, NY, Rep. R-632-57, PIB 560.

Landau, H. J., and D. Slepian, (1971), "Some Computer Experiments in Picture Processing for Bandwidth Reduction," *B. S. T. J.*, pp. 1525–1540, May-June.

Lebedev, D. S., and D. G. Lebedev, (1964), "A New Method of Image Quantization" (in Russian), *Akad. Nauk. Vestnik*, pp. 44–46, Nov.

Lewis, N. W. (1962), "Television Bandwidth and Kell Factor," *Electron. Tech.*, pp. 44–47, Feb.

Lewis, N. W. (1954), "Waveform Response of TV Links," *Proc. Inst. Elec. Engrs.*, **101**, pt. III, pp. 258–270, July.

Limb, J. O., R. F. W. Pease, and K. A. Walsh, (1974), "Combining Intraframe and Frame-to-Frame Coding for TV," *B. S. T. J.*, **53**, pp. 1137–1173, July-Aug.

Limb, J. O. (1967), "Source Receiver Encoding for TV Signals," *Proc. IEEE*, **55**, pp. 364–379, Mar.

Limb, J. O., and F. W. Mounts, (1969), "Digital Differential Quantizer for TV," *B. S. T. J.*, pp. 2583–2599, Jan.

Limb, J. O., and R. F. W. Pease, (1971), "Exchange Spatial and Temporal Resolution in TV Coding," - *B. S. T. J.*, **50**, pp. 191–200, Jan.

Limb, J. O. (1972), "Buffering of Data Generated by Coding of Moving Images," *B. S. T. J.*, **51**, pp. 239–259, Jan.

Limb, J. O. (1966), "Vision Oriented Coding of Visual Signals," Ph. D. dissertation, Univ. Western Australia, Nedlands, Australia.

Limb, J. O. (1969), "Design of Dither Waveforms for Quantized Visual Signals," *B. S. T. J.*, **48**, pp. 2555–2582, Sept.

Limb, J. O. (1967), "Coarse Quantization of TV Signals," *Australian Telecommun.*, Res. 1, nos. 1 and 2, pp. 32–42, Nov.

Limb, J. O., C. B. Rubinstein, and K. A. Walsh, (1971), "Digital Coding of Color Picturephone Signals by Element-Differential Quantization," *IEEE Trans. Commun. Tech.*, **19**, pp. 992–1006, Dec.

Limb, J. O., and R. F. W. Pease, (1971), "A Simple Interframe Coder For Video Telephone," *B. S. T. J.*, **50**, pp. 1877–1888, July-Aug.

Limb, J. O. (1970), "Efficiency of Variable Length Encodings," *UMR Mervin J. Kelly Commun. Conf.*, pp. 13.3.1–13.3.9.

Limb, J. O. (1973), "*Picture Coding Algorithm for the Merli Scan,*" *IEEE Trans. Commun.*, **21**, pp. 300–305, Apr.

Limb, J. O. (1973), "Picture Coding: The use of a Viewer Model in Source Coding," *B. S. T. J.*, **52**, pp. 1271–1302, Oct.

Limb, J. O. (1969), "Adaptive Encoding of Picture Signals," MIT Conf. on Bandwidth Compression, Apr.

Limb, J. O. (1972), "Buffering of Data Generated by the Coding of Moving Images," *B. S. T. J.*, **51**, pp. 239–259, Jan.

Limb, J. O. (1968), "Entropy of Quantized TV Signals," *Proc. IEE*, **115**, no. 1, pp. 16–20, Jan.

Lindgren, N. (1965), "Machine Recognition of Human Language—Part II, Theoretical Models of Speech Perception and Language," *IEEE Spectrum*, **2**, pp. 45–59, Apr.

Lippel, B. (1967), "Improvement of a Coarse Picture Quantizer by Means of Ordered Dither," M. S. thesis, Newark College of Engineering, Newark, NJ, June.

Lippel, B., M. Kurland, and A. H. Marsh, (1971), "Ordered Dither Patterns for Coarse Quantization in Pictures," *Proc. IEEE,* **59**, pp. 429–431, Mar.

Lippel, B. (1969), "Experiments with a New Message Format for Digital Encoding of Pictorial Information," ECOM Report 3083, (AD 685 819), Jan.

Lippel, B. (1963), "Bit Maps and Digital TV," Special Report, U. S. Army Electronics Res. and Dev. Lab., Ft. Monmouth, NJ, July.

Lippel, B., and M. Kurland, (1971), "The Effect of Dither on Luminance Quantization of Pictures," *IEEE Trans. Commun. Tech.,* **19**, pp. 879–947, Dec.

Lippman, R. (1973), "A Technique for Channel Error Correction in DPCM Picture Transmission," *Proc. Int. Conf. Commun.,* II, Seattle, WA, pp. 48.12–48.18.

Lord, A. V. (1953), "Standards Conversion for International TV," *Electronics,* p. 144.

Lowry, E. M., and J. J. DePalma, (1961), "Sine Wave Response of the Visual System I; The Mach Phenomenon," *J. Opt. Soc. Amer.* **51**, pp. 740–746, July.

Lyche, T., and L. Schumaker, (1974), *"Computation of Smoothing and Interpolating Natural Splines via Local Bases,"* - Center for Numerical Analysis, Univ. Texas, Austin, Texas, Report 17, April.

Mackay, D. M. (1953), *"Operational Aspects of Some Fundamental Concepts of Human Communication"* - Synthese, **9**, pp. 182–198.

Macovski, A., and E. D. Jones, (1971), *"Facsimile Bandwidth Compression Using Nonlinear Analog Processing,"* IEEE Trans. Commun. Tech., **19**, pp. 1110–1116, Dec.

Mancill, C. E. (1975), "Digital Color Image Restoration," Univ. So. California Image Processing Inst. USCIPI Report 630, Aug.

Martin, R. S., and J. H. Wilkinson, (1965), "Symmetric Decomposition of a Positive Definite Matrix," *Numer. Math. 7,* pp. 362–383.

Martinson, L. W., and W. K. Smith, (1975), "Digital Matched Filtering with Pipelined Floating-Point Fast Fourier Transforms," *IEEE Trans. Acoust., Speech, and Signal Processing,* **23**, pp. 222–234, Apr.

Martinson, L. W. (1978), "A 10 MHz Image Bandwidth Compression Model," Proc. IEEE Computer Soc. Conf. on Pattern Recognition and Signal Processing, Chicago, pp. 132–136, May/June.

Mascarenhas, N. D. (1974),"Digital Image Restoration Under a Regression Model; The Unconstrained, Linear Equality and Inequality Constrained Approaches," Univ. So. California, Los Angeles, USCIPI Report 520, Jan.

Mascarenhas, N. D. A., and W. K. Pratt, (1975), "Digital Image Restoration Under a Regression Model," *IEEE Trans. Circuits and Systems,* **22**, pp. 252–266, Mar.

Massa, R. J. (1964), "Visual Data Transmission," Airforce Cambridge Res. Labs., Bedford, MA, Physical Sci. Paper No. 13, AFCLR-64-323, Apr.

Mather, R. L. (1957), "Gamma-ray Collimator Penetration and Scattering Effects," *J. Appl. Phys.,* **28**, pp. 1200–1207, Oct.

Maurice, R. D. A., M. Gilbert, G. F. Newell, and J. G. Spencer, (1955), "The Visibility of Noise in TV," British Broadcasting Corp., London, Monograph 3, Oct.

Max, J. (1960), "Quantizing for Minimum Distortion," *IRE Trans. Inform. Theory,* **6**, pp. 7–12, Mar.

Mayo, J. S. (1962), "A Bipolar Repeater for PCM Signals," *B. S. T. J.,* **41**, pp. 25–97, Jan.

McAdam, D. P., (1969), "Digital Image Restoration by Constrained Deconvolution," *J. Opt. Soc. Amer.,* **59**, pp. 748–752.

McDonald, R. A. (1966), "Signal-to-Noise and Idle Channel Performance of DPCM Systems—Particular Applications to Voice Signals," *B. S. T. J.,* **45**, pp. 1123–1151, Sept.

McDonald, R. A., and P. M. Schultheis, (1964), "Information Rates of Gaussian Signals Under Criteria Constraining the Error Spectrum," *Proc. IEEE,* **52**, pp. 415–416, Apr.

McDonald, R. A. (1974), *Finite Rings with Identity,* Marcel Dekker, New York.

McGlamery, B. L. (1967), "Restoration of Turbulence Degraded Images," *J. Opt. Soc. Amer.,* **57,** pp. 293–297, Mar.

Means, R. W., H. J. Whitehouse, and J. M. Speiser, (1974), "Television Encoding Using a Hybrid Discrete Cosine Transform and DPCM in Real Time," *Proc. Nat. Telecommun. Conf.,* San Diego, CA, pp. 61–66, Dec.

Mertz, L. (1965), *Transformation in Optics,* Wiley, New York.

Mertz, L. (1950), "Perception of TV Random Noise," *J. SMPTE,* **54,** pp. 8–34, Jan.

Mertz, L. (1951), "Data on Random Noise Requirements for Theater TV," *J. SMPTE,* **57,** pp. 87–107, Aug.

Meyr, H., *et al.* (1973), "Optimum Run-Length Codes," *Int. Conf. Commun.,* pp. 48.19–48.25.

Michel, W. S. (1957), "A Coded Facsimile System," *IRE Wescon Conv. Rec.,* pp. 84–93.

Millard, J. B., and H. I. Maunsell, (1971), "Digital Encoding of Video Signals," *B. S. T. J.,* **50,** pp. 459–479, Feb.

Miller, K. S. (1964), *Multidimensional Gaussian Distributions,* Wiley, New York, p. 8.

Miller, R. L. (1961), "The Possible Use of Log DPCM for Speech Transmission in Unicom," Globecom Conf., Chicago, IL.

Minsky, M., and S. L. Papert, (1967), "Research on Intelligent Automata," Project MAC, MIT, Status Report II, Sept.

Montgomery, W. D. (1965), "Reconstruction of Pictures from Scanned Records," *IEEE Trans. Inform. Theory,* **11,** pp. 204–206, Apr.

Moon, P., and D. E. Spencer, (1966), "The Visual Effect of Nonuniform Surrounds," *J. Opt. Soc. Amer.,* **56,** pp. 1628–1633, Nov.

Mounts, F. W. (1970), "Conditional Replenishment: A Promising Technique for Video Transmission," *Bell Labs Rec.,* **48,** no. 4, pp. 110–121, Apr.

Mounts, F. W. (1967), "Low Resolution TV: An Experimental Digital System for Evaluating Bandwidth Reduction Techniques," *B. S. T. J.,* **46,** pp. 167–198, Jan.

Mounts, F. W. (1969), "A Video Encoding System Employing Conditional Picture Element Replenishment," *B. S. T. J.,* **48,** pp. 2545–2554, Sept.

Murakami, H., and I. S. Reed, (1977), "Recursive Realization of Finite Impulse Filters Using Finite Arithmetic," *IEEE Trans. Inform. Theory,* **23,** no. 2, pp. 232–242, Mar.

Musmann, H. D., and D. Preuss, (1973), "A Redundancy Reducing Facsimile Coding System," *Nachrichtentech. Z.,* **26,** pp. 91–94, Feb.

Myers, R. H. (1971), *Response Surface Methodology,* Allyn and Bacon, Boston, MA.

Nahi, N. E. (1972), "Role of Recursive Estimation in Statistical Image Enhancement," *Proc. IEEE,* **60,** pp. 872–877, July.

Nahi, N. E. (1969), *Estimation Theory and Applications,* Wiley, New York.

Nahi, N. E., and T. Assefi, (1971), "Bayesian Recursive Image Estimation," *Proc. Two-Dimensional Image Processing,* Columbia, MO, Oct.

Nahi, N. E., and B. M. Schaefer, (1972), "Decision Directed Adaptive Recursive Estimators: Divergence Prevention," *IEEE Trans. Automat. Contr.,* **17,** pp. 61–68, Feb.

Nahi, N. E., and I. M. Weiss, (1971), "Bounding Filters in the Presence of Inexactly Known Parameters," *IEEE Decision and Control Conf.,* Dec.

Naraghi, M. (1975), "An Algorithm Image Estimation Method Applicable to Nonlinear Observations," Univ. South California, Image Processing Inst., USCIPI Report No. 580.

Narendra, A., L. S. Davis, D. L. Milgram, and A. Rosenfeld, (1978), "*Piecewise Approximation of Pictures Using Maximal Neighborhoods,*" *IEEE Trans. Comput.,* **27,** no. 4, pp. 375–379, Apr.

NASA Report, (1976), "Solid State Television Camera, Final Report (CCD Buried Channel)," NASA No. N77-25429, Section III, (Prepared by Fairchild Camera & Instrument Corp.), Dec.

Nathan, R. (1966), "Digital Video Data Handling," Jet Propulsion Lab., Pasadena, CA, Tech. Report, TR-32-877, Jan.

Nathan, R. (1968), "Picture Enhancement for the Moon, Mars, and Man," (1968), *Pictorial Pattern Recognition,* Washington DC, Thompson.

Naval Electronics Systems Command, (1975), "CCD Photosensor Array Development Program," Phase II, Final Report, No. ED-AX-61, AD-A022-881, Section III (Prepared by Fairchild Camera & Instrument Corp.) Apr.

Neuberger, A. (1966), "Characterization of Pictures by Essential Contours," M. Sm. thesis, Dept. Elec. Eng., MIT, Cambridge, MA, Jan.

Newell, G. F., and W. K. E. Geddes, (1963), "The Visibility of Small Luminance Perturbations in TV Displays," BBC Eng. Div. Res. Dept., Report T-106.

Nilsson, N., and B. Raphael, (1967), Preliminary Design of an Intelligent Robot, Computers and Information Sciences II, Academic Press, New York.

Nishikawa, A., R. J. Massa, and J. C. Mott-Smith, (1965), *"Area Properties of TV Pictures," IEEE Trans. Inform. Theory,* 11, pp. 348–353, July.

Nitadori, K. (1965), "Statistical Analysis of Δ-PCM," *Electronics Commun. (Japan),* 48, Feb. 1965.

O'Brien, V. (1958), "Contour Perception, Illusion and Reality," *J. Opt. Sci. Amer.,* 48, pp. 112–119, Feb.

Ohira, T., K. Hayakawa, and K. Shibata, (1973), "Picture Quality of Hadamard Transform Coding Using Nonlinear Quantizing for Color TV Signals," *Symp. Applications of Walsh Functions,* Washington D. C., Apr.

Ohmori, K. *et al.* (1972), "An Application of Cellular Logic for High Speed Decoding of Minimum Redundancy Codes," *Proc. AFIPS Fall Joint Comput. Conf.,* 41, pp. 345–351, Dec.

Ohnsorg, F. R. (1971), "Properties of Complex Walsh Functions," *Proc. IEEE Fall Electronics Conf.,* pp. 383–385.

Ohnsorge, H. (1973), "A Data Compression System for the Transmission of Digitalized Video Signals," *Conf., Rec. Int. Commun.,* June.

Oliver, B. N., J. R. Pierce, and C. E. Shannon, (1948), "The Philosophy of PCM," *Proc. IRE,* 36, pp. 1324–1331, Nov.

Oliver, B. N. (1966), "Efficient Coding," *B. S. T. J.,* 45, pp. 724–750, Jan.

O'Neal, J. B. (1966), "Delta Modulation Quantizing Noise Analytical and Computer Simulation Results for Gaussian and TV Input Signals," *B. S. T. J.,* 45, pp. 117–142, Jan.

O'Neal, J. B. (1966), "Predictive Quantization Systems (DPCM) for Transmission of TV Signals," *B. S. T. J.,* pp. 689–721, May/June.

O'Neal, J. B. (1971), "Entropy Coding in Speech and TV DPCM Systems," *IEEE Trans. Info. Theory,* 17, pp. 758–761, Nov.

O'Neal, J. B. (1971), "Bounds on Subjective Performance Measures for Source Encoding Systems," *IEEE Trans. Inform. Theory,* 17, pp. 224–231.

O'Neal, J. B. (1971), "Signal-to-Quantizing-Noise Ratios for DPCM," *IEEE Trans. Commun. Tech.,* 19, pp. 568–570, Aug.

O'Neal, J. B. (1967), *"A Bound on Signal-to-Quantizing-Noise Ratios for Digital Encoding Systems,"* Proc. IEEE, 55, pp. 287–292, Mar.

O'Neill, E. F. (1959), "Tasi: Time Assignment Speech Interpolation," *Bell Labs. Rec.,* 37, no. 3, pp. 83–87, Mar.

Oppenheim, A. V., R. W. Schafer, and T. G. Stockham, Jr. (1968), "Nonlinear Filtering of Multiplied and Covalued Signals," *Proc. IEEE,* 56, pp. 1264–1291, Aug.

Oppenheim, A. V., and C. J. Weinstein, (1972), "Effects of Finite Register Length in Digital Filtering and the Fast Fourier Transform," *Proc. IEEE,* 60, pp. 967–976, Aug.

Palermo, C. J., R. V. Palermo, and H. Horwitz, (1965), "The Use of Data Omission for Redundancy Removal," *Rec. Int. Space Electron. Telemetry Symp.,* pp. 11D1–11D16.

Pan, J. W. (1962), "Reduction of Information Redundancy In Pictures," Sc. D. dissertation, Dept. Elec. Eng., MIT, Cambridge, MA, Sept.

Panter, P. F. and W. Dite, (1951), "Quantization Distortion In Pulse Count Modulation With Nonuniform Spacing of Levels," *Proc. IRE,* pp. 44–48, Jan.

Papoulis, A. (1965), Probability, Random Variables, And Stochastic Processes, New York, McGraw-Hill.

Parkyn, W. A. Jr. (1970), "Digital Image Processing Aspects of the Walsh Transform," Applications of Walsh Functions Symp., Apr.

Pawula, R. F. (1963), "Comments on Statistical Properties of the Contours of Random Surfaces," *IRE Trans. Inform. Theory,* 9, pp. 208–209, July.

Pearl, J., H. C. Andrews, and W. K. Pratt, (1972), "Performance Measures for Transform Data Coding," *IEEE Trans. Commun.,* 20, pp. 411–415, June.

Pearl, J. (1971), "Basis-Restricted Transformations And Performance Measures For Spectral Representations," *Proc. Hawaiian Conf.;* also *IEEE Trans. Inform. Theory,* 17, pp. 751–752, Nov.

Pearl, J. (1970), *"On the Distance Between Representations,"* Symp. Picture Coding, North Carolina State Univ., Raleigh, NC, Sept.

Pease, R. F. W. and J. O. Limb, (1971), "Exchange of Spatial and Temporal Resolution in TV Coding," *B. S. T. J.,* 50, pp. 191–200.

Pease, R. F. W. (1972), "Conditional Vertical Subsampling—A Technique to Assist in the Coding of TV Signals," *B. S. T. J.,* 51, pp. 787–802, Apr.

Pease, R. F. W. and W. G. Scholes, "TV Coding Using Field Difference Quantization," private communication.

Pedlar, C. (1965), "Rods And Cones, A Fresh Approach," *Color Vision Physiology Experimental Psychology,* London, Churchill, pp. 52–83.

Pennington, K. S., and P. M. Will, (1970), "A Grid-Coded Technique for Recording Three-Dimensional Scenes Illuminated with Ambient Light," *Optical Commun.,* 2, pp. 167–169, Sept.

Pennington, K. S., P. M. Will, and G. L. Shelton, (1970), "Grid Coding: A Technique for the Extraction of Differences from Scenes," *Opt. Commun.,* 2, 1970, pp. 113–119, Aug.

Peterson, D. P., and D. Middleton, (1962), "Sampling and Reconstruction of Wavenumber—Limited Functions in N-Dimensional Euclidian Space," *Inform. Control,* 5, pp. 279–323, Dec.

Phillips, D. L. (1964), "A Technique for the Numerical Solution of Certain Integral Equations of the First Kind," *J. Assoc. Comput. Mach.,* 9, pp. 84–97.

Pingle, K. K. (1966), "A Program to Find Objects in a Picture," Artificial Intelligence Project, Stanford Univ., Stanford, CA, Memo no. 30, Jan.

Pollard, J. M. (1971), "The Fast Fourier Transform in a Finite Field," *Math. Comput.,* 25, pp. 365–374, Apr.

Pollard, J. M. (1976), "Implementation of Number-Theoretic Transforms," *Electron. Lett.,* 12, pp. 378–379, July.

Powers, K. H., and H. Staras, (1957), "Some Relations Between TV Picture Redundancy and Bandwidth Requirements," *Commun Electronics,* 32, pp. 492–496, Sept.

Prasada, B. (1963), "Some Possibilities of Picture Signal Bandwidth Compression," *IEEE Trans. Commun. Systems,* 11, pp. 315–328, Sept.

Pratt, W. K., J. Kane, and H. C. Andrews, (1969), "Hadamard Transform Image Coding," *Proc. IEEE,* 57, pp. 58–68, Jan.

Pratt, W. K., L. B. Welch, and W. Chen, (1972), "Slant Transforms for Image Coding," *Proc. Symp. on Walsh Functions,* Mar. 1972.

Pratt, W. K. (1971), "Spatial Transform Coding of Color Images," *IEEE Trans. Commun. Tech.,* 19, pp. 980–992, Dec.

Pratt, W. K. (1962), "Stop-Scan Detection System for Interplanetary TV Transmission," Nat. Symp. Space Electron. and Telemetry, Section 4.3, Oct.

Pratt, W. K., and H. C. Andrews, (1970), "Transform Image Coding," USCEE Report 387, Univ. South California, Los Angeles, Mar.

Pratt, W. K. (1972), "BSC Error Effects on PCM Color Image Transmission," *IEEE Trans. Inform. Theory,* pp. 636–643, Sept.

Pratt, W. K., W. Chen, and L. R. Welch, (1974), "Slant Transform Image Coding," *IEEE Trans. Commun.,* **22,** pp. 1075–1093, Aug.

Pratt, W. K. (1970), "Linear and Nonlinear Filtering in the Walsh Domain," *Proc. Symp. Walsh Functions,* Washington DC, pp. 38–42.

Pratt, W. K. (1974), "Transform Image Coding Spectrum Extrapolation," Honolulu, HI, pp. 7–9, Jan.

Pratt, W. K., and F. Davarian, (1977), "Fast Computational Techniques for Pseudoinverse and Wiener Image Restoration," *IEEE Trans. Comput.,* **26,** pp. 571–580, June.

Pratt, W. K. (1977), "Pseudoinverse Image Restoration Computational Algorithms," *Optical Inform. Processing,* **II,** G. W. Stroke, V. Nesterikhin, and E. S. Barrekette, (Eds.), Plenum Press, New York.

Pratt, W. K. (1967), "A Bibliography on TV Bandwidth Reduction," *IEEE Trans. Inform. Theory,* Jan.

Pratt, W. K. (1972), "Generalized Wiener Filtering Computational Techniques," *IEEE Trans. Comput.,* **21,** pp. 636–641, July.

Pratt, W. K., and H. C. Andrews, (1968), "Fourier Transform Coding of Images," *Proc. Hawaii Int. Conf. System Sciences,* Jan.

Pratt, W. K., and H. C. Andrews, (1969), "Two-Dimensional Transform Coding for Images," IEEE Int. Symp. Inform. Theory, Jan.

Pratt, W. K., and H. C. Andrews, (1969), "Application of Fourier Hadamard Transformation to Bandwidth Compression," MIT Symp. on Picture Bandwidth, Apr.

Pratt, W. K. (1960), "Karhunen-Loeve Transform Coding of Images," IEEE Int. Symp. Info. Theory, June.

Pratt, W. K. (1970), "Application of Transform Coding to Color Images," Symp. Picture Coding, North Carolina State Univ., Raleigh, NC, Sept.

Pratt, W. K. (1970), "A Comparison of Digital Image Transforms," Mervin J. Kelly Commun. Conf., Rolla, MO, Sept.

Pratt, W. K. (1977), "Pseudoinverse Image Restoration Computational Algorithms," in *Optical Inform. Processing,* **II,** G. W. Stroke, Y. Nesterikhin, and E. S. Barrekette, (Eds.), Plenum Press, New York.

Pratt, W. K. (1978), *Digital Image Processing,* Wiley, New York.

Proctor, C. W., and P. A. Wintz, (1971), "Picture Bandwidth Reduction for Noisy Channels," School Elec. Eng., Purdue Univ., Lafayette, IN, Tech. Rept. TR-EE-71-30, Aug.

Prosser, R. D., J. W. Allnat, and N. W. Lewis, (1966), "Quality Grading of TV Pictures," *Proc. Inst. Elec. Eng.,* London, **III,** pp. 491–502.

Protonotarius, E. N. (1967), "Slope Overload Noise in DPCM Modulation Systems," *B. S. T. J.,* pp. 2119–2161, Nov.

Quastler, H. (1956), "Studies in Human Channel Capacity," in *Inform. Theory,* Third London Symp., Butterworth, London, pp. 361–371.

Rabiner, L. R., and B. Gold, (1975), *Theory and Application of Digital Signal Processing,* Prentice-Hall, Englewood Cliffs, NJ, p. 367.

Rabiner L. R., R. W. Schafer, and C. M. Rader, (1969), "The Chirp-Z Transform Algorithm and its Applications," *B. S. T. J.,* Appendix, pp. 1290–1291, June.

Rader, C. M. (1972), *"Discrete Convolution via Mersenne Transforms,"* IEEE Trans. Comput. **21**, pp. 1269–1273, Dec.

Rader, C. M. (1975), "On the Application of Number-Theoretic Transforms for High Speed Convolution to Two-Dimensional Filtering," *IEEE Trans. Circuits Syst.* **22**, June.

Raga, G. L. (1965), "Digital TV Compression—Fact or Myth," Nat. *Telemetering Conf. Proc.*

Raiffa, H., and R. Schlaiffer, (1961), *Applied Statistical Decision Theory,* Harvard Press, Boston, MA.

Rao, K. R., M. A. Narasimhan, and K. Revuluri, (1974), "The Hadamard Haar Transform," *Nat. Electron. Conf.,* pp. 336–339.

Rao, K. R., M. A. Narasimhan, and V. Devarajan, (1977), "Cal-Sal Walsh-Hadamard Transform," *Proc. Midwest Symp. Circuits and Systems*, pp. 705–709.

Ratcliff, F. (1965), *Mach Bands*, Holden-Day, San Francisco, CA.

Rathbone, W. M. (1970), "Output Format Coding for Data Compressors," *Nat. Telemetering Conf. Rec.,* pp. 117–121.

Ray, W. D., and R. M. Driver, (1970), "Further Decomposition of a Stationary Random Process," *IEEE Trans. Inform. Theory,* **16**, no. 6, pp. 470–476, Nov.

Ready, P. J., and P. A. Wintz, (1972), "Multispectral Data Compression Through Transform Coding and Block Quantization," School Elec. Eng., Purdue Univ., Lafayette, IN, Tech. Report TR-EE-72-29, May.

Ready, P. J., P. A. Wintz, S. J. Whitsett, and D. A. Landgrebe, (1971), "Effects of Compression and Random Noise on Multispectral Data," *Proc. 7th Int. Symp. Remote Sensing Environment,* Michigan Univ., May.

Ready, P. J., and P. A. Wintz, and D. A. Landgrebe, (1971), "A Linear Transformation for Data Compression and Feature Selection in Multispectral Imagery," Lab. for the Applications of Remote Sensing, Purdue Univ., Lafayette, IN, Inform. Note 072071, July.

Reed, I. S., and T. K. Truong, (1975), "Use of Finite Field to Compute Convolution," *IEEE Trans. Inform. Theory,* pp. 208–213, Mar.

Reed, I. S., and T. K. Truong, (1975), "Complex Integer Convolutions Over a Direct Sum of Galois Fields," *IEEE Trans. Inform. Theory,* **21**, pp. 657–661, Nov.

Reed, I. S., and T. K. Truong, (1976), "Convolutions Over Residue Classes of Quadrature Images," *IEEE Trans. Inform. Theory,* **22**, pp. 468–475, July.

Reed, I. S., T. K. Truong, Y. S. Kwoh, and E. L. Hall, (1977), "Image Processing by Transforms Over a Finite Field," *IEEE Trans. Comput.,* **26**, pp. 874–881, Sept.

Reinsch, C. H. (1967), *"Smoothing by Spline Functions,"* Numer. Math., **10**, pp. 177–183.

Reiss, R. A. (1965), "Pseudo-Random Scan Applications to Data Processors," *Proc. NEREM Conv. Rec.,* Nov.

Renn, R. L., and R. A. Schaphorst, (1962), "An Image Coding System Which Reduces Redundancy in Two-Dimensions," *Proc. NEC,* Dayton, OH.

Ribonson, G. S. (1971), "Orthogonal Transform Feasibility Study," NASA/MSC by COMSAT, Contract NAS 9 - 11240, Suppl. Reports 1, Aug.; 2, Sept.; 3, Oct.; Final Report, Oct.

Rice, R. (1970), *"The Rice Machine—TV Data Compression,"* General Tech. Document 900-408, Jet Propulsion Lab., Pasadena CA, Sept. 1970.

Rice, R. F., and J. R. Plaunt, (1971), "Adaptive Variable-Length Coding for Efficient Compression of Spacecraft TV Data," *IEEE Trans. Commun. Tech.,* **19**, pp. 889–897, Dec.

Richardson, W. H. (1972), "Bayesian-Based Iterative Methods of Image Restoration," *J. Opt. Soc. Amer.* **62**, pp. 55–59, Jan.

Rindfleisch, T. C., J. A. Dunne, H. J. Frieden, W. D. Stromberg, and R. M. Ruiz, (1971), "Digital Processing of the Mariner 6 and 7 Pictures," *J. Geophys. Res.*, **76**, pp. 394–417, Jan.

Roberts, L. G. (1962), "Picture Coding Using Pseudo-Random Noise," *IRE Trans. Inform. Theory.* **8**, pp. 145–154, Feb.

Roberts, L. G. (1965), "Machine Perception of Three Dimensional Solids," *Optical and Electro-Optical Inform. Proc.,* MIT Press, Cambridge, MA.

Robertson, G. H. (1969), "Computer Study of Quantized Output Spectra," *B. S. T. J.,* **48**, pp. 2391–2403, Sept.

Robinson, G. S. (1972), "Logical Convolution and Discrete Walsh and Fourier Power Spectra," *IEEE Trans. Audio Electroacoust.,* **20**, pp. 271–280, Oct.

Robinson, G. S. (1958), "A Report on Primes of the Form $K \times 2^n + 1$ and on Factors of Fermat Numbers," *Proc. Amer. Math. Soc.,* **9**, Oct.

Robson, J. G. (1966), "Spatial and Temporal Contrast—Sensitivity Functions of the Visual Data System," *J. Opt. Soc. Amer.,* **56**, pp. 1141–1142, Aug.

Rocca, F. (1972), "TV Bandwidth Compression Utilizing Frame-to-Frame Correlation and Movement Correlation and Movement Compensation," Symp. Picture Bandwidth Compression, Gordon & Breach, New York.

Rocca, F., and S. Zanoletti, (1970), "Bandwidth Reduction via Movement Compensation on a Model of the Random Video Process," *Nat. Telemetry Conf.,* pp. 2.10–2.15.

Rogers, W. L., K. S. Han, L. W. Jones, and W. H. Beierwaltes, (1972), "Application of a Fresnel Zone Plate to Gamma-Ray Imaging," *J. Nuclear Medicine,* **13**, p. 612.

Rosenfeld, A. (1969), *"Picture Processing by Computer,"* Comput. Surveys, pp. 147–176, Sept.

Rosenfeld, A. (1973), "Progress in Picture Processing: 1967–71," *ACM Comput. Surveys,* **5**, pp. 81–108, June.

Rosenheck, B. M. (1970), "FASTFAX, A Second Generation Facsimile System Employing Redundancy Reduction Techniques," *IEEE Trans. Commun. Tech.,* **18**, pp. 772–779, Dec.

Ross, I., and L. G. Zuckerman, (1959), "Some Coding Concepts to Conserve Bandwidth," *IRE Telemetering Conf.,* May.

Rubinstein, C. B., and J. O. Limb, (1973), "Color Border Sharpness," *Colour Conf. Int. Colour Assn. (AIC),* York, England.

Rushforth, C. K., and R. W. Harris, (1968), "Restoration, Resolution, and Noise," *J. Opt. Soc. A.,* **58**, pp. 539–545, Apr.

Rust, B. W., and W. R. Burrus, (1972), *Mathematical Programming and the Numerical Solution of Linear Equations,* American Elsevier, New York.

Saaty, T. L., and J. Bram, (1964), *Nonlinear Mathematics,* McGraw-Hill, 1964.

Sakrison, D. J., and V. R. Algazi, (1971), "Comparison of Line-by-Line and Two-Dimensional Encoding of Random Images," *IEEE Trans. Inform. Theory,* **17**, pp. 386–398, July.

Sakrison, D. J. (1968), *Communication Theory: Transmission of Waveforms and Digital Information,* Wiley, New York.

Sakrison, D. J. (1968), "The Rate Distortion Function of a Gaussian Process with a Weighted Square Error Criterion," *IEEE PGIT,* **14**, pp. 506–508, May.

Sakrison, D. J. (1970), "Factors Involved in Applying Rate Distortion Theory to Image Transmission," *IEEE M. J. Kelly Commun. Conv.,* Rolla, MO, Oct.

Salaman, R. G. (1961), "Digital TV Encoding," *7th Nat. Commun. Symp. Rec.,* pp. 274–279.

Sawchuk, A. A. (1972), "Space-Variant Image Motion Degradation and Restoration," *Proc. IEEE,* **60**, pp. 854–961, July.

Schade, O. H. (1956), "Optical and Photoelectric Analog of the Eye," *J. Opt. Soc. Amer.,* **46**, pp. 721–739, Sept.

Schade, O. H. (1958), "On the Quality of Color-Television Images and the Perception of Color Detail," *J. SMPTE,* **67**, pp. 801–819, Dec.

Schaphorst, R. A., and R. L. Remm, (1962), "An Image Coding System Which Reduces Redundancy in Two-Dimensions," *Proc. Nat. Aerospace Electronics Conf.,* pp. 116–127, May.

Schaphorst, R. A. (1972), "Frame-to-Frame Coding of NTSC Color TV," *Symp. Picture Bandwidth Compression,* Gordon & Breach, New York.

Schlesinger, K. (1951), "Dot Arresting Improves TV Picture Quality," *Electronics,* 1951, p. 96.

Schouten, J. F., F. DeJager, and J. A. Greefkes, (1952), "Delta Modulation; A New Modulation System for Telecommunications," *Phillips Tech. Review,* Mar.

Schreiber, W. F. (1967), "Picture Coding," *Proc. IEEE,* **55,** pp. 320–330, Mar.

Schreiber, W. F. (1956), "The Measurements of Third Order Probability Distributions of TV Signals," *IRE Trans. Inform. Theory,* **2,** pp. 91–105, Sept.

Schreiber, W. F. (1963), "The Mathematical Foundation of the Synthetic Highs System," Res. Lab. Elec., MIT Cambridge, MA, Quarterly Report 68, p. 140, Jan.

Schreiber, W. F. *et al.* (1968), "Contour Coding of Images," *Wescon Rec.,* pp. 8.3.1–8.3.5.

Schreiber, W. F. (1953), "Probability Distributions of TV Signals," Ph. D. dissertation, Harvard Univ.

Schreiber, W. F., C. F. Knapp, and N. P. Kay, (1959), "Synthetic Highs an Experimental TV Bandwidth Reduction System," *J. Soc. of Motion Picture Engrs.,* p. 525.

Schtein, V. M. (1956), "Computation of Linear Predistorting and Correcting Systems," *Radiotek.,* A, pp. 60–63.

Schultz, M. H. (1973), *Spline Analysis,* Prentice-Hall, Englewood Cliffs, NJ.

Schwartz, J. W., and R. C. Barker, (1966), "Bit-Plane Encoding: A Technique for Source Encoding," *IEEE Trans. Aerospace Electronics Systems,* July.

Schwartz, M. S. (1959), *Information Transmission: Modulation and Noise,* McGraw-Hill, New York.

Schwartz, P. V. (1970), "Analysis of Fourier Compression of Images," Symp. Picture Coding, Raleigh, NC, Sept.

Schwarz, H. R., H. Rustishauser, and E. Stiefel, (1973), *Numerical Analysis of Symmetric Matrices,* - Prentice-Hall, New York.

Schuchman, L. (1964), "Dither Signals and Their Effects on Quantization Noise," *IEEE Trans. Commun. Tech.,* pp. 162–165, Dec.

Schoenberg, I. J. (1964), "On Interpolation by Spline Functions and its Minimal Properties," *Proc. Conf. Approximation Theory,*" P. L. Butzer, and J. Korevaar, Eds., Verlag Bosel.

Searle, N. H. (1970), "A Logical Walsh-Fourier Transform," *Proc. Applications Walsh Functions Symp.,* pp. 95–98, Apr./Oct.

Seitzer D. (1970), "Instantaneous Priority Multiplexing of Gray Pictures," *IEEE Trans. Commun. Tech.,* p. 690, Oct.

Sekey, A. (1965), "Detail Detection in TV Signals," *Proc. IEEE,* pp. 75–76, Jan.

Selin, I. (1965), *Detection Theory,* Princeton University Press.

Seyler, A. J., and A. Korpel, (1959), "A Voltage Controlled Continuously Variable Low Pass Filter," *Electron. Eng.* pp. 16–22.

Seyler, A. J. (1962), "The Coding of Visual Signals to Reduce Channel Capacity Requirements," *Proc. IEE* (London), pp. 676–684, Sept.

Seyler, A. J., and J. B. Potter, (1960), "*Waveform Testing of TV Transmission Facilities,*" *Proc. IRE* (Australia), **21,** pp. 470–478, July.

Seyler, A. J. (1962), "Visual Communications and Psychophysics of Vision," *Proc. IRE* (Australia), **23,** p. 291, May.

Seyler, A. J., and Z. L. Budrikis, (1959), "Measurements of Temporal Adaptation to Detail Vision," *Nature* (London), **184,** pp. 1215–1217.

Seyler, A. J., and Z. L. Budrikis, (1965), "Detail Perception After Scene Changes in TV Image Presentation," *IEEE Trans. Inform. Theory,* **11,** pp. 31–43, Jan.

Seyler, A. J. (1963), "Real-Time Recording on TV Frame Differences Areas," *Proc. IEEE,* **51,** pp. 478–480, Mar.

Seyler, A. J. (1965), "*Probability Distributions of TV Frame Differences*," *Proc. Inst. Radio Electron. Eng.* (Australia), p. 355, Nov.

Seyler, A. J. (1963), "An Experimental Frame Difference Signal Generator for the Analysis of TV Signals," *IRE Radio Electron. Eng. Conv.* (Australia).

Seyler, A. J. (1955), "Bandwidth Reduction in TV Relaying," *Proc. IRE*, (Australia), pt. I, pp. 218–225; pt. II, pp. 261–276, Aug.

Seyler, A. J., and P. R. Wallace, (1965), "Acquisition of Statistical TV Signals," *Proc. IRE* (Australia), pp. 343–354, Nov.

Shanks, J. L. (1969), "Computation of the Fast Walsh-Fourier Transform," *IEEE Trans. Comput.*, **18**, pp. 457–459, May.

Shannon, C. E. (1960), "Coding Theorems of a Discrete Source with a Fidelity Criterion," *Inform. Decision Processes*, McGraw-Hill, New York.

Shannon, C. E., and W. Weaver, (1953), "The Mathematical Theory of Communications," Univ. of Illinois Press, Urbana, IL, p. 33.

Shannon, C. E. (1968), "A Mathematical Theory of Communications," *B. S. T. J.*, **27**, pp. 379–423, 623–656.

Sheldahl, S. A. (1968), "Channel Identification Coding for Data Compressors," *Eascon Rec.*, pp. 319–325.

Shennum, R. H., and J. R. Gray, (1962), "Performance Limitations of a Practical PCM Terminal," *B. S. T. J.*, **41**, pp. 143–171, Jan.

Shibata, K., T. Ohira, *et al.* (1969), "PCM Terminal Equipment for Bandwidth Compression of TV Signals," *Joint Conv. IECE* (Japan), Paper 2619.

Shulman, A. R. (1970), *Optical Data Processing*, Wiley, New York.

Shurtleff, D. A. (1967), "Studies in TV Legibility—A Review of the Literature," *Inform. Display*, **4**, pp. 40–45.

Sidhu, G. S., and R. T. Boute, (1972), "Property Encoding: Applications in Binary Picture Encoding and Boundary Following," - *IEEE Trans. Comput.*, **21**, pp. 1206–1215, Nov.

Siedman, A. E. (1975), "How to Compute Two Even Fourier Transforms with One Transform Step," *Proc. IEEE*, p. 544, Mar.

Silverman, H. F. (1977), "An Introduction to Programming the Winograd Fourier Transform Algorithm," *IEEE Trans. Acoust. Speech Signal Processing*, **25**, pp. 152–165, Apr.

Singleton, R. C. (1967), "On Computing the Fast Fourier Transform," *Commun. As. Comput. Mach.*, **10**, pp. 647–654, Oct.

Singleton, R. C. (1969), "An Algorithm for Computing the Mixed Radix Fast Fourier Transform," *IEEE Trans. Audio Electroacoust.*, **17**, pp. 93–103, June.

Sjostrand, F. S. (1965), "The Synaptology of the Retina," *Colour Vision Physiology and Experimental Psychology*, Churchill, London, pp. 110–144.

Smith, E. A., and D. R. Phillips, (1972), "Automatid Cloud Tracking Using Precisely Aligned Digital ATS Pictures," *IEEE Trans. Comput.*, **21**, pp. 715–729, July.

Smith, B. (1957), "Instantaneous Companding of Quantized Signals," *B. S. T. J.*, **36**, pp. 653–709, May.

Sneddon, I. N. (1951), *Fourier Transforms* - McGraw-Hill, New York, p. 62.

Som, S. C., and A. K. Gosh, (1966), "Assessment of Sharpness and Overall Picture Quality of Different Psychometric Methods," *J. Opt. Soc. Amer.*, **56**, pp. 44–48, Jan.

Sondhi, M. M. (1972), "Image Restoration: The Removal of Spatially Invariant Degradation," *Proc. IEEE*, pp. 843–853, July.

Spang, H. A., and P. M Schultheis, (1962), "Reduction of Quantizing Noise by the Use of Feedback," *IRE Trans. Commun. Syst.*, p. 373, Dec.

Spencer, D. R., and T. Huang, (1969), "Bit-Plane Encoding of Continuous Tone Pictures," *Proc. Symp. Computer Processing in Commun.*, Polytechnic Inst., Brooklyn, NY, pp. 101–120, Apr.

Sperling, G. (1970), "Model of Visual Adaptation and Contrast Detection," *Perception Psychophysics,* **8**, pp. 143–157, Mar.

Sperling, G. (1965), "Temporal and Spatial Masking, 1. Masking by Impulse Flashes," *J. Opt. Soc. Amer.*, **55**, pp. 541–559, May.

Srihari, S. N., and M. K. Ohanesian, (1979), "An Efficient Algorithm for Determining Hadamard Sequence Vectors," *IEEE Trans. Comput.*, **28**, pp. 243–244, Mar.

Stell, D. E., D. T. Hess, and K. K. Clarke, (1969), "Digital Storage, Measurement, Transformation, and Recovery of Video Information," *Proc. Symp. Computer Processing in Communications,* Brooklyn, NY, June.

Stell, D. E. (1969), "Digital TV Storage and Video Redundancy Reduction," M. Sc. Report, Polytechnic Inst., Brooklyn, NY, June.

Stevens, S. S., C. T. Morgan, and J. Volkman, (1941), "Theory of the Neural Quantum in the Discrimination of Loudness and Pitch," *Am. J. Psychol.*, **54**, pp. 315–335.

Stockham, T. G., Jr. (1966), "High-Speed Convolution and Correlation," *Spring Joint Computer Conf., AFIPS Conf. Proc.*, **28**, Washington DC, Spartan, pp. 229–233.

Stockham, T. G., Jr. (1972), "Image Processing in the Context of a Visual Model," *Proc. IEEE,* **60**, pp. 828–842.

Stockham, T. G., Jr. (1972), "Intra-Frame Encoding for Monochrome Images by Means of a Psychophysical Model Based on Nonlinear Filtering of Multiplied Signals," *Proc. Symp. on Picture Bandwidth Reduction,* Gordon & Breach, New York.

Stoddard, J. C. (1955), "Measurement of Second Order Probability Distributions of Pictures by Digital Means," Mass. Inst. Tech. Res. Lab. Elec., Tech. Report No. 302.

Strintzis, M. (1976), "Comments on Two-Dimensional Bayesian Estimate of Images," *Proc. IEEE,* **64**, pp. 1255–1257, Aug.

Stroh, R. W., and R. R. Boorstijn, (1970), "Optimum and Adaptive DPCM," Tech Report for NASA, NASA Grant NGR 33-006-040, Polytechnic Inst. of Brooklyn, PIB-EE.

Stroud, J. M. (1954), "The Fine Structure of Psychological Time in Information Theory in Psychology," Free Press, Glencoe, IL, pp. 174–207.

Stuki, P. (1970), "Efficient Transmission of Graphics Using Polyomino Filtering at the Receiver," *IEEE Trans. Aerospace Electronics Systems,* **6**, pp. 811–814, Nov.

Stuki, P. (1971), "Limits of Instantaneous Priority Multiplexing Applied to Black-and-White Pictures," *IEE Trans. Commun. Tech.,* **19**, pp. 169–177, Apr.

Sutton, R. N., and E. I. Hall, (1972), "Texture Measures for Automatic Classification of Pulmonary Disease," *IEEE Trans. Comput.,* **21**, pp. 667–676, July.

Swerling, P. (1962), "Statistical Properties of Contours of Random Surfaces," *IRE Trans. Inform. Theory,* **8**, pp. 315–321, July.

Swets, J. A., W. P. Tanner, and T. G. Birdsall, (1961), "Decision Processes in Perception," *Psychol. Rev.,* **68**, pp. 301–340.

Swing, R. E. (1973), "The Optics of Microdensitometry," *Opt. Eng.,* **12**, pp. 185–198, Nov.

Swinton, A. C. (1912), "The Possibilities of TV," *J. Roentgen Soc.*

Sziklai, G. C. (1956), "Some Studies in the Speed of Visual Perception," *IRE Trans. Inform. Theory,* p. 125.

Taki, Y., N. Tanaka, A. Yamazaki, (1960), "Statistical Characteristics of TV Signals and Estimation of a Limit of Bandwidth Reduction," *Tech. J. Jap. Broadcasting,* **12**, p. 26.

Tam, L. D. C., and Y. Goulet, (1972), "On Arithmetical Shift for Walsh Functions," **21**, p. 1451, Dec.

Tasto, M., and P. A. Wintz, (1971), "Image Coding by Adaptive Block Quantization," *IEEE Trans. Commun. Tech.,* **19**, Dec.

Tasto, M., and P. A. Wintz, (1970), "Picture Bandwidth Compression by Adaptive Block Quantization," School Elec. Enc., Purdue, Univ., Lafayette, IN, Report TR-EE-70-14, June.

Tasto, M., and P. A. Wintz, (1970), "An Adaptive Picture Bandwidth Compression System Using Pattern Recognition Techniques," *Proc. 3rd Hawaiian Conf. Syst. Sci.,* pp. 513–515, Jan.

Tasto, M., A. Habibi, and P. A. Wintz, (1970), "Adaptive and Nonadaptive Image Coding by Linear Transformation and Block Quantization," Symp. Picture Coding, North Carolina State Univ., Raleigh, NC, Sept.

Tasto, M., and P. A. Wintz, (1972), "A Bound on the Rate-Distortion Function and Application to Images," *IEEE Trans. Inform. Theory,* **18**, pp. 150–159, Jan.

Tatachar, M., and J. C. Prabhakar, (1977), "Fast Walsh Transform-Another Approach," *Proc. Midwest Symp. Circuits and Systems,* pp. 692–297.

Tatachar, M., and J. C. Prabhakar, (1976), "Design of Digital Walsh Filters," *Region V IEEE Conf.,* Austin, TX, Apr.

Teacher, C. F. (1974), "Digital TV Techniques," Nat. Commun. Symp., Oct.

Teer, K. (1959), "Some Investigations on Redundancy and Possible Bandwidth Compression in TV Transmission," Thesis Technische Hogeschool Delft, The Netherlands, Sept.

Tescher, A. G. (1973), "The Role of Phase in Adaptive Image Coding," Univ. So. California, Los Angeles, USCIPI Report 510, Dec.

Thoma, W. (1973), "Combination of DPCM and Frame Repeating for Picture Coding," *Siemens Res. Devel. Rep.,* **2**, pp. 121–128.

Thoma, W., and A. G. Siemens, (1972), "Video Transmission Network with Intraframe and Optional Interframe Coding," *Nat. Telemetry Conf.,* pp. 39.1–39.6.

Thomas, J. P. (1966), "Brightness Variations in Stimuli With Ramp-Like Contours," *J. Opt. Soc. Amer.,* **56**, pp. 238–241, Feb.

Thompson, J. E. (1967), U. K. Patent 1218015, Mar.

Thompson, J. E. (1971), "A 36 M Bits/S TV Coder Employing Pseudorandom Quantization," *IEE Trans. Commun. Tech.,* **19**, Dec.

Thompson, J. E. (1974), "Differential Encoding of Composite Color TV Signals Using Chrominance Corrected Prediction," *IEEE Trans. Commun.,* pp. 1106–1113, Aug.

Thompson, J. E. (1971), "U. K. Patent Application," No. 40309/71., Aug.

Thompson, J. E. (1972), "Differential Coding for Digital Transmission of PAL Colour TV Signals," Int. Broadcasting Conv., London, (IEE Conf. Publ. 88) pp. 26–32, Sept.

Thompson, J. E., and J. J. Sparkes, (1967), "A Pseudorandom Quantizer for TV Signals," *Proc. IEEE,* **55**, pp. 353–355, Mar.

Thompson, W. E. (1972), "The Synthesis of a Network to Have a Sine Squared impulse Response," *Proc. IEE,* **99**, pt. III, pp. 373–376, Nov.

Tomnovec, F. M., and R. L. Mather, (1957), "Experimental Gamma-Ray Collimator Sensitivity Patterns," *J. Applied Phys.,* **28**, pp. 1208–1211, Oct.

Toothill, G. C. (1960), "The Use of Cyclic Permuted Chain Codes for Digitizers," Inform. Processing Conf., Paris, France, UNESCO.

Totty, R. E., and G. C. Clark, (1967), "Reconstruction Error in Waveform Transmission," *IEEE Trans. Inform. Theory,* **13**, pp. 336–338, Apr.

Totty, R. E., and J. C. Hancock, (1963), "On Optimum Finite Dimensional Signal Representation," *Proc. 1st Allerton Conf. Circuits and Systems Theory.*

Treitel, S., and J. L. Shanks, (1971), "The Design of Multistage Separable Planar Filters," *IEEE Trans. Geosci. Electron.,* **9**, pp. 10–27, Jan.

Trench, W. F. (1960), "The Generation of Pseudorandom Sequences for Use in Testing Damp Data Processing Techniques," *Damp Data Procedures,* **2**, Sec. 10.10, RCA Tech. Report, Jan.

Ttetiak, O. J. (1966), "Scanner Displays," Res. Lab. Elec., MIT, Cambridge, MA, Quarterly Report 83, p. 129, Oct.

Tsukkerman, I. I., (1958), "The Transmission of the Coordinates of a TV Image," *Radio Eng.,* **13**, pp. 93–105, July.

Tsybakov, B. S. (1962), "Linear Coding of Images," *Radio Eng. Electron. Physics,* **7**, pp. 351–360, Mar.

Twomey, S. (1963), "On The Numerical Solution of Fredholm Integral Equation of the First Kind by the Inversion of the Linear System Produced by Quadrature," *J. Assoc. Comput. Mach.,* **10**, pp. 97–101.

Tyler, V. M. (1969), "Two Hardcopy Terminals for PCM Communications of Meteorological Products," *1969 Int. Conf. Commun.,* pp. 11.21–11.28.

Urquhard, and D. I. Pullen, "Improvements Related to Multiplied Transmission Systems," British Patent No. 1057024.

U. S. Patent No. 337423, "Reduced Bandwidth Binary Picture Transmission," or U. K. Patent Specification No. 1057024, "Improvements Relating to Binary Picture Transmission."

Vanderlugt, A. (1968), "A Review of Optical Data Processing Techniques," *Opt. Acta.,* **15**, pp. 1–33.

Van De-Weg, H. (1953), "Quantizing Noise of a Single Integration Delta Modulation System With an N-Digit Code," *Phillips Res.* Reports, **8**, pp. 367–385.

Van Trees, H. L. (1968), *Detection, Estimation, and Modulation Theory,* Wiley, New York.

Varah, J. M. (1973), "On the Numerical Solution of Ill-Conditioned Linear Systems with Applications to Ill-Posed Problems," *SIAM J. Numerical Analysis,* No. 2, **10**, p. 257.

Vegh, E., and L. M. Leibowitz, (1976), "Fast Complex Convolution in Finite Rings," *IEEE Trans. Acoust., Speech, Signal Processing,* **24**, pp. 343–344, Aug.

Viere, B. J. (1971), "Image Quality in Facsimile Systems," *Int. Conf. Commun.,* pp. 7.17–7.20.

Watanabe, S., "The Loeve-Karhunen Expansion as a Means of Information Compression for Classification of Continuous Signals," IBM Watson Res. Center, Yorktown Heights, NY, Tech. Report, AM RL-TR-65-114.

Wecksung, G. W., and K. Campbell, (1974), "Digital Image Processing at EG & G," **7**, pp. 63–71, May.

Wee, E., and S. Hsieh, (1975), "An Application of Projection Transform Technique in Image Transmission," *Proc. Nat. Telecommun. Conf.,* pp. 22-1–22-6.

Weiner, N. (1949), *Extrapolation, Interpolation, and Smoothing of Stationary Time Series,* MIT Press, Cambridge, MA.

Weiss, I. M. (1970), "A Survey of Discrete Kalman-Bucy Filtering with Unknown Noise Covariances," AIAA Guidance and Control Conf., Santa Barbara, CA.

Wendland, B., and F. May, (1973), "Ein Adaptiver Intraframe-Codierer fur Fernsehsignale," *Intern. Elektron. Rundschau,* **27**, pp. 12–18, Jan.

Wendland, B. (1972), "An Adaptive Source Coder For TV Signals," *Dig. Conf. Digital Processing of Signals Commun.,* Loughborough, England.

Wendt, H. (1973), "Interframe-Codierung Fur Videosignale," *Int. Elektron. Rundschan,* **27**, pp. 2–6, Jan.

Whelan, J. W. (1965), "Digital TV Bandwidth Reduction Techniques as Applied to Spacecraft TV," *AIAA Unmanned Spacecraft Meeting Proc.,* pp. 5–22, Mar.

Whelan, J. W., "Analog FM vs Digital-PSK," *IEEE Trans. Commun. Tech.,* to be published.

Whelan, J. W. (1965), "Digital TV Bandwidth Reduction Techniques as Applied to Spacecraft TV," AIAA Unmanned Spacecraft Meeting, Los Angeles, CA, CP-12, Mar.

Whelchel, J. E. Jr. and D. F. Guinn, (1968), "The Fast Fourier-Hadamard Transform And Its Use In Signal Representation And Classification," *EASCON Conv. Rec.,* pp. 561–573.

White, H. E. *et al.* (1972), "Dictionary Look-up Encoding Of Graphics Data," in *Picture Bandwidth Compression,* New York, Gordon and Breach, pp. 267–281.

Wholey, J. S. (1961), "The Coding of Pictorial Data," *IRE Trans. Inform. Theory,* 7, pp. 99–104, Apr.

Wichman, T. (1963), "Optimum Coding Of Pictorial Data," *9th Nat. Commun. Symp. Rec.,* pp. 18–27.

Widrow, B. (1961), "Statistical Analysis Of Amplitude Quantized Sampled Data Systems," *Trans. AIEE (Applications and Industry),* pt. II, **79**, pp. 555–568, Jan.

Widrow, B. (1956), "A Study of Rough Amplitude Quantization by Means of Nyquist Sampling Theory," *IRE Trans. Circuit Theory,* 3, pp. 266–276, Dec.

Wilkins, L. C., and P. A. Wintz, (1970), "Studies On Data Compression, Part I Picture Coding By Contours, Part II Error Analysis of Run-Length Codes," School Elec. Eng., Purdue Univ., Lafayette, IN, Tech. Report TR-EE-70-17, Sept.

Will, P. M. and K. S. Pennington, (1972), "Grid Coding: A Novel Technique for Image Processing," *Proc. IEEE,* **60**, pp. 669–680, June.

Williams, C. S. and O. A. Becklund, (1972), *Optics* Wiley, New York.

Winkler, M. R. (1965), "Pictorial Transmission with HIDM," *IEEE Int. Conv. Rec.,* pt. 1, pp. 285–291.

Winkler, M. R. (1964), "Digital TV: Shrinking Bulky Bandwidths," *Electronics,* Dec.

Winkler, M. R. (1953), "Discussion on the Impact of Information Theory on TV," *J. Brit. IRE,* **13**, pp. 590–591, Dec.

Winkler, M. R. (1953), "Saving TV Bandwidth," *Wireless World,* **59**, pp. 158–162, Apr.

Winkler, M. R. (1959), "TV Bandwidth Reduction," *Wireless World,* **65**, p. 402, Sept.

Winograd, S. (1976), "On Computing the Discrete Fourier Transform," *Proc. Nat. Acad. Sci.,* **73**, no. 4, pp. 1005–1006, Apr.

Wintz, P. A. (1972), "Transform Picture Coding," *Proc. IEEE,* **60**, pp. 809–820, July.

Wintz, P. A., and A. J. Kurtenbach, (1968), "Waveform Error Control in PCM Telemetry," *IEEE Trans. Inform. Theory,* **14**, Sept.

Wintz, P. A., and A. J. Kurtenbach, (1967), "Analysis and Minimization of Message Error in PCM Telemetry Systems," School Elec. Eng., Purdue Univ., TR-EE-67-19.

Woods, J. W. (1972), "Two-Dimensional Discrete Markovian Fields," *IEEE Trans. Inform. Theory,* **18**, Mar.

Wood, R. C. (1969), "On Optimum Quantization," *IEEE Trans. Inform. Theory,* **15**, pp. 248–252, Mar.

Woods, J. W., and T. S. Huang, (1972), "Picture Bandwidth Compression by Linear Transformation and Block Quantization," in *Picture Bandwidth Compression,* Gordon and Breach, New York, pp. 555–573.

Woods, J. W. (1960), "On Optimum Quantization," *Video Bandwidth Compression by Linear Transformation"* MIT QPR No. 91, pp. 219–224, Oct.

Wyle, H., T. Erb, and R. Banow, (1961), "Reduced-Time Facsimile Transmission by Digital Coding," *IRE Trans. Commun. Syst,* 9, pp. 215–222, Sept.

Wyszecki, G., and W. S. Stiles, (1967), *Color Science,* Wiley, New York.

Young, T. Y., and W. H. Huggins, (1963), "On The Representation of Electrocardiographs," *IEEE Trans. Bio-Medical Electronics,* **10**, pp. 86–95, July.

Young, I. T., and J. Mott-Smith, (1965), "On Weighted PCM," *IEEE Trans. Inform. Theory,* **11**, pp. 596–597, Oct.

Youngblood, W. A. (1958), "Estimation of the Channel Capacity Required for Picture Transmission," Sc. D. thesis, MIT, Cambridge, MA.

Youngblood, W. A. (1958), "Picture Processing," Quarterly Progress Rep. 54, MIT Res. Lab. Elec., p. 138.

Yuen, C. K. (1972), "Remarks on the Ordering of the Walsh Functions," *IEEE Trans. Comput.* p. 1452, Dec.

Zamperoni, P. (1972), "An Improved Run-Length Code for Typewritten Text and Drawings," *Nachrichtentech Z.,* **25,** Apr.

Zetterburg, L. (1955), "A Comparison Between Delta and Pulse Code Modulation," *Ericsson Technics*, **2,** pp. 95–154.

Zworykin, V. K., and E. G. Ramberg, (1952), "Standards Conversion of TV Signals," *Electronics*, p. 86.

2   ERROR-CORRECTION CODING

Abend, K., T. J. Hartley, Jr., B. D. Fritchman, and C. Gumacos, (1968), "On Optimum Receivers for Channels Having Memory," *IEEE Trans. Inform. Theory,* **14,** pp. 819–820, Nov.

Abend, K., and B. D. Fritchman, (1970), "Statistical Detection for Communication Channels with Intersymbol Interference," *Proc. IEEE.,* **58,** pp. 779–785, May.

Abramson, N. M. (1969), "Encoding and Decoding Cyclic Code Group," The Aloha System, Honolulu, HI, A 69-8, Oct.

Abramson, N. M. (1958), "A Class of Systematic Codes for Non-Independent Errors," Elec. Lab. Stanford Univ., Stanford, CA, Tech. Report No. 51, Dec.

Alexander, A. A., R. M. Gryb, and D. W. Nast, (1960), "Capabilities of Telephone Network for Data Transmission," *B. S. T. J.,* **39,** pp. 431–476, May.

Anderson, B. D. O., and J. B. Moore, (1968), "State Estimation via The Whitening Filter," Michigan Univ. *JACC*, pp. 123–129.

Artin, E. (1970), *Galois Theory*, Notre Dame, IN, Notre Dame Univ. Press.

Ash, R. (1965), *Information Theory,* Wiley, New York, p. 176.

Assmus, E. F., and H. F. Mattson Jr., (1969), "New 5-Designs," *J. Comb. Theory,* **6,** pp. 122–151.

Bahl, L. R., J. Cocke, F. Jelinek, and J. Raviv, (1972), "Optimal Decoding of Linear Codes for Minimizing Symbol Error Rate," *Int. Symp. Inform. Theory,* p. 90, Jan.

Bahl, L. R., C. D. Cullum, W. D. Frazer, and F. Jelinek, (1972), "An Efficient Algorithm for Computing Free Distance," *IEEE Trans. Inform. Theory,* **18,** pp. 437–439, May.

Bahl, L. R. and R. T. Chien, (1971), "Single - and Multiple - Burst - Correcting Properties of a Class of Cyclic Product Codes," *IEEE Trans. Inform. Theory,* **17,** pp. 594–600, Sept.

Bahl, L. R. and F. Jelinek, (1972), "On the Structure of Rate I/N Convolutional Codes," *IEEE Trans. Inform. Theory,* **18,** pp. 192–196, Jan.

Bahl, L. R. and F. Jelinek, (1971), "Rate 1/2 Convolutional Codes with Complementary Generators," *IEEE Trans. Inform. Theory*, **17,** pp. 718–727, Nov.

Balakrishnan, A. V., and I. J. Abrams, (1960), "Detection of Levels and Error Rates in PCM Telemetry Systems," *Nat. Telemetry Conf.,* pp. 37–55.

Barritt, P. R., and E. J. Habib, (1970), "Tracking and Data Relay Satellite System: An Overview," *AIAA 7th Annual Meeting and Technical Display,* Paper No. 70-1305, Oct.

Bartee, T. C. (1960), *Digital Computer Fundamentals,* McGraw-Hill, New York.

Bartee, T. C., and D. L. Schneider, (1962), "An Electronic Decoder for Bose-Chaudhuri-Hocqenghem Error Correcting Codes," *IRE Trans. Inform. Theory,* **8,** pp. 17–24.

Bartee, T. C., and P. Wood, (1961), *"Coding for Tracking Radar Ranging,"* Lincoln Labs., MIT Report.

Bartee, T. C., and C. K. Ray-Chaudhuri, (1960), "On A Class of Error Correcting Binary Group Codes," *Inform. Contr.,* 3, pp. 68–79.

Bartee, T. C., and D. I. Schneider, (1963), "Computation with Finite Fields," *Inform Contr.,* 6, pp. 79–98, Mar.

Behn, T. D., and Y. C. Ho, (1968), "On A Class of Linear Stochastic Differential Games," *IEEE Trans. Automat. Contr.,* 13, pp. 227–240, June.

Bellman, R. (1970), *Introduction to Matrix Analysis,* McGraw-Hill, New York, pp. 117–118.

Berlekamp, E. R. (1968), *Algebraic Coding Theory,* McGraw-Hill, New York.

Berlekamp, E. R. *Note on Recurrent Codes, IEEE Trans. Inform. Theory,* 10, p. 257, July.

Berlekamp, E. R. (1965), "On Decoding Binary Bose-Chaudhuri-Hocquenghem Codes," *IEEE Trans. Inform. Theory,* pp. 577–579, Oct.

Berlekamp, E. R. (1963), "A Class of Convolutional Codes," *Inform Contr.,* 6, pp. 1–13.

Berman, S. D. (1967), "On the Theory of Group Codes," *Kibernetika,* 3, pp. 31–39.

Birkhoff, G., and S. MacLane, (1965), *A Survey of Modern Algebra,* 3rd ed., MacMillan, New York, p. 71.

Blasbalg, H., D. Freeman, and R. Keeler, (1964), "Random-Access Communications Using Frequency Shifted PN (Pseudo-Noise) Signals," *IEEE Int. Conv.,* New York City, NY, pp. 1–34, Mar.

Blizard, R. B. (1969), "Convolutional Coding for Data Compression," Martin Marietta Corp., Denver, CO. Report R-69-17.

Bose, R. C. and D. K. Ray-Chaudhuri, (1960), "On a Class of Error Correcting Binary Group Codes," *Inform. Contr.,* 3, pp. 68–79, Mar.

Bose, R. C., and J. G. Caldwell, (1967), "Synchronizable Error-Correcting Codes," *Inform. Contr.,* 10, pp. 616–630.

Bowen, R. R. (1969), "Bayesian Decision Procedures for Interfering Digital Signals," *IEEE Trans. Inform. Theory,* 15, pp. 506–507, July.

Bower, E. K., and S. J. Dwyer, III (1969), "A Strengthened Asymptotic Gilbert Bound for Convolutional Codes," *IEEE Trans. Inform. Theory,* 15, pp. 433–435, May.

Brockett, R. W., and M. D. Mesarovic, (1965), "The Reproducibility of Multivariable Systems," *J. Math. Anal. and Appl.,* 11, pp. 548–563.

Brown, A. B., and S. T. Meyers, (1958), "Evaluation of Some Error Correction Methods Applicable to Digital Data Transmission," *IRE Conv. Rec.,* p. 37.

Bucher, E. A., and J. A. Heller, (1960), "Error Probability Bounds of Systematic Convolutional Codes," *IEEE Trans. Inform. Theory,* 16, pp. 219–224, Mar.

Bucher, E. A. (1968), "Error Mechanisms for Convolutional Codes," Ph. D. dissertation, Dept. Elec. Eng., MIT, Cambridge, MA.

Burton, H., and E. J. Weldon, Jr, (1965), "Cyclic Products Codes," *IEEE Trans. Inform. Theory,* 11, pp. 433–439, July.

Busacker, R., and T. Saaty, (1965), *Finite Graphs and Networks: An Introduction with Applications,* McGraw-Hill, New York.

Bussgang, J. J. (1965), "Some Properties of Binary Convolutional Code Generators," *IEEE Trans. Inform. Theory,* 11, pp. 90-100, Jan.

Byron, E., and J. S. McIntyre, (1973), "Support Study for Command Link-Uplink Signal Design," Report No. CSC-0-212, John Hopkins Univ., Applied Physics Lab., Feb.

Cain, J. B. (1972), "Utilization of Low-Redundancy Convolutional Codes," *Nat. Telemetry Conf. Rec.,* Houston, TX, pp. 21B1–21B8, Nov.

Carlyle, J. W. (1963), "Equivalent Stochastic Sequential Machines," Univ. Calif., Berkeley, CA, Tech. Report Ser. 60.

Carlyle, J. W. (1963), "Reduced Forms for Stochastic Sequented Machines," *J. Math. Anal. and Appl.,* 7, pp. 167–175.

Chang, R. W. and J. C. Hacock, "On Receiver Structures for Channels Having Memory," *IEEE Trans. Inform. Theory,* 12, pp. 463–468, Oct.

Chen, C. L. (1970), "Computer Results on the Minimum Distance of Some Binary Cyclic Codes," *IEEE Trans. Inform. Theory,* 16, pp. 359–360, May.

Chen, C. L., C. A. Desoer, A. Niederlinski, and R. E. Kalman, (1966), "Simplified Conditions for Controllability and Observability of Linear Time-Invariant Systems," *IEEE Trans. Automat. Contr.,* 11, pp. 613–614, July.

Chien, R. T. (1964), "Cyclic Decoding Procedures for Bose-Chaudhuri-Hocquenghem Codes," *IEEE Trans. Inform. Theory,* 10, pp. 357–363, Oct.

Chien, R. T., and D. T. Tang, (1964), "On Detecting Errors After Correction," *Proc. IEEE,* 62, p. 974, Aug.

Clark, G. C. (1971), "Implementation of Maximum Likelihood Decoders for Convolutional Codes," *Proc. Int. Telemetering Conf.,* Washington DC.

Clark, G. C., and R. C. Davis, (1971), "Two Recent Applications of Error Correction Coding to Communications System Design," *IEEE Trans. Commun. Tech.,* 19, pp. 856–863, Oct.

Clark, G. C., and R. C. Davis (1927), "Reliability-of-Decoding Indicators for Maximum Likelihood Decoders," *Proc. 5th Hawaii Int. Conf. System Science,* Honolulu, HI, Western Periodicals, pp. 447–450.

Clark, G. C., R. C. Davis, J. C. Herndon, and D. D. McRae, (1969), "Interim Report on Convolutional Coding Research," Advanced System Operation, Radiation Inc., Melbourne, FL, Memo Report 38.

Clark, G. C., and F. A. Perkins, (1972), "Performance of a Digital TV System-Part II: Channel Encoding," *Nat. Telecommun. Conf. Rec.,* Houston, TX, pp. 13E-1–13E-7, Dec.

Codex Corp. (   ), "Coding System Design for Advanced Solar Missions," NASA Ames Res. Center, Moffett Field, CA, Final Report Contract NAS 2-3637.

Codex Corp. (1968), "Final Report High-Speed Sequential Decoder Study," U. S. Army Satellite Communications Agency, Apr.

Cohen, A. R., J. A. Heller, and A. J. Viterbi, (1971), "A New Coding Technique for Asynchronous Multiple Access Communications," *IEEE Trans. Commun. Tech.,* 19, pp. 849–855, Oct.

Cohn, M. (1962), "Controllability in a Linear Sequential Networks," *IRE Trans. Circuit Theory,* 9, pp. 74–78, Mar.

Costello, D. J. (1969), "A Construction Technique For Random-Error-Correcting Convolutional Codes," *IEEE Trans. Inform. Theory,* 15, pp. 631–636, Sept.

Costello, D. J. (1969), "Construction of Convolutional Codes for Sequential Decoding," Univ. Notre Dame, Notre Dame, IN, Tech. Report EE-692, Aug.

Costello, D. J., and T. N. Morrissey, (1971), "Strengthened Lower-Bound on Definite Decoding Minimum Distance for Periodic Convolutional Codes," *IEEE Trans. Inform. Theory,* 17, pp. 212–214, Mar.

Costello, D. J. (1974), "Free Distance Bounds for Convolutional Codes," *IEEE Trans. Inform. Theory,* 20, pp. 356–365, May.

Cowles, J. E. and G. I. Davida, (1972), "Decoding of Triple-Error-Correcting Codes," *Electron. Lett.,* 8.

Cruz, J. B. and W. R. Perkins, (1966), "Conditions for Signal and Parameter Invariance in Dynamical Systems," *IEEE TRans. Automatic Control,* 11, pp. 614–615, July.

Curtis, C. W. and I. Reiner, (1962), *Representation Theory of Finite Groups and Associative Algebras,* Wiley, New York, pp. 94–96.

Davida, G. I. (1971), "Decoding of BCH Codes," *Electron. Lett.,* **7**.

Davida, G. I. and S. M. Reddy, (1972), "Forward-Error Correction with Decision Feedback," *Inform. Contr.,* **21**, pp. 117–133.

Davida, G. I. and J. W. Cowles, (1975), "A New Error-Locating Polynomial for Decoding of BCH Codes," *IEEE Trans. Inform. Theory,* pp. 235–236, Mar.

DeBuda, R. (1972), "Cohenent Demodulation of Frequency-Shift Keying with Low Deviation Ratio," *IEEE Trans. Commun. Tech.,* **20**, pp. 429–435, June.

Devore, A. R., J. A. Heller, and A. J. Viterbi, (1971), "A New Coding Technique for Asynchronous Multiple Access Communications," *IEEE Trans. Commun. Tech.,* **19**, pp. 849–855, Oct.

Ebert, P. M., and S. Y. Tong, (1969), "Convolutional Reed-Solomon Codes," *B. S. T. J.,* pp. 729–743, Mar.

Elias, P. (1961), *Lectures on Communications System Theory,* McGraw-Hill, New York, ch. 13.

Elias, P. (1955), "Coding for Noisy Channels," *IRE Conv. Rec.,* pt. 4.

Elias, P. (1954), "Error Free Coding," *IRE Trans. Inform. Theory,* **4**, pp. 29–37, Sept.

Elspas, B. (1959), "The Theory of Autonomous Linear Sequential Networks," *IRE Trans. Circuit Theory,* **6**, pp. 45–60, Mar.

Elspas, B. (1961), "Design and Instrumentation of Error-Correcting Codes," Stanford Res. Inst., Menlo Park, CA, Project No. 3318, Oct.

Epstein, M. A. (1958), "Coding for the Binary Erasure Channel," Sc. D. dissertation, MIT, Cambridge, MA.

Epstein, M. A. (1959), "Construction of Convolutional Codes by Suboptimalization," MIT Res. Lab. of Elec., Cambridge, MA, Tech. Report No. 341.

Epstein, M. A. (1958), "Algebraic Decoding for a Binary Erasure Channel," *IRE Nat. Conv. Rec.,* **VI**, pp. 56–69.

Everitt, W. L. (Ed.) (1964), *Digital Communications,* Prentice-Hall, Englewood Cliffs, NJ, p. 117.

Fairchild Imaging Systems (1976), "Solid-State Television Camera (CCD-Buried Channel," NASA Report, NAS 9-14844, Dec.

Falb, P. L., and W. A. Wolovich, (1967), "On the Decoupling of Multivariate Systems," *Joint Automat. Control Conf.,* Philadelphia, PA, pp. 791–796, June.

Falconer, D. D. (1967), "A Hybrid Sequential and Algebraic Decoding Scheme," Ph. D. dissertation, Dept. Elec. Eng., MIT, Cambridge, MA.

Falconer, D. D. (1966), "An Upper Bound on the Distribution of Computation for Sequential Decoding with Rate Above R (Comp)," MIT Res. Lab. Elec., Quarterly Progress Report 81, pp. 174–179, Mar.

Falconer, D. D. and C. W. Niessen, (1966), "Simulation of Sequential Decoding for a Telemetry Channel," MIT Res. Lab. Elec., Quarterly Progress Report 80, pp. 183–193, Jan.

Fano, R. M. (1953), "Communication in the Presence of Additive Gaussian Noise," in *Communication Theory,* London, Butterworths, pp. 169–182.

Fano, R. M. (1963), "A Heuristic Discussion of Probabilistic Decoding," *IEEE Trans. Inform. Theory,* **9**, pp. 64–74, Apr.

Fano, R. M. (1964), *Transmission of Information,* Cambridge, MA, MIT Press, and New York, Wiley.

Feinstein, A. (1958), *Foundations of Information Theory,* New York, McGraw-Hill, Sec. 33.

Feinstein, A. (1954), "A New Basic Theorem of Information Theory," *IRE Trans. Inform. Theory,* pp. 2–22.

Feller, W. (1950), *An Introduction of Probability Theory and its Applications,* New York, Wiley.

Ferguson, M. J. (1972), "Optimal Reception for Binary Partial Response Channels," *B. S. T. J.,* **51**, pp. 493–505, Feb.

Fire, P. (1959), "A Class of Multiple-Error Correcting Binary Codes for Non-Independent Errors," Sylvania Electronics Systems, Report RSL-E-2, Mar.

Forney, G. D., Jr. (1972), "Maximum-Likelihood Sequence Estimation of Digital Sequences in the Presence of Intersymbol Interference," *IEEE Trans. Inform. Theory,* **18**, pp. 363–378, May.

Forney, G. D. Jr. and E. K. Bower, (1971), "A High-Speed Sequential Decoder: Prototype Design and test," *IEEE Trans. Commun. Tech.,* **19**, pp. 821–835, Oct.

Forney, G. D., Jr. (1967), "Coding System Design for Advanced Solar Missions," NASA Codex Corp., Watertown, MA, Contract NAS 2-3637, Final Report, Dec.

Forney, G. D., Jr. (1968), "Final Report on a Study of a Sample Sequentail Decoder," Appendix A, Codex Corp., Watertown, MA, U. S. Army Satellite Communication Agency Contract DAA B 07-68-C-0093, Apr.

Forney, G. D., Jr. and R. M. Langelier, (1968), "A High-Speed Sequential Decoder for Satellite Communications," *Conf. Rec. Int. Conf. Commun.,* Boulder, CO, pp. 39.9–39.17, June.

Forney, G. D., Jr. (1965), "On Decoding BCH Codes," *IEEE Trans. Inform. Theory,* p. 549, Oct.

Forney, G. D., Jr. (1970), "Convolutional Codes I: Algebraic Structure," *IEEE Trans. Inform. Theory,* **16**, pp. 720–738, Nov.

Forney, G. D., Jr. (1970), "Use of Sequential Decoders to Analyze Convolutional Code Structure," *IEEE Trans. Inform. Theory,* **16**, pp. 793–795, Nov.

Forney, G. D., Jr. (1966), *Concatenated Codes,* Cambridge, Ma, MIT Press.

Forney, G. D., Jr. (1972), "Lower Bounds on Error Probability in the Presence of Large Intersymbol Interference," *IEEE Trans. Commun. Tech.,* **20**, pp. 76–77, Feb.

Forney, G. D., Jr. (1972), "Convolutional Codes II: Maximum Likelihood Decoding," Stanford Elec. Labs., Stanford, CA, Tech. Report 7004-1, June.

Forney, G. D., Jr. (1967), "Review of Random Tree Codes," NASA Ames Res. Center, Moffett Field, CA, Contract NAS2-3637, NASA CR 73176, Final Report, Appendix A, Dec.

Forney, G. D., Jr. (1973), "The Viterbi Algorithm," *Proc. IEEE,* **61**, pp. 268–278, Mar.

Frasco, L. A. (1967), "Analysis and Simulation of Semi-Definite Decoding," S. M. thesis, MIT, Cambridge, MA, June.

Freiman, C. V., J. P. Robinson, and A. D. Wyner, (1964), "Redundancy Requirements for the Correction of Finite State Error Patterns," *Proc. Nat. Electronics Conf.,* pp. 687–692.

Freiman, C. V. and J. P. Robinson, (1965), "A Comparison of Block and Recurrent Codes for the Correction of Independent Errors," *IEEE Trans. Inform. Theory,* pp. 445–449, July.

Fritchman, B. D. and J. C. Mixsell, (1974), "Comments on Real-Time Minimal-Bit-Error Probability Decoding for Convolutional Codes," *IEEE Trans. Commun.,* Nov.

Gallager, R. G. (1963), *Low Density Parity Check Codes,* Cambridge, MA, MIT Press.

Gallager, R. G. (1965), "A Simple Derivation of the Coding Theorem and Some Applications," *IEEE Trans. Inform. Theory,* **11**, pp. 3–18, Jan.

Gallager, R. G. (1968), *Information Theory and Reliable Communication,* New York, Wiley, pp. 263–273.

Gallager, R. G. (1964), "Sequential Decoding for Binary Channels with Noise and Synchronization Errors," Lincoln Lab., MIT, Lexington, MA, Rept. No 25G-2.

Geist, J. (1970), "Algorithmic Aspects of Sequential Decoding," Ph. D. dissertation, Dept. Elec. Eng., Univ. Notre Dame, Notre Dame, IN, Aug.

General Electric (1970), "Earth Resources Technology Satellite Spacecraft System Design Studies Final Report, Book 1 of 2, Refer to Section 4.4.1.4," Goddard Space Flight Center, CR-112499, Document No. 705D 4207, Feb. 11.

Gilhousen, K. S. *et al.* (1971), "Coding Systems Study for High Data Rate Telemetry Links," Linkabit Corp., Final Report. Contract No. NAS2-6024, NASA Ames. Res. Center, Jan.

Gilhousen, K. S. and D. R. Lumb, (1973), "A Very High Speed Hard Decision Sequential Decoder," *Nat. Telecommun. Conf. Rec.,* Houston, TX, pp. 21A1–21A7, Nov.

Gilbert, E. N. (1952), "A Comparison of Signaling Alphabets," *B. S. T. J.,* **31**, pp. 504–522.

Gilbert, E. N. (1960), "Capacity of a Burst-Noise Channel," *B. S. T. J.,* **39**, pp. 1253–1265, Sept.

Gill, A. (1966), *Linear Sequential Circuits,* New York, McGraw-Hill, pp. 45–46.

Gill, A. (1962), *Finite State Machines,* New York, McGraw-Hill.

Goddard Space Flight Center (1969), "Tracking and Data Relay Satellite (TDRS) System Concept-Phase A Study Final Report," vols. 1 and 2, Nov.

Goethals, J. M. (1967), "Factorization of Cyclic Codes," *IEEE Trans. Inform. Theory,* **13**, pp. 242–246, Apr.

Golay, M. J. E. (1954), "Binary Coding," *IRE Trans. Inform. Theory,* p. 23.

Golay, M. J. E. (1949), "Note on the Theoretical Efficiency of Information Reception with PPM," *Proc. IRE,* **37**, p. 1031.

Golding, L. (1968), "Digital Television Transmission Systems for Satellite Communication Links," *Proc. Int. Broadcasting Conf.,* Publ. 46.

Gore, W. C. (1970), "Further Results on Product Codes," *IEEE Trans. Inform. Theory,* **16**, pp. 446–451, July.

Gorenstein, D. and N. Zierler, (1961), "A Class of Cyclic Error-Correcting Codes in p^m Symbols," *J. SIAM,* **9**, pp. 207–214, June.

Green, J. H., Jr., and R. L. San Soucie, (1958), "An Error-Correcting Encoder and Decoder of High Efficiency," *Proc. IRE,* **46**, p. 1741.

Green, D. M., and T. S. Birdsall, (1958), "The Effect of Vocabulary Size on Articulation Score," AD 146759.

Haccoun, D. and M. J. Ferguson, (1975), "Generalized Stack Algorithms for Decoding Convolutional Codes," *IEEE Trans. Inform. Theory,* **21**, pp. 638–651, Nov.

Haccoun, D. and M. J. Ferguson, (1973), "Adaptive Sequential Decoding," *Proc. IEEE Int. Symp. Inform. Theory.*

Haccoun, D. (1974), "Multiple-Path Stack Algorithms for Decoding Convolutional Codes," Ph. D. dissertation, Dept. Elec. Eng., McGill Univ., Montreal, Canada, June.

Hagelbarger, D. W. (1959), "Recurrent Codes: Easily Merchanized Burst-Correcting Binary Codes," *B. S. T. J.,* **38**, pp. 969–984, July.

Hagelbarger, D. W. (1967), "Recurrent Codes for the Binary Symmetric Channel," *Lecture Notes Theory of Codes,* Univ. Mich. Summer Conference, Ann Arbor, pp. 18–29, June.

Hall, M. Jr. (1956), "A Survey of Difference Sets," *Proc. Amer. Math. Soc.,* **7**, pp. 975–986.

Hamming, R. W. (1950), "Error Detecting and Error Correcting Codes," *B. S. T. J.,* **26**, pp. 147–160.

Hardy, G. H., J. E. Littlewood, and G. Polya, (1959), *Inequalities,* London, Cambridge Univ. Press.

Hartmanis, J. (1962), "Maximal Autonomous Clocks of Sequential Machines," *IRE Trans. Elec. Comput.,* pp. 83–86, Feb.

Hartmanis, J. and R. E. Stearns, (1963), "A Study of Feedback and Errors in Sequential Machines," *IEEE Trans. Elec. Comput.,* **12**, pp. 223–232, June.

Heller, J. A. (1968), "Short Constraint Length Convolutional Codes," Space Program Summary 37-51, Jet Propulsion Labs., **III**, pp. 171–177, Oct-Nov.

Heller, J. A. (1969), "Improved Performance of Short Constraint Length Convolutional Codes," Space Program Summary 37-56, **III**, Jet Propulsion Labs., pp. 83–84, Feb.-Mar.

Heller, J. A. and I. M. Jacobs, (1971), "Viterbi Decoding for Satellite and Space Communication," *IEEE Trans. Commun. Tech.,* **19**, pp. 835–847, Oct.

Heller, J. A. (1969), "Description and Operation of a Sequential Decoder Simulation Program," Jet Propulsion Lab., Space Programs Summary 37-58, **3**, pp. 36–42, Aug.

Heller, J. A. (1968), "Sequential Decoding: Short Constraint Length Convolutional Codes," Jet Propulsion Lab., Space Programs Summary 37-54, **3** pp. 171–174, Dec.

Heller, J. A. (1974), "On Using Natural Redundancy for Error-Detection," *IEEE Trans. Commun. Tech.,* **22**, pp. 1690–1693, Oct.

Hellman, M. E. (1975), "Convolutional Source Encoding," *IEEE Trans. Inform. Theory,* **21**, pp. 651–656, Nov.

Hilborn, C. G. Jr. (1970), "Applications of Unsupervised Learning to Problems of Digital Communications," *Proc. 9th IEEE Symp. Adaptive Processes, Decision and Control,* Dec.

Hocquenghem, A. (1959), "Codes Correcteurs d'erreurs," *Chiffres,* **2**, pp. 147–156, Sept.

Hoffman, K. and R. Kunze, (1961), *Linear Algebra,* Englewood Cliffs, NJ, Prentice-Hall.

Huffman, D. A. (1956), "The Synthesis of Linear Sequential Coding Networks," in *Information Theory,* C. Cherry, (Ed.), New York, Academic Press.

Huffman, D. A. (1959), "Canonical Forms for Information-Lossless Finite State Logical Machines," *IRE Trans. Circuit Theory,* **6**, pp. 41–59, May.

Huffman, D. A. (1956), "Linear Circuit Viewpoint on Error-Correcting Codes," *IRE Trans. Inform. Theory,* **2**, pp. 20–28.

Huffman, D. A. (1962), "A Method for the Construction of Minimum-Redundancy Codes," *Proc. IRE,* **40**, pp. 1098–1101, Sept.

Huggins, W. H. (1957), "Signal Flow Graphs and Random Signals," *Proc. IRE,* **45**, pp. 74–86, Jan.

Huth, G. K. (1973), "Command Channel Coding for Shuttle Communications," Report No. R7307-1, Axiomatrix, NASA Contract NAS 9-13467, July.

Huth, G. K. (1973), "A Command Encoding Scheme for a Multiplexed Space Communication Link," *Nat. Telecommun. Conf. Rec.* Atlanta, pp. 21A1–21A7, Nov.

Huth G. K. (1972), "Performance Versus Complexity of Viterbi and Sequential Coding," *Nat. Telecommun. Conf. Rec.* Houston, pp. 13B1–13B2, Dec.

Iwadare, Y. (1968), "On Type-B1 Burst-Error-Correcting Convolutional Codes," *IEEE Trans. Inform. Theory,* **14**, no. 4, pp. 577–584.

Jacobs, I. M., and E. R. Berlekamp, (1967), "A Lower to the Distribution of Computation for Sequential Decoding," *IEEE Trans. Inform. Theory,* **13**, pp. 167–174, Apr.

Jacobs, I. M. (1967), "Sequential Decoding for Efficient Communication from Deep Space," *IEEE Trans. Commun. Theory,* **15**, pp. 492–501, Aug.

Jacobs, I. M., and J. A. Heller, (1971), "Performance Study of Viterbi Decoding as Related to Space Communication," Final Tech. Report to U. S. Army Satellite Comm. Agency, Aug.

Jacobs, I. M., and R. J. Sims, (1972), "Configuring a TDMA Satellite Communication System with Coding," *Proc. 5th Hawaii Int. Conf. Systems Science,* Honolulu, HI, Western Periodicals, pp. 443–446.

Jelinek, F. (1969), "A Fast Sequential Decoding Algorithm Using a Stack," *IBM J. Res. Develop.,* **13**, pp. 675–685, Nov.

Jelinek, F. (1968), *Probabalistic Information Theory,* McGraw-Hill, New York.

Jelinek, F. (1969), "Tree Encoding of Memoryless Time-Discrete Sources with a Fidelity Criterion," *IEEE Trans. Inform. Theory*, **15**, pp. 584–590, Sept.

Jelinek, F. (1969), "An Upper Bound on Moments of Sequential Decoding Effort," *IEEE Trans. Inform. Theory*, **15**, pp. 140–149, June.

Jones, D. M., W. R. Wadden, and J. J. Bussgang, (1962), "A Comparative Evaluation of Sequential Algotithms," *Wescon Conv.*, Paper 16.3.

Jordon, K. L. (1966), "The Performance of Sequential Decoding in Connection with Efficient Modulation," *IEEE Trans. Commun. Tech.*, **14**, pp. 283–297, June.

Justesen, J. (1972), "A Class of Constructive Asymptotically Good Algebraic Codes," *IEEE Trans. Inform. Theory*, **18**, pp. 652–656, Sept.

Justesen, J. (1973), "New Convolutional Code Constructions and a Class of Asymptotically Good Time-Varying Codes," *IEEE Trans. Inform. Theory*, **19**, pp. 220–225, Mar.

Kalman, R. E., Y. C. Ho, and K. S. Narendra, (1962), Controllability of Linear Dynamical Systems," *Contrib. Differential Equations*, **1**, pp. 189–213.

Kalman, R. E. (1965), "Irreducible Representations and the Degree of a Rational Matrix," *J. SIAM Contr.*, **13**, pp. 520–544.

Kalman, R. E., P. L. Falb, and M. A. Arbib, (1969), *Topics in Mathematical System Theory*, New York, McGraw-Hill, ch. 10.

Kautz, W. H. (1958), "A Class of Multiple-Error-Correcting Codes," Stanford Res. Inst.

Kilmer, W. L. (1961), "Linear-Recurrent Binary Error-Correcting Codes for Memoryless Channels," *IRE Trans. Inform. Theory*, **7**, pp. 7–13, Jan.

Kilmer, W. L. (1960), "Some Results on Best Recurrent-Type Binary-Error-Correcting Codes," *IRE Int. Conv. Rec.*, pp. 135–147.

Klieber, E. J. (1970), "Some Difference Triangles for Constructing Self-Orthogonal Codes," *IEEE Trans. Inform. Theory*, pp. 237–238, Mar.

Kobayashi, H. (1971), "A Survey of Coding Schemes for Tramsmission or Recording of Data," *IEEE Trans. Commun. Tech.*, **19**, pp. 1087–1100, Dec.

Kobayashi, H. (1971), "Applications of Probabilistic Decoding to Digital Magnetic Recording Systems," *IBM J. Res. Devel.*, **15**, pp. 64–74, Jan.

Kobayashi, H. (1971), "Correlative Level Coding and Maximum Likelihood Decoding," *IEEE Trans. Inform. Theory*, **17**, pp. 586–594, Sept.

Kohavi, Z. (1964), "Secondary State Assignment for Sequential Machines," *IEEE Trans. Electr. Comput.*, **13**, pp. 193–203, June.

Kolor, R. W. (1967), "A Gilbert Bound for Convolutional Codes," M. S. thesis, MIT, Cambridge, MA, Aug.

Kreindler, E., and P. E. Sarachik, (1964), "On the Concepts of Controllability and Observability of Linear Systems," *IEEE Trans. Automat. Contr.*, **9**, Apr.

Koshelev, V. N. (1973), "Dirent Sequential Encoding and Decoding for Discrete Sources," *IEEE Trans. Inform. Theory*, **19**, pp. 340–343, May.

Landon, V. D. (1948), "Theoretical Analysis of Various Systems of Multiplex Transmissions," *RCA Review*, pp. 433–482, July.

Langelier, R. M., D. Quagliato, and J. Quigley, (1968), "Characterization and Coding of DCSP Satellite Communications Channels," *EASCON Conv. Rec.*, pp. 9–13, Sept.

Layland, J. W., and W. A. Lushbaugh, (1971), "A Flexible High-Speed Sequential Decoder for Deep Space Channels," *IEEE Trans. Commun. Tech.*, pp. 813–820, Oct.

Layland, J. W. (1970), "Performance of Short Constraint Length Convolutional Codes and a Heuristic Code Construction Algorithm," Jet Propulsion Lab., Pasadena, CA, Space Programs Summary 37-64, pp. 41–44, Aug.

Layland, J. W., and R. J. McEliece, (1970), "An Upper Bound on the Free Distance of a Tree Code," Jet Propulsion Lab., Pasadena, CA, Space Programs Summary 37-62, pp. 63–64, Apr.

Layland, J. W. (1970), "Synchronizability of Convolutional Codes," Jet Propulsion Lab., Pasadena, CA, Space Programs Summary, 37-64, pp. 44–50, Aug.

Lebow, I. L., P. G. McHugh, A. C. Parker, P. Rosen, and J. M. Wozencraft, (1963), "Application of Sequential Decoding to High-Rate Data Communication on Telephone Lines," *IEEE Trans. Inform. Theory*, Apr.

Lee, Lin-Nan, (1974), "Real-Time Minimal-Bit-Error Probability Decoding of Convolutional Codes," *IEEE Trans. Commun.*, **22**, pp. 146–151, Feb.

Levy, J. E. (1966), "Self-Synchronizing Codes Derived From Binary Cyclic Codes," *IEEE Trans. Inform. Theory*, **12**, pp. 286–290, July.

Lin, S. (1965), "An Investigation of Codes and Decoding Procedures for Sequential Decoding Systems," Ph. D. dissertation, Rice Univ., Houston TX.

Lin, S. and H. Lyne, (1967), "Some Results on Binary Convolutional Code Generators," *IEEE Trans. Inform. Theory*, **13**, pp. 134–140, Jan.

Lin, S. and E. J. Weldon, Jr. (1970), "Further Results on Cyclic Product Codes," *IEEE Trans. Inform. Theory*, **16**, pp. 451–459, July.

Linabit Corp. (1972), "LV 7026 Convolutional Encoder-Viterbi Decoder Acceptance Test," Jan.

Linabit Corp. (1971), "LV 7026 Convolutional Encoder-Viterbi Decoder Instruction Manual," Contract No. DAAB 07-71-C-0148, USASATCOMA, Fort Monmouth, NJ.

Lindsey, W. C. (1969), "Block Coding for Space Communications Systems," *IEEE Trans. Commun. Tech.*, **17**, pp. 217–225, Apr.

Lindsey, W. C. (1967), "Design of Block-Coded Communication Systems," *IEEE Trans. Commun. Tech.*, **15**, pp. 525–534, Aug.

Linkabit Corp. (1971), "Coding Systems Study for High-Speed Data Rate Telemetry Links (Final Report)," Contract NAS 2-6024, NASA Ames. Res. Center, Moffett Field, CA, NASA CR-114278.

Lloyd, S. P. (1957), "Binary Block Coding," *B. S. T. J.,* **36**, p. 517.

Lumb, D. R. and L. B. Hofman, (1970), "An Efficient Coding System for Deep Space Probes with Specific Application to Pioneer Missions," NASA Tech. Note, NASA TN-D-4105, Aug.

Lumb, D. R. and A. J. Viterbi, (1971), "High Data Rate Coding for the Space Station Telemetry Links," *Int. Telemetering Conf. Proc.,* pp. 101–106.

Lyne, H. (1965), "Digital Computer Simulation of a Sequential Decoding System," Ph. D. dissertation, Rice Univ., Houston, TX.

Mackechnie, L. K. (1973), "Maximum Likelihood Receivers for Channels Having Memory," Ph. D. dissertation, Dept. Elec. Eng., Univ. Notre Dame, Notre Dame, IN, Jan.

Macon, N. and A. Spitzbart, (1958), "Inverses of Vandermonde Matrices," *Amer. Math. Monthly,* **65**, pp. 95–100.

MacWilliams, J. (1965), "The Structure and Properties of Binary Cyclic Alphabets," *B. S. T. J.,* **44**, pp. 303–332.

Macy, J. R. (1963), "*Theory of Serial Codes,*" Ph. D. dissertation, Stevens Inst. Tech., Hoboken, NJ.

Magee, F. R. Jr., and J. C. Proakis, (1973), "Adaptive Maximum Likelihood Sequence Estimation for Digital Signalling in the Presence of Intersymbol Interference," *IEEE Trans. Inform. Theory,* **19**, pp. 120–123, Jan.

Mandelbaum, D. (1971), "On Decoding of Reed-Solomon Codes," *IEEE Trans. Inform. Theory,* **17**, pp. 707–712, Nov.

Mandelbaum, D. (1968), "A Note on Synchronizable Error-Correcting Codes," *Inform. Contr.* **13**, pp. 429–432, Nov.

Mandelbaum, D. (1970), "On Synchronization of Convolutional Codes," *IEEE Trans. Commun. Tech.*, pp. 817–821, Dec.

Marcowitz, A. B. (1961), "On Inverses and Quasi-Inverses of Linear Time-Varying Discrete Systems," *J. Franklyn Inst.*, **272**, pp. 23–44, July.

Marcowitz, A. B. (1967), "Codes, Automata and Continuous Systems: Explicit Interconnections," *IEEE Trans. Automat. Contr.*, **12**, pp. 644–650, Dec.

Marcus, M., and H. Minc, (1964), *A Survey of Matrix Theory and Matrix Inequalities*, Allyn and Bacon, Boston, MA.

Massey, J. L., and R. W. Liu, (1964), "Applications of Lypunov's Direct Method to the Error-Propagation Effect in Convolutional Codes," *IEEE Trans. Inform. Theory*, pp. 248–250, July.

Massey, J. L., and M. K. Sain, (1967), "Codes, Automata, and Continuous Systems: Explicit Interconnections," *IEEE Trans. Automat. Contr.*, pp. 664–650, Dec.

Massey, J. L. (1966), "Uniform Codes," *IEEE Trans. Inform. Theory*, pp. 132–134, Apr.

Massey, J. L., and M. K. Sain, (1968), "Inverse of Linear Sequential Circuits," *IEEE Trans. Comput.*, **17**, pp. 330–337, Apr.

Massey, J. L. (1965), "Step-by-Step Decoding of the Bose-Chaudhuri, Hocquenghem Codes," *IEEE Trans. Inform. Theory*, pp. 580–585, Oct.

Massey, J. L. (1963), *Threshold Decoding*, MIT Press, Cambridge, MA, and Wiley, New York.

Massey, J. L., and D. J. Costello, (1971), "Nonsystematic Convolutional Codes for Sequential Decoding in Space Applications," *IEEE Trans. Commun. Tech.*, **19**, pp. 806–813, Oct.

Massey, J. L. (1964), "Reversible Codes," *Inform. Contr.*, **7**, pp. 369–380.

Massey, J. L. (1968), "Some Algebraic and Distance Properties of Convolutional Codes," in *Error-Correcting Codes*, New York, Wiley, p. 90.

Massey, J. L. (1963), "Error Correcting Codes Applied to Computer Technology," *Proc. Nat. Electron. Conf.*, **19**, pp. 142–147.

Massey, J. L. (1965), "Implementation of Burst-Correcting Convolutional Codes," *IEEE Trans. Inform. Theory*, **11**, pp. 416–422, July.

Massey, J. L., M. K. Sain, and J. M. Geist, (1972), "Certain Infinite Markov Chains and Sequential Decoding," *Discrete Math.*, **3**, pp. 163–175, Sept.

Massey, J. L., D. J. Costello, and J. Justesen, (1973), "Polynomial Weights and Code Constructions," *IEEE Trans. Inform. Theory*, **19**, pp. 101–110, Jan.

Massey, J. L. (1962), "Majority Decoding of Convolutional Codes," Res. Lab. Electron. Quarterly Prog. Report 64, MIT, Cambridge, pp. 183–188, Jan.

Massey, J. L. and M. K. Sain, (1967), "Inverse Problems in Coding Automata, and Continuous Systems," *Proc. 8th Annual Symp. on Switching and Automata Theory*, Austin, TX, pp. 226–232, Oct.

Massey, J. L. (1970), "Nonsystematic Convolutional Codes for Sequential Decoding," First NASA Coded Commun. Conf., Feb.

Massey, J. L., "Shift Register Synthesis and BCH Decoding," *IEEE Trans. Inform. Theory*, to be published.

Maxwell, M. and J. Pandelides, (1971), "The Telecommunication System for the Earth Resources Technology Satellite (ERTS) A and B," *Nat. Telemetering Conf. Conv. Rec.*, pp. 137–147.

McAdam, P. L., L. R. Welch, and C. L. Weber, (1972), "M. A. P. Bit Decoding of Convolutional Codes," Int. Symp. Inform. Theory, Asilomar, CA, Jan.

McCluskey, E. J. and T. C. Bartee, (1962), *A Survey of Switching Circuit Theory*, McGraw-Hill, New York.

McEliece, R. and H. C. Rumsey, (1968), "Capabilities of Convolutional Codes," Jet Propulsion Lab., Pasadena, CA, Space Programs Summary 37-50, **3**, pp. 248–251, Apr.

McEliece, R. and J. W. Layland, (1970), "An Upper Bound on the Free-Distance of a Tree Code," Jet Propulsion Lab., Pasadena, CA, Space Programs Summary 37-62, **3** pp. 63–64, Apr.

McMillan, B. (1952), "Introduction to Formal Realizability Theory," *B. S. T. J.*, **31**, pp. 217–279, 541–600.

Meggitt, J. E. (1961), "Error Correcting Codes and Their Implementation for Data Transmission Systems," *IRE Trans. Inform. Theory*, **7**, pp. 234–244, Oct.

Melas, C. M. (1960), "A New Group of Codes for Correction of Dependent Errors in Data Transmission," *IBM J. Res. and Dev.*, **4**, pp. 58–65, Jan.

Merrill, H. M. and G. D. Thompson, (1972), "Design of Communication Systems Using Short-Constraint-Length Convolutional Codes," *Nat. Telecommun. Conf. Rec*, Houston, TX, pp. 13A1–13A6, Dec.

Mertz, P. (1961), "Statistics of Hyperbolic Error Distributions in Data Transmission," *IRE Trans. Commun. Systems*, **9**, pp. 377–382, Dec.

Metzner, J. J. and K. C. Morgan, (1960), "Coded Feedback Communication Systems," *Nat. Electronics Conf.*, Chicago, IL, Oct.

Miczo, A., and L. D. Rudolph, (1970), "A Note on the Free Distance of a Convolutional Code," *IEEE Trans. Inform. Theory*, **16**, pp. 646–648, Sept.

Minty, G. J. (1957), "A Comment on the Shortest-Route Problem," *Oper. Res.*, **5**, p. 724, Oct.

Moore, J. B., and R. E. Mathes, U. S. Patent 2, 183, 147.

Morrissey, T. N. (1968), "Analysis of Decoders for Convolutional Codes by Stochastic Sequential Machine Methods," Univ. Notre Dame, Notre Dame, IN, Tech. Rept. EE-682, pp. 16–19, 23–25, 28–30, May.

Morrissey, T. N. (1970), "Analysis of Decoders for Convolutional Codes by Stochastic Sequential Machine Methods," *IEEE Trans. Inform. Theory*, pp. 460–469, July.

Muller, D. E. "Metric Properties of Boolean Algebra and Their Application to Switching Circuits," Univ. Illinois, Digital Computer Lab. Report, No. 46.

Muller, D. E. (1954), "Application of Boolean Algebra to Switching Circuit Design and to Error Detection," *IRE Trans. Elect. Comput.*, **3**, p. 6.

Neuman, B. (1968), "*Distance Properties of Convolutional Codes,*" M. S. thesis, Dept. Elec. Eng., MIT, Cambridge, MA, Aug.

Neuman, F. and D. Lumb, (1968), "Performance of Several Convolutional and Block Codes with Threshold Decoding," NASA Tech. Notes, TND-4402, Mar.

Neumann, P. G. (1964), "Error-Limiting Coding Using Information-Lossless Sequential Machines," *IEEE Trans. Inform. Theory*, **10**, pp. 108–115, Apr.

North American Rockwell, Space Division, (1972), "Modular Space Station Phase B Extension, Preliminary System Design Volume IV: Subsystems Analyses, Section 7.4," Manned Spacecraft Center, Contract NAS9-9953. Jan.

Nuttall, A. H. (1962), "Error Probabilities for Equicorrelated M-ary Signals," *IRE Trans. Inform. Theory*, pp. 305–314, July.

Odenwalder, J. P. (1970), "*Optimal Decoding of Convolutional Codes,*" Ph. D. dissertation, Dept. Elec. Eng., UCLA, Los Angeles, Jan.

Omura, J. K. (1969), "On the Viterbi Decoding Algorithm," *IEEE Trans. Inform. Theory*, **15**, pp. 177–179, Jan.

Omura, J. K. (1971), "Optimal Receiver Design for Convolutional Codes and Channels with Memory via Control Theoretical Concepts," *Inform. Sci.*, **3**, pp. 243–266, July.

Omura, J. K. (1972), "On the Viterbi Algorithm for Source Coding," *IEEE Int. Symp. Inform. Theory*, Pacific Grove, CA, p. 21, Jan.

Pelchat, M. G., R. C. Davis, and M. B. Luntz, (1971), "Coherent Demodulation of Continuous-Phase-Binary FSK Signals," *Proc. Int. Telemetry Conf.* Washington, DC, pp. 181–190.

Pennington, R. H. (1965), *Introductory Computer Methods and Numerical Analysis,* New York, MacMillan, pp. 281–283.

Perry, K. E. and J. M. Wozencraft, (1962), "SECO: A Self-Regulating Error Correcting Coder-Decoder," *IRE Trans. Inform. Theory,* pp. 128–135, Sept.

Peterson, W. W. (1961), *Error Correcting Codes,* Cambridge, MA, MIT Press.

Peterson, W. W. and D. T. Brown, (1961), "Cyclic Codes for Error Detection," *Proc. IRE,* **49**, pp. 228–235, Jan.

Peterson, W. W. (1963), *Error Correcting Codes,* Wiley, New York, pp. 137–160.

Peterson, W. W. and D. T. Brown, (1960), "Encoding and Error-Correction Procedures for the Bose-Chaudhuri Codes," *IRE Trans. Inform. Theory,* **6**, pp. 459–470, Sept.

Peterson, W. W. and E. J. Weldon, Jr., (1972), *Error Correcting Codes,* MIT Press, pp. 285–290, 427, 435.

Peterson, W. W., T. G. Birdsall, and W. C. Fox, (1954), "Theory of Signal Detectability," *IRE Trans. Professional Group on Inform. Theory, Symp. on Inform. Theory,* pp. 15–17, Sept.

Pinsker, M. S. (1965), "Complexity of the Decoding Process," *Problems of Information Transmission,* **1**, pp. 113–116.

Piret, P. (1975), "On a Class of Alternating Cyclic Convolutional Codes," *IEEE Trans. Inform. Theory,* **21**, pp. 64–69, Jan.

Piret, P. (1972), "Convolutional Codes and Irreducible Ideals," *Philips Res. Report,* **27**, pp. 257–271.

Plotkin, M. (1951), "Binary Codes with Specified Minimum Distance," Univ. Penna., Moore School of Res. Div. Report 51-20.

Pollack, M, and W. Wiebenson, (1960), "Solutions of the Shortest-Route Problem-A Review," *Oper. Res.,* **8**, pp. 224–230, Mar.

Prange, E. (1959), "The Use of Coset Equivalence in the Analysis and Decoding of Group Codes," USAF Cambridge Res. Cen., Bedford, Mass. Tech. Report AFCRC-TR-59-164, June.

Preparata, F. P. (1970), "An Improved Upper-Bound to Free Distance for Rate I/N Convolutional Codes," *Proc. UMR-Mervin J. Kelly Communications Conf.,* Rolla, MO, Oct.

Preparata, F. P. (1964), "Systemative Construction of Optimal Linear Recurrent Codes for Burst Error protection," *Calcolo,* **1**, pp. 147–153.

Preparata, F. P. (1964), "State Logic Relations for Autonomous Sequential Networks," *IEEE Trans. Electron. Comput.,* **13**, pp. 542–548, Oct.

Preparata, F. P. and S. R. Ray, (1972), "An Approach to Artificial Nonsymbolic Cognition," *Inform. Sci.,* **4**, pp. 65–86, Jan.

Puente, J. G., W. G. Schmidt, and A. M. Werth, (1971), "Multiple-Access Techniques for Commercial Satellites," *Proc. IEEE,* **59**, pp. 218–229, Feb.

Quagliato, D. N. (1970), "Advanced Coding Techniques for Satellite Communications," CADPL Report No. 167, Sept.

Quagliato, D. N. (1972), "Error Correcting Codes Applied to Satellite Channels," *IEEE Int. Conf. Commun.,* Philadelphia, PA, pp. 15.13–15.18.

Qureshi, S., and F. Newhall, (1973), "An Adaptive Receiver for Data Transmission Over Time-Dispersive Channels," *IEEE Trans. Inform. Theory,* **19**, pp. 448–457, July.

Ramsey, J. L. (1970), "Realization of Optimum Interleavers," *IEEE Trans. Inform. Theory,* **16**, pp. 338–345, May.

Ramsey, J. L. (1970), "Cascaded Tree Codes," MIT Res. Lab. Electron., Cambridge, MA, Report 478, Sept.

Raviv, J. (1967), "Decision Making in Markov Chains Applied to the Problem of Pattern Recognition," *IEEE Trans. Inform. Theory*, **13**, pp. 536–551, Oct.

Reed, I. S. (1954), "*A Class of Multiple Error-Correcting Codes and the Decoding Scheme*," *IRE Trans. Inform. Theory*, **4**, pp. 38–49.

Reed, I. S., and G. Solomon, (1960), "Polynomial Codes Over Certain Finite Fields," *J. Soc. Industrial Applied Math.*, **8**, pp. 300–304, June.

Reiffen, B. (1960), "Sequential Encoding and Decoding for the Discrete Memoryless Channel," Ph. D. dissertation, Dept. Elec. Eng., MIT, Cambridge, MA, Aug.

Reiffen, B., W. G. Schmidt, and H. L. Yudkin, (1961), "The Design of an Error-Free-Data Transmission System for Telephone Circuits," *AIEE, Commun. and Electronics*, pt. D, pp. 224–231, July.

Rice, S. O. (1950), "Communications in the Presence of Noise-Probability of Error of Two Encoding Schemes," *B. S. T. J.*, **29**, pp. 60–93.

Robinson, J. P., and A. J. Bernstein, (1967), "A Class of Binary Recurrent Codes with Limited Error Propagation," *IEEE Trans. Inform. Theory*, **13**, pp. 106–113, Apr.

Robinson, J. P. (1967), "The Construction of All Linear Uniform Codes," *Princeton Conf. Inform. Sci. and Systems*, Princeton, NJ, pp. 217–220.

Robinson, J. P. (1963), "Error Propagation in Decoding Recurrent Codes," IBM Watson Res. Lab., Yorktown Heights, NY, Res. Note NC-324, Nov.

Robinson, J. P. (1965), "An Upper Bound on the Minimum Distance of a Convolutional Code," *IEEE Trans. Inform. Theory*, **11**, pp. 567–571, Oct.

Robinson, J. P. (1964), "Note on Optimal Recurrent Codes," IBM Watson Res. Lab., Yorktown Heights, NY, Res. Paper RC-1270, Aug.

Robinson, J. P. (1968), "Error Propagation and Definite Decoding of Convolutional Codes," *IEEE Trans. Inform. Theory*, **14**, pp. 121–128, Jan.

Robinson, J. P. (1968), "Definite Decoding and Error Propagation of Convolutional Codes," *IEEE Trans. Inform. Theory*, **14**, pp. 121–128, Jan.

Robinson, J. P., and A. J. Bernstein, (1965), "Self Orthogonal Codes," Tech. Report 43, Dept. Elec. Eng., Digital Systems Lab., Princeton Univ., Princeton, NJ.

Robinson, J. P. (1968), "Reversible Convolutional Codes," *IEEE Trans. Inform. Theory*, pp. 609–610, July.

Rocher, E. Y. and R. L. Pickholtz, (1970), "An Analysis of the Effectiveness of Hybrid Transmission Schemes," *IBM J. Res. Devel.*, **14**, pp. 426–433, July.

Rosenberg, W. J. (1971), "Structural Properties of Convolutional Codes," Ph. D. dissertation, Dept. Elec. Eng., Univ. California, Los Angeles, ch. 4.

Sacks, G. E. (1958), "Multiple Error Correction by Means of Parity Checks," *IRE Trans. Inform. Theory*, **4**, pp. 145–147, Dec.

Sain, M. K. and J. L. Massey, (1969), "Invertibility of Linear Time-Invariant Dynamical Systems," *IEEE Trans. Automat. Contr.*, **14**, pp. 141–149, Apr.

Sain, M. K. (1967), "Functional Reproducibility and the Existence of Classical Sensitivity Matrices," *IEEE Trans. Automat. Contr.*, **12**, p. 458, Aug.

Sain, M. K. (1967), "On A Useful Matrix Inversion Formula and Its Applications," *Proc. IEEE*, **55**, p. 1753, Oct.

Savage, J. E. (1965), "The Computation Problem with Sequential Decoding," MIT Lincoln Lab., Cambridge, MA, Tech. Report 371, Feb.

Savage, J. E. (1966), "Sequential Decoding: The Computation Problem," *B. S. T. J.*, **45**, pp. 149–175, Jan.

Savage, J. E. (1966), "The Distribution of Sequential Decoding Computation Time," *IEEE Trans. Inform. Theory*, **12**, pp. 143–147, Apr.

Schalkwijk, J. P. M., and A. J. Vinck, (1975), "Syndrome Decoding of Convolutional Codes," *IEEE Trans. Commun.*, pp. 789–792, July.

Sellers, F. F., Jr. (1962), "Bit Loss and Gain Correction Code," *IRE Trans. Inform. Theory*, **8**, pp. 35–38, Jan.

Shannon, C. E., R. G. Gallager, and E. R. Berlekamp, (1967), "Lower Bounds to Error Probability for Coding on Discrete Memoryless Channels," *Inform. Contr.*, **10**, pt. I, pp. 65–103; pt. II, pp. 522–552.

Shannon, C. E. and W. Weaver, (1949), *The Mathematical Theory of Communications*, Urbana, IL, Univ. Illinois Press.

Shannon, C. E. (1951), "Prediction and Entropy of Printed English," *B. S. T. J.*, **30**, pp. 50–64, Jan.

Shannon, C. E. (1948), "A Mathematical Theory of Communications," *B. S. T. J.*, **27**, pp. 379–423, 623–656.

Shannon, C. E. (1949), "Communication in the Presence of Noise," *Proc. IRE*, **37**, pp. 10–21.

Shannon, C. E. (1949), "Communication Theory of Secrecy Systems," *B. S. T. J.*, **28**, pp. 656–715.

Silverman, L. M. (1968), "Properties and Application of Inverse Systems," *IEEE Trans. Automat. Contr.*, **13**, pp. 436–437, Aug.

Silverman, L. M. (1969), "Inversion of Multivariable Linear Systems," *IEEE Trans. Automat. Contr.*, **14**, June.

Singer, J. (1938), "A Theorem in Finite Projective Geometry and Some Applications to Number Theory," *Trans. Amer. Math. Soc.*, **43**, pp. 377–385.

Sittler, R. W. (1956), "Systems Analysis of Discrete Markov Processes," *IRE Trans. Circuit Theory*, **3**, pp. 257–266, Dec.

Slepian, D. (1956), "A Class of Binary Signaling Alphabets," *B. S. T. J.*, **35**, p. 203.

Slepian, D. (1969), "Some Further Theory of Group Codes," *B. S. T. J.*, **39**, pp. 1219–1251, Sept.

Sloane, N. J., S. M. Reddy, and C. L. Chen, (1972), "New Binary Codes," *IEEE Trans. Inform. Theory*, **18**, pp. 503–510, July.

Solomon, G., And J. J. Stiffler, (1964), "Algebraicly Punctured Cyclic Codes," *Inform. Contr.* **8**, pp. 170–179.

Sommer, R. C. (1968), "High Efficiency Multiple Access Communications Through a Signal Processing Repeater," *IEEE Trans. Commun. Tech.*, **16**, pp. 222–232, Apr.

Souza, C. R., and R. J. Leake, (1967), "Stochastic Finite-State Systems: Four Models and Their Relationships," *Proc. 5th Annual Allerton Conf. on Circuit and System Theory*, Monticello, IL, Oct.

Springett, J. C., and M. K. Simon, (1971), "An Analysis of the Phase Coherent/Incoherent Output of the Bandpass Limiter," *IEEE Trans. Commun. Tech.*, **19**, pp. 42–49, Feb.

Stiffler, J. J. (1965), "Comma-Free Error-Correcting Codes," *IEEE Trans. Inform. Theory*, **11**, pp. 107–112, Jan.

Stone, J. J. (1963), "Multiple-Burst Error Correction with the Chinese Remainder Theorem," *J. Soc. Ind. Appl. Math.*, **11**, pp. 74–81.

Storch, P. (1934), "Evolution of Long Distance Type-Printing Traffic by Wire and Radio," *Elek. Z.*, **55**, pp. 109–141.

Sullivan, D. D. (1967), "Error Propagation Properties of Uniform Codes," *Int. Symp. Inform. Theory*, San Remo, Italy, Sept.

Sullivan, D. D. (1966), "Control of Error Propagation in Convolutional Codes," Univ. Notre Dame, Notre Dame, Ind. Tech. Report EE-667, Nov.

Swain, P. H., and K. S. Fu, (1966), "On Relationships Between the Theories of Discrete-State and Continuous-State Linear Time-Invariant Systems," *Proc. 4th Annual Allerton Conf. on Circuit and Systems Theory*, pp. 794–804, Oct.

Taqvi, S. Z. H., and W. H. Ng, (1970), "The Simulation of Error Correction Codes for Space Communications Channels," HASD 642D-823239, Lockheed Electronics Co.

Taqvi, S. Z. H. (1970), "Performance Prediction for the Viterbi Decoder," OB 2059, Lockheed Electric Co., Sept.

Taqvi, S. Z. H. (1970), "Computer Analysis of Noise for Coding Simulations Applications," HASD 642C-823-246, Lockheed Electric Co.

Taqvi, S. Z. H. (1971), "Some Interesting Simulation Results of the Viterbi Decoder," *Nat. Telemetry Conf.*, pp. 240–246.

Timor, U. (1971), "Sequential Ranging with the Viterbi Algorithm," Jet Propulsion Lab., Pasadena, CA, JPL Tech. Report 32-1526, II, pp. 75–79, Jan.

Tippett, L. C. H. (1925), "On the Extreme Individuals and the Range of Samples Taken From a Normal Population," *Biometrics*, **17**, p. 364.

Tong, S. Y. (1966), "Synchronization Recovery Techniques for Binary Cyclic Codes," *B. S. T. J.*, pp. 515–522, 561–595, Apr.

Tong, S. Y. (1968), "Systematic Construction of Self-Orthogonal Diffuse Codes," Bell Telephone Labs., Tech. Memo, Holmdel, NJ.

Tong, S. Y. (1968), "Burst Trapping Techniques for a Compound Channel," Bell Telephone Labs., Tech. Memo, Holmdel, NJ.

Tong, S. Y. (1968), "Correction of Synchronization Errors with Burst Error Correcting Cyclic Codes," Bell Telephone Labs., Holmdel, NJ.

Trench, W. F (1963), "An Algorithm for the Inversion of the Finite Toeplitz Matrices," *J. SIAM*, **12**, pp. 515–522, Sept.

Ungerboeck, G. (1971), "Nonlinear Equalization of Binary Signals in Gaussian Noise," *IEEE Trans. Commun. Tech.*, **19**, pp. 1128–1137, Dec.

U.S. Dept. Commerce, N.T.I.S. (1975), "CCD Photosensor Array Development Program (Phase II)," AD A022 881, Apr.

Van de Meeberg, L. (1974), "A Tightened Upper Bound on the Error Probability of Binary Convolutional Codes with Viterbi Decoding," *IEEE Trans. Inform. Theory*, **20**, pp. 389–391, May.

Viterbi, A. J. (1961), "On Coded Phase Coherent Communications," *IRE Trans. Space Electronics and Telemetry*, **7**, pp. 3–14, Mar.

Viterbi, A. J. (1964), "Lower Bounds of Maximum Signal-to-Noise Ratios for Digital Communications Over the Gaussian Channel," *IEEE Trans. Commun.*, **12**, pp. 10–17, Mar.

Viterbi, A. J. (1967), "Error Bounds for Convolutional Codes and an Asymptotically Optimum Decoding Algorithm," *IEEE Trans. Inform. Theory*, **13**, pp. 260–269, Apr.

Viterbi, A. J. (1971), "Convolutional Codes and Their Performance in Communication Systems," *IEEE Trans. Commun. Tech.*, **19**, pp. 751–772, Oct.

Viterbi, A. J., and J. P. Odenwalder, (1969), "Further Results on Optimal Decoding of the Convolutional Codes," *IEEE Trans. Inform. Theory*, **15**, pp. 732–734, Nov.

Viterbi, A. J. (1967), "Orthogonal Tree Codes for Communication in the Presence of White Gaussian Noise," *IEEE Trans. Commun. Tech.*, **15**, pp. 238–242, Apr.

Viterbi, A. J. (1966), *Principles of Coherent Communication*, McGraw-Hill, New York.

Viterbi, A. J. (1967), "The State Diagram Approach to Optimal Decoding," Jet Propulsion Labs., Pasadena, CA, Space Programs Summary, 37-58, pp. 50–55.

Viterbi, A. J. (1962), "Classification and Evaluation of Coherent Synchronous Sampled-Data Telemetry Systems," *IRE Trans. Space Electronics and Telemetry*, SET-8, pp. 13–22, Mar.

Wadden W. R., D. M. Jones, and J. C. Pennypacker, (1961), "An Investigation of Sequential Decoding," *RCA Review*, **22**, Sept.

Wagner, T. J. (1968), "A Gilbert Bound for Periodical Binary Convolutional Codes," *IEEE Trans. Inform. Theory,* **14**, pp. 752–755, Sept.

Warner, B., and R. Loffredo, (1972), "Design and Performance of the Convolutional Codes for the DITEC Projects," COMSAT Lab. Tech. Memo CL-28-72.

Weldon, E. J., Jr., (1968), "A Note on Synchronization Recovery with Extended Cyclic Codes," *Inform. Contr.,* **13**, pp. 354–356, Oct.

Wintz, P. A. (1972), "Transform Picture Coding," *Proc. IEEE,* **60**, pp. 809–820, July.

Wozencraft, J. M., and B. Reiffen, (1961), *Sequential Decoding,* MIT Press, Cambridge, MA, and Wiley, New York.

Wozencraft, J. M., and I. M. Jacobs, (1965), *Principles of Communication Engineering,* Wiley, New York.

Wozencraft, J. M., and M. Horstein, (1961), "Coding for Two-Way Channels," in *Information Theory,* C. Cherry, (Ed.), Butterworth, London.

Wozencraft, J. M. (1957), "Sequential Decoding for Reliable Communication," Sc. D. dissertation, Dep. Elec. Eng. MIT, Cambridge, MA, June.

Wozencraft, J. M. (1957), "Sequential Decoding for Reliable Communications," Tech. Report No. 325, MIT Res. Lab. Elec., Cambridge, MA, Aug.

Wu, W. W. (1975), "New Convolutional Codes - Part I," *IEEE Trans. Commun.,* **23**, pp. 942–956, Sept.

Wu, W. W. (1971), "Applications of Error-Coding Techniques to Satellite Communication," *COMSAT Tech, Rev.*

Wyner, A. D., and R. B. Ash, (1963), "Analysis of Recurrent Codes," *IEEE Trans. Inform. Theory,* **9**, pp. 143–156, July.

Wyner, A. D. (1963), "Some Results on Burst-Correcting Recurrent Codes," *IEEE Conv. Rec.,* pp. 139–152.

Wyner, A. D. (1965), "Capabilities of Bounded Discrepance Decoding," *B. S. T. J.,* **44**, pp. 1064–1122, July-Aug.

Wyner, A. D. (1963), "Gilbert Bounds for Recurrent Codes," IBM Watson Res. Center, Yorktown Heights, NY, Res. Paper, July.

Yau, S. S. (1965), "Autonomous Clocks in Sequential Machines," *IEEE Trans. Electr. Comput.,* pp. 467–472, June.

Yudkin, H. L. (1964), "Channel State Testing in Information Decoding," Ph.D. dissertation, Dept. Elec. Eng., MIT, Cambridge, MA, Sept.

Zadeh, L. A., and C. A. Desoer, (1963), "*Linear System Theory,*" McGraw-Hill, New York, pp. 302–306.

Zeoli, G. W. (1971), "Coupled Decoding of Block-Convolutional Concatenated Codes," Ph.D. dissertation, Dept. Elec. Eng., UCLA.

Zigangirov, K. (1966), "Some Sequential Decoding Procedures," *Probl. Peredach. Inform.,* pp. 13–25.

Zierler, N. (1959), "Linear Recurring Sequences," *J. Soc. Ind. Appl. Math.,* **7**, pp. 31–48.

Zierler, N. (1955), "Several Binary Sequence Generators," Lincoln Labs., MIT Tech. Report 95.

Zierler, N., and D. Gorenstein, (1961), "A Class of Error Correcting Codes in p^m Symbols," *J. Soc. Ind. Appl. Math.,* **9**, pp. 207–214.

Ziv, J. (1967), "Asymptotic Performance and Complexity of a Coding Scheme for Memoryless Channels," *IEEE Trans. Inform. Theory,* **13**, pp. 356–359, July.

Ziv, J. (1966), "Further Results on the Asymptotic Complexity of an Iterative Coding Scheme," *IEEE Trans. Inform. Theory,* **12**, pp. 168–171, Apr.

Ziv, J. (1963), "Successive Decoding Scheme for Memoryless Channels," *IEEE Trans. Inform. Theory,* **9**, pp. 97–105, Apr.

Symbols and Abbreviations

Name	Definition	Page
A	Amplifier	155
A	Normalization constant	20
A	Signal sample amplitude	5
A/D	Analog to digital	148
A_f	Ratio of DPCM process autocorrelation to variance	39
AGC	Automatic gain control	163
ALC	Automatic light control	191
AMP	Amplifier	155
$A-S$	Subscript denoting dither subtraction	28
ATTN	Attenuator	155
AU	Arithmetic unit	96
B	Number of bits per code word	15
B	Number of bits per sample	15
B	Constant	20
B	Filter bandwidth	5
BAL	Balanced	155
BCH	Bose-Chaudhuri-Hocquenghen	138
BLANK	Blanking	154
BPF	Band-pass filter	155
BSC	Binary symmetrical channel	256
BW	Bandwidth	199

Name	Definition	Page
CCD	Charge-coupled device	175
cm	Centimeter	187
COMP	Composite	191
Corr	Correlation	60
COUPL	Coupler	163
C/S	Clutter to signal	158
CSOC	Canonical self-orthogonal codes	283
D	Dither level	26
D	Deviation of the mean output	22
dB	Decibel	9
dc	Direct current	96
DCT	Discrete cosine transform	91
DD	Drains of differential amplifier	187
DDA	Dual-gate differential amplifier	187
DET	Detector	154
DFGA	Distributed floating gate amplifier	187
d(kt)	Discrete time function output	39
DIFF	Differential	145
DIM	Dimensional	134
DPCM	Differential pulse code modulation	46
E	Mean square error	20
E	Signal energy	9
E_k	Output mean square error	20
$\bar{e}_n^2$	RMS error	4
E/N_0	Signal-to-noise ratio (S/N)	9
eV	Electron volts	187
exp	Exponential	156
f_B	Bit rate in pps	9
FF	Flip-flop circuit; also divide by two	146
FFT	Fast Fourier Transform	79
FGA	Floating-gate amplifier	187
FGA1	Floating-gate amplifier No. 1	187
FGA2	Floating-gate amplifier No. 2	187
FM	Frequency-modulation	4
f_S	Sampling frequency	9
$f(t)$	Continuous time function of the input	39
GEN	Generator	236

Name	Definition	Page
GHz	Gigahertz (frequency)	163
HORIZ	Horizontal	154
Hz	Hertz (frequency)	199
I & D	Integrate and dump	155
IDG	Input differential gate	187
IF	Intermediate frequency	10
ILT	In-line transfer	297
INV	Inverter	143
kHz	Kilohertz (frequency)	144
MIS	Metal-insulator-semiconductor	176
MIX	Mixer	155
MOD	Modulator	145
mod 7	Modulo seven (Cyclical numbering systems)	268
MU	Memory unit	96
mV	Millivolt	187
MV	Multivibrator	153
N	Noise power at the output of a band-pass filter	6
n	Number of bits per word	20
$\bar{n}^2$	Noise power	15
nA	Nanoamperes	187
N_0	Noise power in a unit bandwidth	10
NRZ	Non-return zero	240
NTSC	National Television System Committee	47
OSC	Oscillator	19
OUT	Output	187
P	Probability of error	15
P	Pulse	166
PAL	Color TV system of the Phase Alternation type	41
PAM	Pulse amplitude modulation	4
PAM/FM	PAM plus FM	4
P_b	Probability of bitt errors	7
PCM	Pulse code modulation	2
PCM/FM	PCM plus FM	3
PDM	Pulse duration modulation	4
PDM/FM	PDM plus FM	4
$p(i)$	Probability of an ith level error	7
P_k	Codeword probability	20

Name	Definition	Page
$P(0)$	Probability of zero level being in error	7
$P(E)$	Probability of error	36
PRN	Pseudorandom noise coding	143
$p(x)$	Probability of event x occurring	20
$P(y, x)$	Joint Probability of y and x	20
$P(y\|x)$	Probability of y; given x	20
q	Quantal step size	26
$Q_A(i, i+1)$	Quantization of binary levels for slicing between levels i and $i+1$	26
R	Reset	238
R	Resistor	29
R	Channel error rate	15
RD	Reset drain	187
REG	Register	145
RG	Reset ground	187
RL	Run length coding	66
rms	Root mean square	4
RS	Reed-Solomon	287
S	Received signal power	6
S	Set	238
SAT	Saturation	187
SC	Source of compensation amplifier	187
SD	Signal output drain	187
SH	Shot	146
SIPSF	Space invariant point spread function	108
$\bar{s}_n^2$	Mean-square signal	5
$\bar{s}_n^2/\bar{e}_n^2$	Mean-square signal-to-error ratio	4
SNR	Signal-to-noise ratio	24
SVD	Singular value decomposition	105
SVPSF	Space variant point spread function	108
SW	Switch	96
SYNC	Synchronization	155
T	Matrix transpose when used as a superscript	75
T	Sample transmission time period	6
T	Time period of 1 bit	26
t	Time in general (a variable)	6
TES	Trailing edgeshaper	146
TRANSF	Transformations	134

Name	Definition	Page
TV	Television	1
TWT	Traveling wave tube	236
V	Variance	20
VCO	Voltage controlled oscillator	155
VERT	Vertical	154
W	Word transmission time period	6
$W(\omega)$	Power spectral weighting function	28
x_k	Input point	20
x_n	Sample number	6
XTAL	Crystal	236
XVCO	Crystal voltage controlled oscillator	163
x, y, z	Binary signals with weights $\frac{1}{2}$, $\frac{1}{4}$, and $\frac{1}{8}$	26
y_k	Signal output	20
$\hat{y}(kt)$	Expected value of discrete time function output	39
$Y(t)$	Sampled output as a function of time	39
$\bar{y}_x$	Mean value of output	20
α	Arbitrary constant	34
α	Digital multiplier in feedback loop of Fig. 4.4	37
α	Peak-to-peak noise amplitude	24
α_x	Digital impulse response of the DPCM linear predictor, for $x = 1, 2, \ldots, N$	41
$\phi_{ff}(0)$	Autocorrelation function when $T = 0$ (Variance)	39
$\delta(x)$	Unit delta function of x	6
ρ	Ratio equal to A_f	41
σ^2	Variance	6
ω	Radian carrier frequency	6
$\oplus$	Exclusive OR gate	206
$(\bar{x})$	Superscript bar denotes average value of x	39

Index